Jan-Philipp Weiß

Numerical Analysis of Lattice Boltzmann Methods for the Heat Equation on a Bounded Interval

Numerical Analysis of Lattice Boltzmann Methods for the Heat Equation on a Bounded Interval

von
Jan-Philipp Weiß

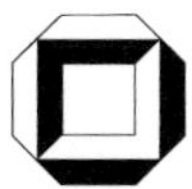

universitätsverlag karlsruhe

Dissertation, Universität Karlsruhe (TH)
Fakultät für Mathematik, 2006

Impressum

Universitätsverlag Karlsruhe
c/o Universitätsbibliothek
Straße am Forum 2
D-76131 Karlsruhe
www.uvka.de

Universitätsverlag Karlsruhe 2006
Print on Demand

ISBN-13: 978-3-86644-069-2
ISBN-10: 3-86644-069-3

Vorwort

Diese Dissertationsschrift entstand im Rahmen meiner Tätigkeit am Institut für Angewandte und Numerische Mathematik, Lehrstuhl II, der Universität Karlsruhe (TH). Der Anstoß zu dieser Arbeit fand sich während meiner Zeit am Fachbereich Mathematik der Universität Kaiserslautern. In Zusammenarbeit mit dem Fraunhofer Institut für Techno- und Wirtschaftsmathematik (ITWM) in Kaiserslautern nahm ich dort an der Bearbeitung eines Projektes des Bundesministeriums für Bildung und Forschung (BMBF) mit dem Titel *Adaptive Gittersteuerung für Lattice-Boltzmann Verfahren zur Simulation von Füllprozessen im Gießereibereich* teil. Ich möchte all jenen danken, die während dieser Zeit zum Gelingen meiner Arbeit beigetragen haben.

Mein besonderer Dank gilt Herrn Prof. Dr. Willy Dörfler für die vielen wertvollen und hilfreichen Anregungen und die immerwährende Unterstützung in sämtlichen Angelegenheiten. Mein weiterer Dank gilt Herrn Prof. Dr. Vincent Heuveline für die nützlichen Hinweise und die fruchtbaren Gespräche. Den Mitarbeitern des Institutes, insbesondere unserer Sekretärin Frau Kerstin Dick, danke ich für die angenehme Atmosphäre und die hervorragende Zusammenarbeit.

Meine Familie, meine Freunde und meine Freundin Steffi möchte ich für die seelische, geistige und moralische Unterstützung, die Geduld und die wunderbare Zeit, die ich mit Ihnen verbringen durfte, an dieser Stelle in besonderem Maße würdigen.

Herzlichen Dank Euch allen!

Karlsruhe, August 2006   Jan-Philipp Weiß

Contents

Introduction

The solution of fluid-dynamic equations from the viewpoint of analytical, numerical and algorithmic aspects is a challenging task. In recent years, lattice Boltzmann methods have been developed to face this problem. The performance of the proposed lattice Boltzmann methods renders convincing results, but less is known about the mathematical analysis.

Fluid-dynamic equations, for example the Navier-Stokes or the Euler equations, deal with macroscopic quantities like velocity, pressure or temperature. Lattice Boltzmann methods follow a different approach. They treat distribution functions that stem from a particle kinetic framework. Both approaches rely on conservation principles, the first one on a macroscopic level, the second one on a microscopic level. Since only macroscopic quantities, that is, physically measurable quantities, are in the scope of the researcher's interest, the distribution functions have to be averaged appropriately to achieve interpretable results. With respect to specific scalings, it turns out that the averaged quantities approximate solutions of the fluid-dynamic equations in a certain limit; see for example Ref. [**24**].

It is an open question whether these different approaches can be related to each other rigorously or if they represent two independent models. Not only owing to the complexity of the limiting equations, the analysis of the lattice Boltzmann methods in the multi-dimensional case is equipped with formidable difficulties. Up to now there are only a few answers concerning essential issues like stability or convergence of the used schemes. But the convincing results achieved so far are encouraging to advance these methods and their analysis.

By the end of the 1980s, lattice Boltzmann methods were introduced by engineers and physicists. Many papers have been written since then, but the mathematical background is still obscure in many fields. Areas of application of lattice Boltzmann methods are the simulation of incompressible flows in complex geometries, for example the flow of blood in vessels, multiphase and multicomponent fluids, free surface problems, moving boundaries, fluid-structure interactions, chemical reactions, flow through porous media, suspension flows, magnetohydrodynamics, semiconductor simulations, non-Newtonian fluids, large eddy and turbulence simulations in aerodynamics, and many more.

The limiting fluid-dynamic equations are determined by the scaling and the choice of the collision operator, where various models are possible. For this reason, lattice Boltzmann methods are applicable to a great variety of different problems. Although lattice Boltzmann methods are universally acclaimed for the applicability to complex geometries and interfacial dynamics, severe problems appear in the case of boundary conditions.

The advantages of lattice Boltzmann methods find expression in a comparably simple explicit algorithm on uniform grids with only local interactions. The parallelization of the algorithms for the speed-up of the computations is straightforward. A great advantage is the gain of differentiated quantities without performing numerical differentiations.

The convection step of the lattice Boltzmann method is a linear operation that is often combined with a relaxation process. This feature allows for a multiscale analysis. The incompressible Navier-Stokes equations can be attained in the nearly incompressible limit. The pressure is computed by the evaluation of an equation of state, whereas the direct numerical simulation requires the solution of a Poisson equation for the pressure.

The lattice Boltzmann methods only utilize a minimal set of velocities in the phase space. This greatly simplifies the kinetic processes motivated by the full Boltzmann equation. In spite of the simplicity of the discrete models, the convergence often shows a better behavior than for their continuous counterparts.

Although the lattice Boltzmann equations can be viewed as a finite difference discretization of the fluid-dynamic equations in combination with an explicit Euler method, they preserve much more physics than any direct discretization of the macroscopic equations. This is owed to the higher order moments that are not matched by the fluid-dynamic equations.

In the diffusive scaling of the lattice Boltzmann equations, a coupling of the grid size h and the time step τ of the form $\tau \sim h^2$ is required. Hence, many time steps have to be performed and the computational effort grows immensely when refining the grid. Since the lattice Boltzmann schemes are designed for uniform grids, a local refinement of the grid is a nontrivial task. As in the boundary case, severe problems appear at the grid intersection zones. Sophisticated algorithms have to be applied. In order to minimize the additional effort, it is desirable to develop a posteriori error estimation techniques to gain information, whether the grid has to be refined globally or locally or not at all.

In this work, we consider a one-dimensional model problem with the aim to answer some of the open questions concerning the numerical analysis of lattice Boltzmann methods. This includes a detailed analysis of the continuous limit equations.

In the first chapter of this work, we outline the passage from the full Boltzmann equation to the lattice Boltzmann methods. As an intermediate step, we pay attention to the discrete velocity models that can be seen as an interface.

In Chapter 2, we set up a linear model problem in one dimension. This model problem consists of three parts: the macroscopic limit equation, the velocity discrete system and the discrete lattice Boltzmann equations. In this work, we examine the convergence of the solutions of the velocity discrete system towards solutions of the macroscopic limit equation, as well as the convergence of the lattice Boltzmann solutions. The chosen model enables a deepened analysis of the initial boundary-value problem for a set of different boundary conditions. In spite of its systematic simplicity, most of the features of general lattice Boltzmann methods are covered.

The limiting fluid-dynamic equation of the model problem is the heat equation on a bounded interval equipped with several kinds of boundary conditions. The basic properties of the corresponding solutions are examined. The regularity of the solutions of the limiting equations plays a decisive role for all types of convergence processes. For this reason, special care is taken on this topic. Necessary prerequisites of the data are worked out that ensure the suitable smoothness of the solutions. These results are important for the ensuing convergence analysis.

The interface between the heat equation and the lattice Boltzmann schemes is formed by the discrete velocity model presented in Chapter 3, namely the Goldstein-Taylor model. We examine the behavior of the solutions in the limiting case. It turns out that we are faced with the solution of telegraph equations that present a singularly perturbed problem for the heat equation. The corresponding solutions are dealt with in Chapter 4. With the acquired knowledge we can put up

Fourier solutions for the Goldstein-Taylor model. The perturbation parameter of the equations is in close correspondence with the grid size of the lattice Boltzmann schemes.

In Chapter 5, we set up the lattice Boltzmann schemes as discretizations of the Goldstein-Taylor model. Several types of boundary conditions are introduced. The discrete schemes are interpreted in a finite difference as well as in a finite volume context. We prove discrete stability estimates that are in accordance with the continuous case. Furthermore, we present discrete Fourier solutions that give a deep insight into the asymptotic behavior of the solutions, when the grid size h approaches zero. The coupling of the discrete density and the discrete flux is revealed. As observed in many applications, the choice of the initial data for the flux is of subordinate importance. The investigation of the eigenvalue distribution of the time evolution operator leads to the insight that problems may appear, if the provided data hurt the smoothness prerequisites. These observations are confirmed by numerical results reported in Chapter 7.

By following the concept of consistency and stability discussed in Chapter 6, we prove second order convergence in terms of the discrete $L^{\infty}(L^2)$-norms with respect to the grid size h for fixed end times and for all types of the presented boundary conditions. Second order convergence is also proved for the flux in the discrete $L^2(L^2)$-norms. Further investigations lead to the conjecture that the convergence orders can be improved for vanishing source terms and a specific choice of the parameters.

In Chapter 7, we perform numerical experiments to confirm the theoretical results of the preceding chapters. The experiments show second order convergence for the density and the flux. We examine the influence of the boundary and the initial conditions and the dependence on the parameters. We find fourth order convergence for vanishing source terms and a specific choice for the relaxation parameter. The lattice Boltzmann methods are compared to the explicit and implicit Euler methods and to the Crank-Nicolson scheme.

Oscillations in the lattice Boltzmann solutions occur for nonsmooth initial data. We point out that this is a consequence of the eigenvalue distribution of the discrete time evolution operator and that there is no remedy for this disadvantage. In this point, a discretization via the explicit Euler method appears to be superior. Similar observations are found for the lattice Boltzmann methods for the advection-diffusion equation. At the end of this chapter, we consider lattice Boltzmann solutions for the viscous Burgers equation.

Lattice Boltzmann methods are designed for uniform grids. The symmetry of the stencils is a necessary fundament for many effects. In Chapter 8, we present algorithms for the coupling of grids with different grid sizes. The performance of these algorithms is investigated. The grid coupling is a basic ingredient for the construction of adaptive methods that need to be tackled in future works.

List of Abbreviations

LB	Lattice Boltzmann
FD	Finite difference
FV	Finite volume
VC	Vertex centered
CC	Cell centered
VCFD	Vertex centered finite difference
VCFV	Vertex centered finite volume
CCFD	Cell centered finite difference
CCFV	Cell centered finite volume
EOC	Experimental order of convergence

CHAPTER 1

From Boltzmann to Lattice Boltzmann

In this introductory chapter we outline the relationship between the Boltzmann equation and the discrete velocity models and their physical properties. For the one-dimensional discrete velocity models, we present the equations resulting in the fluid-dynamic limits. Lattice Boltzmann schemes are derived from the discrete velocity models by applying discretizations with respect to the characteristic directions on regular lattices.

1.1. The Boltzmann Equation

In a microscopic picture, we consider moving particles subject to binary collisions. In a mathematical framework, the particles are modeled by a distribution function $f = f(t, \boldsymbol{x}, \boldsymbol{v})$ that describes particles at the time t at the spatial point $\boldsymbol{x} = [x_1; x_2; x_3] \in \mathbb{R}^3$ having the velocity $\boldsymbol{v} = [v_1; v_2; v_3] \in \mathbb{R}^3$. The time evolution of the distribution function for particles with mass m subject to an external force $\boldsymbol{K}$ is modeled by the *Boltzmann equation* (Ref. [9])

$$\frac{d}{dt} f(t, \boldsymbol{x}, \boldsymbol{v}) = \left(\frac{\partial}{\partial t} + \boldsymbol{v} \cdot \boldsymbol{\nabla}_{\boldsymbol{x}} + \frac{\boldsymbol{K}}{m} \cdot \boldsymbol{\nabla}_{\boldsymbol{v}} \right) f(t, \boldsymbol{x}, \boldsymbol{v}) = Q(f)(t, \boldsymbol{x}, \boldsymbol{v}). \qquad (1.1)$$

This equation describes the behavior of a rarefied gas. The derivatives have to be considered with respect to time, space and velocity. We use the operators $\boldsymbol{\nabla}_{\boldsymbol{x}} := [\partial/\partial x_1; \partial/\partial x_2; \partial/\partial x_3]$ and $\boldsymbol{\nabla}_{\boldsymbol{v}} := [\partial/\partial v_1; \partial/\partial v_2; \partial/\partial v_3]$. The binary collisions of the particles are determined by the collision operator

$$Q(f)(t, \boldsymbol{x}, \boldsymbol{v}) := \int_{\mathbb{R}^3} \int_{S^2} B(\boldsymbol{v}, \boldsymbol{w}, \boldsymbol{e}) \left[f(\boldsymbol{v}') f(\boldsymbol{w}') - f(\boldsymbol{v}) f(\boldsymbol{w}) \right] d\boldsymbol{e} \, d\boldsymbol{w}$$

with the collision kernel B. The domain of integration is the velocity space $\mathbb{R}^3$ and the unit sphere S^2. Here, $\boldsymbol{v}$ and $\boldsymbol{w}$ denote the velocities before the collision and $\boldsymbol{v}'$ and $\boldsymbol{w}'$ are the velocities after the collisions. The collision kernel describes the probability that a particle moving with velocity $\boldsymbol{v}$ and colliding with a particle with velocity $\boldsymbol{w}$ is reflected into the direction $\boldsymbol{e} := (\boldsymbol{v} - \boldsymbol{v}')/|\boldsymbol{v} - \boldsymbol{v}'|$. The velocities after the collision are determined by the relations

$$\boldsymbol{v}' = \boldsymbol{v} + \boldsymbol{e}\boldsymbol{e} \cdot (\boldsymbol{w} - \boldsymbol{v}),$$
$$\boldsymbol{w}' = \boldsymbol{w} - \boldsymbol{e}\boldsymbol{e} \cdot (\boldsymbol{w} - \boldsymbol{v})$$

that express the conservation of energy and momentum.

Physical effects of the rarefied gas are described by macroscopic quantities that are obtained by averaging the weighted particle distribution function. The mass density ρ is defined by

$$\rho(t, \boldsymbol{x}) := \int_{\mathbb{R}^3} f(t, \boldsymbol{x}, \boldsymbol{v}) \, d\boldsymbol{v}.$$

The mean velocity $\boldsymbol{u}$ of the fluid is gained by setting

$$\boldsymbol{u}(t, \boldsymbol{x}) := \frac{1}{\rho} \int_{\mathbb{R}^3} \boldsymbol{v} f(t, \boldsymbol{x}, \boldsymbol{v}) \, d\boldsymbol{v},$$

1

and the local temperature T is given by

$$T(t, \boldsymbol{x}) := \frac{1}{\rho} \int_{\mathbb{R}^3} |\boldsymbol{v} - \boldsymbol{u}|^2 f(t, \boldsymbol{x}, \boldsymbol{v}) \, d\boldsymbol{v}.$$

We now disregard external forces and rewrite the Boltzmann equation in scaled form

$$\kappa \frac{\partial}{\partial t} f + \boldsymbol{v} \cdot \boldsymbol{\nabla}_{\boldsymbol{x}} f = \frac{1}{\epsilon} Q(f),$$

where κ is the *Mach number* and ϵ is the *Knudsen number*. The regime $\kappa = 1$, $\epsilon \to 0$ corresponds to the so-called Euler limit, whereas the scaling $\kappa = \epsilon \to 0$ yields the diffusive or Navier-Stokes limit. In the corresponding limits, the macroscopic quantities fulfill equations of continuum theory up to a given order. The fluid-dynamic limits of the Boltzmann equation are examined in Ref. [21], [2], [3] and [12]. In both scalings we find $Q(f) = 0$ in the limit ϵ to zero. A distribution function f that fulfills $Q(f) = 0$ is called a *Maxwellian*.

A function $\varphi : \mathbb{R}^3 \to \mathbb{R}$ is called a *collisional invariant*, if it is a solution of

$$\int_{\mathbb{R}^3} Q(f)(\boldsymbol{v}) \varphi(\boldsymbol{v}) \, d\boldsymbol{v} = 0.$$

By a manipulation of the collision operator, one finds the necessary condition

$$\varphi(\boldsymbol{v}) + \varphi(\boldsymbol{w}) = \varphi(\boldsymbol{v}') + \varphi(\boldsymbol{w}'),$$

and φ has to be of the form

$$\varphi(\boldsymbol{v}) = a + \boldsymbol{b} \cdot \boldsymbol{v} + |\boldsymbol{v}|^2$$

for $a \in \mathbb{R}$ and $\boldsymbol{b} \in \mathbb{R}^3$. This choice corresponds to the conservation of mass, momentum and energy (Ref. [10]).

1.2. A General Discrete Velocity Model

The Boltzmann equation is a seven-dimensional integro-differential equation that is difficult to handle. For this reason, *discrete velocity models* are introduced, where the particles can only possess a finite number of velocities. The basis is a gas with identical particles. Binary particle collisions result in changes of the velocity. We follow the description in Ref. [19] and in Ref. [36].

We consider the finite velocity space

$$\mathcal{F} := \{\boldsymbol{u}_0, \ldots, \boldsymbol{u}_{p-1} \mid \boldsymbol{u}_i \in \mathbb{R}^d, \, i = 0, \ldots, p-1\}$$

with p velocities for $d \in \{1, 2, 3\}$. The corresponding distribution functions are

$$f_i := f_i(t, \boldsymbol{x}) := f(t, \boldsymbol{x}, \boldsymbol{u}_i) \quad \text{for } \boldsymbol{x} \in \mathbb{R}^d, \, t \in \mathbb{R}, \, i = 0, \ldots, p-1.$$

Other labels for the distribution functions are particle densities or populations.

For $d \in \{2, 3\}$, we only allow for binary collisions obeying conservation of momentum and energy. Let $\boldsymbol{u}_i$ and $\boldsymbol{u}_j$ be the precollisional velocities and $\boldsymbol{u}_k$ and $\boldsymbol{u}_l$ be the postcollisional velocities with $i, j, k, l \in \{0, \ldots, p-1\}$. The velocities have to obey

$$\boldsymbol{u}_i + \boldsymbol{u}_j = \boldsymbol{u}_k + \boldsymbol{u}_l, \quad \boldsymbol{u}_i^2 + \boldsymbol{u}_j^2 = \boldsymbol{u}_k^2 + \boldsymbol{u}_l^2.$$

External forces are neglected. Mechanical considerations lead to the equations for the ensemble $\boldsymbol{F} := [f_0; \ldots; f_{p-1}]$

$$\left(\frac{\partial}{\partial t} + \boldsymbol{u}_i \cdot \boldsymbol{\nabla}_{\boldsymbol{x}} \right) f_i = Q_i(\boldsymbol{F}) := \frac{1}{2} \sum_{j,k,l} p_{ij}^{kl} |\boldsymbol{u}_i - \boldsymbol{u}_j| (f_k f_l - f_i f_j) \quad \text{for } i = 0, \ldots, p-1.$$

$$(1.2)$$

The collisions are determined by the transition probabilities p_{ij}^{kl}. The analogy to the Boltzmann equation (1.1) is obvious. We rewrite (1.2) in matrix form by

$$\left(\boldsymbol{Id}\frac{\partial}{\partial t} + \boldsymbol{V} \cdot \boldsymbol{\nabla}_{\boldsymbol{x}} \right) \boldsymbol{F} = \boldsymbol{Q}(\boldsymbol{F}),$$

where $\boldsymbol{Id}$ is the unit matrix and $\boldsymbol{V}$ is a diagonal vector matrix with coefficients $\boldsymbol{V}_{ii} := \boldsymbol{u}_{i-1}$ for $i = 1, \ldots, p$. The vector $\boldsymbol{Q}(\boldsymbol{F})$ has the components $Q_{i-1}(\boldsymbol{F})$ for $i = 1, \ldots, p$.

For a function $\varphi : \mathcal{F} \to \mathbb{R}$, we define $\boldsymbol{\Phi} : \mathcal{F} \to \mathbb{R}^p$ with the components $\boldsymbol{\Phi}_i := \varphi(\boldsymbol{u}_{i-1})$ for $i = 1, \ldots, p$. For a given set of distribution functions, we define the density

$$\rho := \sum_{i=0}^{p-1} f_i$$

and the moments

$$\overline{\boldsymbol{\Phi}} := \frac{1}{\rho} \sum_{i=0}^{p-1} \boldsymbol{\Phi}_{i+1} f_i \, .$$

Multiplication of (1.2) by $\boldsymbol{\Phi}_{i+1}$ and the summation over $i = 0, \ldots, p-1$ lead to the transport equation

$$\frac{\partial}{\partial t}(\boldsymbol{\Phi}, \boldsymbol{F}) + (\boldsymbol{\Phi}, \boldsymbol{V} \cdot \boldsymbol{\nabla}_{\boldsymbol{x}} \boldsymbol{F}) = (\boldsymbol{\Phi}, \boldsymbol{Q}(\boldsymbol{F})),$$

where we employ the scalar product $(\cdot, \cdot)$ in $\mathbb{R}^p$. The right hand side can be transformed to

$$(\boldsymbol{\Phi}, \boldsymbol{Q}(\boldsymbol{F})) = \frac{1}{8} \sum_{j,k,l} p_{ij}^{kl} |\boldsymbol{u}_i - \boldsymbol{u}_j| (f_k f_l - f_i f_j)(\boldsymbol{\Phi}_{i+1} + \boldsymbol{\Phi}_{j+1} - \boldsymbol{\Phi}_{k+1} - \boldsymbol{\Phi}_{l+1}).$$

In analogy to the case with continuous velocities, we define the collisional invariants by the relation

$$\boldsymbol{\Phi}_i + \boldsymbol{\Phi}_j - \boldsymbol{\Phi}_k - \boldsymbol{\Phi}_l = 0.$$

For collisional invariants we obtain a conservation law in well-known form

$$\frac{\partial}{\partial t}(\boldsymbol{\Phi}, \boldsymbol{F}) + (\boldsymbol{\Phi}, \boldsymbol{V} \cdot \boldsymbol{\nabla}_{\boldsymbol{x}} \boldsymbol{F}) = 0.$$

Typical collisional invariants are the mass with $\varphi(\boldsymbol{u}_i) := 1$, the components of the velocity with $\varphi(\boldsymbol{u}_i) := \boldsymbol{u}_i \cdot \boldsymbol{e}_j$ for $j \in \{1, \ldots, d\}$, where $\boldsymbol{e}_j$ is the jth unit vector in $\mathbb{R}^d$, and the energy with $\varphi(\boldsymbol{u}_i) := |\boldsymbol{u}_i|^2$ for $i = 0, \ldots, p-1$.

By defining the entropy

$$H(t) := \sum_{i=0}^{p-1} f_i \log(f_i), \quad \text{if } f_i > 0 \text{ for } i = 0, \ldots, p-1,$$

one can find a stability result in the form of the *H-Theorem*, which essentially states that H is a monotone decreasing function in time, that is,

$$\frac{d}{dt}H(t) \le 0,$$

and that $H(t)$ is bounded from below; see Ref. [**19**].

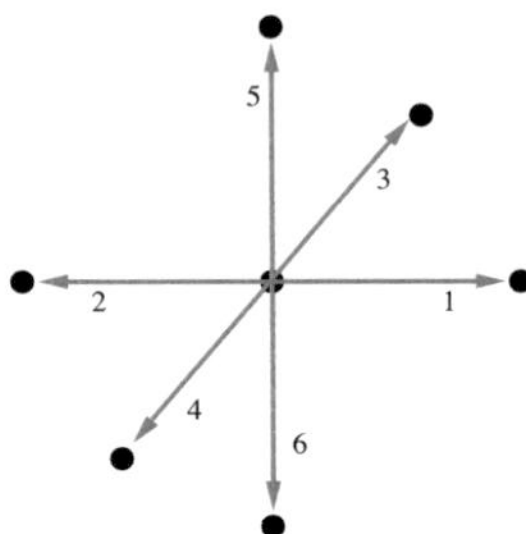

FIGURE 1.1. D3Q6-model with six velocities in three dimensions.

1.3. Examples of Discrete Velocity Models

As a first example of a three-dimensional discrete velocity model we give the *Broadwell model* (Ref. [5], [6]). The particles are allowed to move along the six directions defined by the x-, y- and z-axis according to Figure 1.1. This regular lattice is known as the D3Q6-model with six velocities in three dimensions. In unit scales, the equations of the Broadwell model read

$$\partial_t f_1 + \partial_x f_1 = \sigma(f_3 f_4 + f_5 f_6 - 2 f_1 f_2),$$
$$\partial_t f_2 - \partial_x f_2 = \sigma(f_3 f_4 + f_5 f_6 - 2 f_1 f_2),$$
$$\partial_t f_3 + \partial_y f_3 = \sigma(f_1 f_2 + f_5 f_6 - 2 f_3 f_4),$$
$$\partial_t f_4 - \partial_y f_4 = \sigma(f_1 f_2 + f_5 f_6 - 2 f_3 f_4),$$
$$\partial_t f_5 + \partial_z f_5 = \sigma(f_1 f_2 + f_3 f_4 - 2 f_5 f_6),$$
$$\partial_t f_6 - \partial_z f_6 = \sigma(f_1 f_2 + f_3 f_4 - 2 f_5 f_6),$$

where σ is the collision frequency. Furthermore, we abbreviate $\partial_t := \partial/\partial t$, $\partial_x := \partial/\partial x$, $\partial_y := \partial/\partial y$ and $\partial_z := \partial/\partial z$. We find conservation of mass, momentum and energy. By imposing the restriction that there is no dependence on the variables y and z and setting $f_3 = f_4 = f_5 = f_6$, we find the one-dimensional reduced Broadwell model with the equations

$$\partial_t f_1 + \partial_x f_1 = 2\sigma(f_3^2 - f_1 f_2),$$
$$\partial_t f_2 - \partial_x f_2 = 2\sigma(f_3^2 - f_1 f_2),$$
$$\partial_t f_3 = \sigma(f_1 f_2 - f_3^2).$$

As a consequence of the reduction, we do not have conservation of energy anymore. Conserved quantities are the density $\rho := f_1 + f_2 + 4f_3$ and the flux $j := f_1 - f_2$. The fluid-dynamic limit of the reduced Broadwell model is examined in Ref. [7].

In many applications the quadratic collision terms are replaced by linearized relaxation-type collision operators, where the deviations from an equilibrium state are considered. We want to introduce these models by means of the D2Q9-model in two dimensions with nine velocities that is displayed in Figure 1.2. We restrict to a two-dimensional square lattice, where each node has eight neighbors: a left and a right neighbor, an upper and a lower neighbor, and four neighbors that can be reached over the diagonals. Including the zero velocity case, we have to consider nine different velocities $\boldsymbol{u_0}, \ldots, \boldsymbol{u_8}$ and the corresponding distribution functions

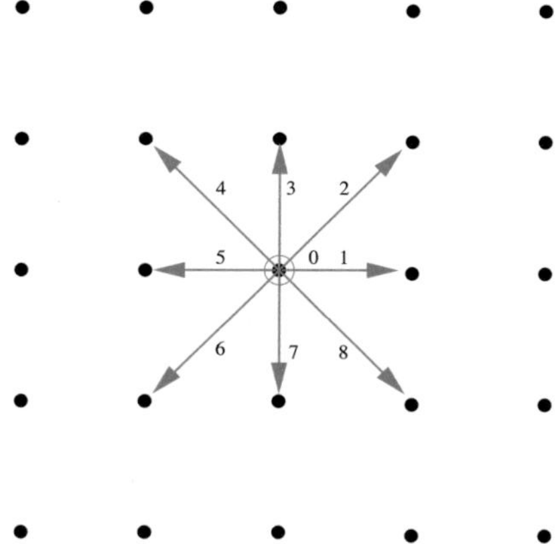

FIGURE 1.2. D2Q9-model with nine velocities in two dimensions.

$f_0, \ldots, f_8$ on this square lattice. The equations of the corresponding discrete velocity model are written as

$$\frac{\partial f_i}{\partial t} + \boldsymbol{u}_i \cdot \boldsymbol{\nabla}_{\boldsymbol{x}} f_i = \frac{1}{\tau} \sum_{j=0}^{8} A_{ij}(f_j^{\text{eq}} - f_j) \quad \text{for } i = 0, \ldots, 8. \qquad (1.3)$$

The collision matrix $\boldsymbol{A} = [A_{ij}]_{i,j=0,\ldots,8}$ is chosen in a way that the total mass and the total momentum are preserved. The collision frequency is expressed in the form $1/\tau$. Details on the collision operator and the equilibrium distributions can be found in Ref. [30]. Due to the nonlinear computation of the equilibrium distribution functions $f_j^{\text{eq}} = f_j^{\text{eq}}(f_0, \ldots, f_8)$, $j = 0, \ldots, 8$, the equations in (1.3) form a nonlinear system.

In the limit ϵ to zero we distinguish two kinds of scalings. The Euler scaling

$$\frac{\partial f_i}{\partial t} + \boldsymbol{u}_i \cdot \boldsymbol{\nabla}_{\boldsymbol{x}} f_i = \frac{1}{\epsilon} Q_i \quad \text{for } i = 0, \ldots, 8,$$

and the diffusive or Navier-Stokes scaling

$$\frac{\partial f_i}{\partial t} + \frac{\boldsymbol{u}_i}{\epsilon} \cdot \boldsymbol{\nabla}_{\boldsymbol{x}} f_i = \frac{1}{\epsilon^2} Q_i \quad \text{for } i = 0, \ldots, 8.$$

Asymptotic and multiscale expansions like the Hilbert or the Chapman-Enskog expansions are used to find the compressible and the incompressible Navier-Stokes equations in the diffusive limit (Ref. [2], [10], [24]).

The distribution functions can be described by a set of moments and the discrete velocity model can be rewritten as a set of hierarchical equations for the moments. For certain choices of the moments, the equations for the moments yield the fluid-dynamic equations up to a certain order. The number of moments has to be chosen equal to the number of discrete velocities; see Ref. [30]. Since the physical meaning of the moments of higher order is unclear in some situations, the resulting equations are cryptic to some extent.

In the following we restrict to one-dimensional discrete velocity models. The goal is to find appropriate models that cover the properties of multi-dimensional models. In one dimension we can only postulate conservation of mass and momentum. Conservation of energy in addition is impossible for these models. The two

FIGURE 1.3. D1Q2-model with two velocities in one dimension.

velocity case with $u_1 := 1$ and $u_2 := -1$ is depicted in Figure 1.3. For this specific model there is only conservation of mass. This corresponds to a nonphysical choice of the collision operator. Hence, we are treating a fictitious gas. The general one-dimensional scaled discrete velocity model reads

$$\partial_t f_1 + \partial_x f_1 = q(f_1, f_2),$$
$$\partial_t f_2 - \partial_x f_2 = -q(f_1, f_2).$$

We define the mass density $\rho := f_1 + f_2$ and the flux $j := f_1 - f_2$. Conservation of mass is given by the equation

$$\partial_t \rho + \partial_x j = 0,$$

where j turns out to be a function of ρ and $\partial_x \rho$ in the limiting case.

In the *Goldstein-Taylor model* (Ref. [20], [48]), we choose the collision term

$$q(f_1, f_2) := \sigma(f_2 - f_1).$$

This model can be seen in the framework of the relaxation-type schemes with the equilibrium distribution $f^{\text{eq}} := (f_1 + f_2)/2$. The *Carleman model* (Ref. [8]) uses

$$q(f_1, f_2) := \sigma(f_2^2 - f_1^2).$$

The *linear Ruijgrok-Wu model* (Ref. [44]) employs

$$q(f_1, f_2) := \alpha f_2 - \beta f_1 \quad \text{for } \alpha \neq \beta.$$

The *nonlinear Ruijgrok-Wu model* (Ref. [44]) uses a collision operator of the form

$$q(f_1, f_2) := \alpha f_2 - \beta f_1 + \gamma f_1 f_2.$$

An overview over the presented models and their properties can be found in Ref. [28].

The hyperbolic scaling

$$\partial_t f_1 + \partial_x f_1 = \frac{1}{\epsilon} q(f_1, f_2),$$
$$\partial_t f_2 - \partial_x f_2 = -\frac{1}{\epsilon} q(f_1, f_2).$$

in the Euler limit ϵ to zero leads to conservation laws of the form

$$\partial_t \rho + \partial_x F(\rho) = 0,$$

where F depends on the choice of the model. Macroscopic variables are the density $\rho := f_1 + f_2$ and the flux $j := f_1 - f_2$. In Ref. [18], the relation

$$j = F(\rho) = \sqrt{\rho^2 + 1} - 1$$

is proved for the nonlinear Ruijgrok-Wu model in the limit ϵ to zero for the choice $\gamma = 2\alpha = 2\beta$. Entropy bounds and BV-estimates on unbounded domains are employed to prove the convergence. One gets

$$f_1 = \frac{1}{2}(\rho + F(\rho)), \quad f_2 = \frac{1}{2}(\rho - F(\rho)).$$

For small ρ we have

$$F(\rho) \approx \frac{1}{2}\rho^2.$$

Further results can be found in Ref. [27].

More interesting are the Navier-Stokes limits of the equations

$$\partial_t f_1 + \frac{1}{\epsilon}\partial_x f_1 = \frac{1}{\epsilon^2}q(f_1, f_2),$$

$$\partial_t f_2 - \frac{1}{\epsilon}\partial_x f_2 = -\frac{1}{\epsilon^2}q(f_1, f_2)$$

in the diffusive scaling. As transformed variables we consider the density $\rho := f_1 + f_2$ and the flux $j := (f_1 - f_2)/\epsilon$, where we now have

$$f_1 = \frac{1}{2}(\rho + \epsilon j), \quad f_2 = \frac{1}{2}(\rho - \epsilon j).$$

The diffusive limit ϵ to zero gives rise to the relation

$$j = F(\rho, \partial_x \rho).$$

The Goldstein-Taylor model yields the heat equation (Ref. [**32**], [**35**], [**34**])

$$\partial_t \rho = \frac{1}{2\sigma}\partial_x^2 \rho$$

in the fluid-dynamic limit with the flux $j := -\partial_x\rho/(2\sigma)$. The Carleman model (Ref. [**32**], [**35**]) renders

$$\partial_t \rho = \frac{1}{4}\partial_x^2 \log(\rho).$$

For the linear Ruijgrok-Wu model the choice $q(f_1, f_2) = \sigma\big(f_2 - f_1 + \epsilon A(f_1 + f_2)\big)$ renders the advection-diffusion equation

$$\partial_t \rho + A\partial_x \rho = \frac{1}{2\sigma}\partial_x^2 \rho$$

with the flux $j := A\rho - \partial_x\rho/(2\sigma)$. For the nonlinear Ruijgrok-Wu model (Ref. [**44**], [**18**]) we find

$$j = \frac{1}{2}\rho^2 - \frac{1}{2\sigma}\partial_x \rho,$$

where ρ is a solution of the viscous Burgers equation

$$\partial_t \rho + \frac{1}{2}\partial_x \rho^2 = \frac{1}{2\sigma}\partial_x^2 \rho.$$

Most of the results in the literature are obtained in the absence of boundary conditions; see for example Ref. [**18**]. In Ref. [**45**] the diffusive limit for an initial-boundary value problem for the Goldstein-Taylor model is investigated without giving error estimates and precise orders of convergence.

1.4. Lattice Boltzmann Methods

The general aim of the study of lattice Boltzmann methods is the solution of incompressible Navier-Stokes like problems. The basic idea in this approach is to model the fluid flow by a simplified time evolution of microscopic particles. The particles in lattice Boltzmann models can only stay in the discrete nodes of a regular lattice. In each time step the particles can move to neighboring nodes or stay where they are. Depending on the underlying grid, only a limited number of possibilities for the movement is given. As in the case of velocity discrete Boltzmann models, the time evolution of the particles is described by the distribution functions f_i, $i = 0, \ldots, p - 1$, each one belonging to one selected velocity $\boldsymbol{u}_i$, $i = 0, \ldots, p - 1$. The simplicity of lattice Boltzmann models is constituted in the discrete nature of the particle positions, the particle velocities and the discrete time.

Macroscopic quantities are determined by an averaging of the distribution functions. The discrete density ρ and the discrete average velocity $\boldsymbol{u}$ are defined by

$$\rho := \sum_{i=0}^{p-1} f_i, \quad \boldsymbol{u} := \frac{1}{\rho} \sum_{i=0}^{p-1} f_i \boldsymbol{u}_i.$$

The time evolution consists of two steps: in the collision step there are interactions between the particles in the nodes, which result in changed velocities and a redistribution of the densities f_i, $i = 0, \ldots, p-1$. Invariant quantities like the total mass and the total momentum are unchanged by the collisions in the dimension $d = 2$ or $d = 3$. In the propagation step, the particles move to the neighboring nodes. The combination of both steps can be written in the simple form

$$f_i(t + \tau, \boldsymbol{x} + \tau \boldsymbol{u}_i) = f_i(t, \boldsymbol{x}) + \sum_{j=0}^{p-1} A_{ij}(f_j^{\text{eq}}(t, \boldsymbol{x}) - f_j(t, \boldsymbol{x})) \quad \text{for } i = 0, \ldots, p-1.$$

$$(1.4)$$

These are the *lattice Boltzmann equations* for this specific collision model. The evaluation is carried out only in the discrete points $\boldsymbol{x}$ of the lattice. The collision operator involves the collision matrix $\boldsymbol{A} = [A_{ij}]_{i,j=0,\ldots,p-1}$ and the local equilibrium distributions f_j^{eq} for $j = 0, \ldots, p-1$. In most applications the grid step and the time step are scaled to one. But this procedure somehow obscures the inherent coupling of space and time that is determined by the choice of the discrete velocities. Details for the D2Q9-model with $p = 9$ can be found in Ref. [**30**]. In the same paper and in Ref. [**29**], it is proved that the lattice Boltzmann equations are a second order finite difference approximation in space and a first order explicit Euler discretization in time for the incompressible Navier-Stokes equations.

In the historical development, the lattice Boltzmann models originate from the theory of *cellular lattice gas automata*, where only Boolean states for the distribution functions are considered (Ref. [**17**], [**16**], [**50**], [**4**]). But it is clear that the lattice Boltzmann equations (1.4) can be seen as a discretization of the discrete velocity model (1.3) with respect to the characteristic directions determined by the p velocities.

Severe problems for the lattice Boltzmann equations occur in the presence of boundary conditions. At the nodes close to the boundary it is unclear how to extract the inflow velocity distribution functions. Since the boundary conditions are expressed in terms of macroscopic quantities, one has to provide conversion formulas between the macroscopic quantities and the discrete distribution functions. In most of the models, the number of distribution functions exceeds the number of macroscopic variables. Hence, the conversion formulas do not form a closed system. Higher order moments of the distribution functions have to be involved, although their physical meaning is not clear in many situations. The most commonly used boundary conditions are of periodic type or the so-called *bounce-back* boundary conditions, where the outflow values are used to determine the inflow values. In the physical interpretation this matches the case of solid walls. Review articles of lattice Boltzmann methods can be found in Ref. [**11**] and in Ref. [**4**].

The lattice Boltzmann equations for the one-dimensional models read

$$f_1(t + \tau, x + h) = f_1(t, x) + q(f_1, f_2)(t, x),$$
$$f_2(t + \tau, x - h) = f_2(t, x) - q(f_1, f_2)(t, x).$$

Boundary values have to be provided for f_1 at the left boundary and f_2 at the right boundary.

CHAPTER 2

The Heat Equation

As pointed out in the introduction, the analysis of lattice Boltzmann methods is a challenging task that is difficult to cope with. For this reason, we restrict our considerations to a model problem that allows for a deepened analysis. In spite of the simplicity of the model problem, it preserves the most fundamental features of more complex velocity models and lattice Boltzmann schemes. We focus on the linear Goldstein-Taylor model, whose limit equation is the heat equation. At the end of this work, we present extensions of this model, where the limiting equations are the advection-diffusion equation and the nonlinear Burgers equation.

The fluid-dynamic limit equation of our linear model problem is the heat equation on a bounded interval. For the analysis of our model problem and a deep understanding of the continuous and the discrete convergence processes, we need to gain information on the behavior of the solutions of the heat equation subject to different boundary conditions.

We start with a collection of helpful definitions. Then we give an overview on the passage from the kinetic velocity model to the heat equation in the fluid-dynamic limit. For the analysis, the underlying advection system is transformed into two telegraph equations.

For the continuous and the discrete convergence processes, the regularity of the limit equation is of particular importance. We examine the necessities for the data to achieve the required regularity for the solutions of the heat equation. Four different types of boundary conditions are proposed and the reduction of the boundary values is described even for higher order derivatives. In the space-time corners specific compatibility conditions need to be fulfilled. Solutions of the heat equation are investigated in the classical as well as in the weak sense. Fourier series of the solutions in terms of orthogonal eigenfunctions are presented. A priori estimates are derived from the Fourier representations and from energy principles.

2.1. Definitions

Let $\Omega := (x_L, x_R) \subseteq \mathbb{R}$ be a bounded open domain and let $|\Omega| := x_R - x_L$ be its length. For $0 < T < \infty$, time-dependent partial differential equations are considered on the space-time domain $\Omega_T := (0, T] \times \Omega$. Let the parabolic boundary be defined by

$$\Gamma_T := \{(t, x) \in \overline{\Omega_T} : t = 0 \text{ or } x = x_L \text{ or } x = x_R\}.$$

Partial derivatives with respect to space and time are denoted by

$$\partial_x := \frac{\partial}{\partial x}, \quad \partial_x^k := \frac{\partial^k}{\partial x^k} \quad \text{for } k \in \mathbb{N},$$

$$\partial_t := \frac{\partial}{\partial t}, \quad \partial_t^k := \frac{\partial^k}{\partial t^k} \quad \text{for } k \in \mathbb{N},$$

in the classical as well as in the weak sense. The total derivative with respect to time is given by d/dt.

Classical solutions of the partial differential equations can be found in the spaces of k-times continuously differentiable functions

$$C^k(\Omega) := \left\{ u : \Omega \to \mathbb{R} \ : \ \partial_x^l u \text{ is continuous for } 0 \leq l \leq k \right\} \quad \text{for } k \in \mathbb{N}_0,$$
$$C^\infty(\Omega) := \cap_{k \in \mathbb{N}_0} C^k(\Omega).$$

In the same manner, the spaces $C^k(\overline{\Omega})$, $C^k([0,T])$ and $C^0(\Omega_T)$ are defined. In $C^k(\overline{\Omega})$, $k < \infty$, we define the norm

$$\|u\|_{C^k(\overline{\Omega})} := \max_{0 \leq l \leq k} \sup_{x \in \Omega} |\partial_x^l u(x)|.$$

The space of test functions with compact support is defined by

$$C_0^\infty(\Omega) := \left\{ u \in C^\infty(\Omega) \ : \ \overline{\{x \in \Omega \ : \ u(x) \neq 0\}} \subset \Omega \right\}.$$

For time-dependent functions, we define for $k, l \in \mathbb{N}_0$

$$C^{k,l}(\Omega_T) := \left\{ u : \Omega_T \to \mathbb{R} \ : \ u, \partial_x u, \ldots, \partial_x^k u, \partial_t u, \ldots, \partial_t^l u \in C^0(\Omega_T) \right\}.$$

For $\alpha \in (0,1)$ and $k \in \mathbb{N}_0$, we define the spaces of Hölder continuous functions

$$C^\alpha(\overline{\Omega}) := \left\{ u : \overline{\Omega} \to \mathbb{R} \ : \ \|u\|_{C^\alpha(\overline{\Omega})} < \infty \right\},$$
$$C^{k+\alpha}(\overline{\Omega}) := \left\{ u : \overline{\Omega} \to \mathbb{R} \ : \ \partial_x^l u \in C^\alpha(\overline{\Omega}) \text{ for } 0 \leq l \leq k \right\},$$

where the norms are given by

$$\|u\|_{C^\alpha(\overline{\Omega})} := \sup_{x \in \overline{\Omega}} |u(x)| + \sup_{\substack{x,x' \in \overline{\Omega} \\ x \neq x'}} \frac{|u(x) - u(x')|}{|x - x'|^\alpha},$$

$$\|u\|_{C^{k+\alpha}(\overline{\Omega})} := \sum_{l=0}^{k} \|\partial_x^l u\|_{C^\alpha(\overline{\Omega})}.$$

Similarly, we define the Hölder spaces $C^\alpha([0,T])$ and $C^{k+\alpha}([0,T])$. For time-dependent functions defined on Ω_T, we use

$$C^\alpha(\overline{\Omega_T}) := \left\{ u \in C^0(\overline{\Omega_T}) \ : \ \|u\|_\alpha < \infty \right\},$$
$$C^{k+\alpha, l+\alpha}(\overline{\Omega_T}) := \left\{ u \in C^{k,l}(\overline{\Omega_T}) \ : \ \|u\|_{k+\alpha, l+\alpha} < \infty \right\}.$$

The corresponding norms are given by

$$\|u\|_\alpha := \sup_{(t,x) \in \Omega_T} |u(t,x)| + \sup_{\substack{(t,x),(t',x') \in \Omega_T \\ (t,x) \neq (t',x')}} \frac{|u(t,x) - u(t',x')|}{\operatorname{dist}((t,x),(t',x'))^\alpha},$$

$$\|u\|_{k+\alpha, l+\alpha} := \|u\|_\alpha + \sum_{i=1}^{k} \|\partial_x^i u\|_\alpha + \sum_{i=1}^{l} \|\partial_t^i u\|_\alpha,$$

with $\operatorname{dist}((t,x),(t',x')) := \left(|t - t'| + |x - x'|^2 \right)^{1/2}$. See Ref. [15, Chapter 3.2].

The Lebesgue spaces $L^p(\Omega)$ with respect to the Lebesgue measure are defined for $1 \leq p \leq \infty$ by

$$L^p(\Omega) := \left\{ u : \Omega \to \mathbb{R} \ : \ \|u\|_{L^p(\Omega)} < \infty \right\},$$

with the norms

$$\|u\|_{L^p(\Omega)} := \left(\int_\Omega |u|^p \right)^{1/p} \quad \text{for } 1 \leq p < \infty,$$
$$\|u\|_{L^\infty(\Omega)} := \sup_{x \in \Omega} |u(x)|.$$

More precisely, $L^p(\Omega)$ is the space of classes of equivalence of Lebesgue measurable functions, satisfying the above norm definitions with respect to the equivalence relation: $u \equiv v$ if u and v are different only on a subset of Ω with Lebesgue measure zero. For $p = 2$, we get the Hilbert space $L^2(\Omega)$ endowed with the scalar product $(\cdot, \cdot)$ defined by

$$(u, v) := \int_\Omega uv \quad \text{for all } u, v \in L^2(\Omega).$$

The Lebesgue space $L^2(\Omega_T)$ and $L^2((0, T))$ are defined in an analogous way.

For the variational formulations of the partial differential equations under examination, we define the usual Sobolev spaces for $k \in \mathbb{N}_0$

$$H^k(\Omega) := \{u \in L^2(\Omega) : \partial_x^l u \in L^2(\Omega) \text{ for } 0 \leq l \leq k\},$$
$$H_0^1(\Omega) := \{u \in H^1(\Omega) : u(x_L) = u(x_R) = 0\},$$

where the derivatives $\partial_x u$ are defined in the following weak sense. A function $u \in L^2(\Omega)$ has a weak derivative $v =: \partial_x u \in L^2(\Omega)$, if

$$(u, \partial_x \psi) = -(v, \psi) \quad \text{for all } \psi \in C_0^\infty(\Omega).$$

The spaces $H^k(\Omega)$ are Hilbert spaces with respect to the scalar products

$$(u, v)_k := \sum_{l=0}^{k} |\Omega|^{2l} (\partial_x^l u, \partial_x^l v).$$

Further on, we define the symmetric bilinear forms

$$\langle u, v \rangle_k := (\partial_x^k u, \partial_x^k v) \quad \text{for all } u, v \in H^k(\Omega).$$

The associated norms and semi-norms on H^k are defined by

$$\|u\|_k := (u, u)_k^{1/2} \quad \text{for } u \in H^k(\Omega),$$
$$|u|_k := \langle u, u \rangle_k^{1/2} \quad \text{for } u \in H^k(\Omega).$$

In $H_0^1(\Omega)$, the symmetric bilinear form $\langle \cdot, \cdot \rangle_1$ is a scalar product due to Poincaré's inequality

$$\|u\|_0 \leq \frac{|\Omega|}{\pi} |u|_1 \quad \text{for all } u \in H_0^1(\Omega).$$

We note the conformity $\| \cdot \|_0 = | \cdot |_0 = \| \cdot \|_{L^2(\Omega)}$.

In a Hilbert space V with the scalar product $(\cdot, \cdot)_V$ and the associated norm $\| \cdot \|_V$, the Cauchy-Schwarz inequality

$$(u, v)_V \leq \|u\|_V \|v\|_V \quad \text{for all } u, v \in V$$

is valid. The dual space V^* of V is defined by

$$V^* := \{f : V \to \mathbb{R} : v \mapsto \langle f, v \rangle_{V^*} := f(v), f \text{ is linear and continuous}\}.$$

For $f \in V$, we can define $f^* \in V^*$ by $\langle f^*, v \rangle_{V^*} := (f, v)_V$ for $v \in V$. The norm in V^* is defined by

$$\|f\|_{V^*} := \sup_{0 \neq \varphi \in V} \frac{\langle f, \varphi \rangle_{V^*}}{\|\varphi\|_V}.$$

We get $\|f\|_{V^*} = \|f\|_V$ for $f \in V$.

When considering space-time functions $u : \Omega_T \to \mathbb{R}$, we employ the spaces

$$L^\infty((0,T); L^2(\Omega)) := \left\{ u : (0,T) \to L^2(\Omega) \; : \; \sup_{t \in (0,T)} \|u(t)\|_{L^2(\Omega)} < \infty \right\},$$

$$L^2((0,T); V) := \left\{ u : (0,T) \to V \; : \; \int_0^T \|u(t)\|_V^2 \, dt < \infty \right\},$$

$$C^0([0,T]; V) := \left\{ u : [0,T] \to V \; : \; u \text{ continuous} \right\},$$

where V is a Banach space with norm $\| \cdot \|_V$. See Ref. [22] and Ref. [40] for further details. Let V be a closed subspace of $H^1(\Omega)$ with $H_0^1(\Omega) \subseteq V \subseteq H^1(\Omega)$ and the corresponding norm $\| \cdot \|_V$. We define the space

$$H^1((0,T); V) := \left\{ u \in L^2((0,T); V) \; : \; \partial_t u \in L^2((0,T); V^*) \right\}.$$

The derivative $\partial_t u$ in the above definition is defined in the following weak sense. A function $u \in L^2((0,T); V)$ has a weak derivative $v =: \partial_t u \in L^2((0,T); V^*)$, if

$$\int_0^T u \partial_t \psi = - \int_0^T \langle v, \psi \rangle_{V^*} \quad \text{for all } \psi \in C_0^\infty(0,T).$$

The above time integral is understood as a Bochner integral over functions with values in the function space V. A scalar product in $H^1((0,T); V)$ is given by

$$\langle u, v \rangle_{H^1((0,T);V)} := \int_0^T \left((u,v)_V + \langle \partial_t u, \partial_t v \rangle_{V^*} \right) \quad \text{for all } u, v \in H^1((0,T); V).$$

For further reference, we refer to Ref. [49, IV.25] and Ref. [51, §23].

For a function $r \in C^0\left(\overline{\Omega_T}\right)$ and $N \in \mathbb{N}$, we define the discrete grid vector $\boldsymbol{R} = \boldsymbol{R}(t) := [R_l(t)]_{l=1,\ldots,N} := [r(t, x_L + lh]_{l=1,\ldots,N} \in \mathbb{R}^N$ with $h := |\Omega|/N$. Then we define the discrete L^2-norm of $\boldsymbol{R}$ by

$$\|\boldsymbol{R}(t)\|_2 := \left(\sum_{l=1}^N h |R_l(t)|^2 \right)^{1/2}.$$

In the applications we use slight changes of the definition at the boundaries.

For a prescribed time step τ, we introduce the discrete times $\{t_k\}_{k=0,\ldots,M}$ with $t_k := k\tau$ for $k = 0, \ldots, M$. With this definitions we introduce the discrete $L^\infty(L^2)$-norm by setting

$$\|\boldsymbol{R}\|_{L^\infty(L^2)} := \max_{k=0,\ldots,M} \|\boldsymbol{R}(t_k)\|_2,$$

and the discrete $L^2(L^2)$-norm by

$$\|\boldsymbol{R}\|_{L^2(L^2)} := \left(\sum_{k=0}^M \tau \|\boldsymbol{R}(t_k)\|_2^2 \right)^{1/2}.$$

We say that function f depending on a small parameter $\epsilon > 0$ is of order $\mathcal{O}(\epsilon^p)$ in the limit ϵ to zero for $p \in \mathbb{R}$, if we have

$$\lim_{\epsilon \to 0} \frac{|f|}{\epsilon^p} < \infty.$$

In the same manner, we say that a function g depending on the large value $N \gg 0$ is of the order $\mathcal{O}(N^k)$ in the limit N to infinity for $k \in \mathbb{R}$, if we have

$$\lim_{N \to \infty} \frac{|g|}{N^k} < \infty.$$

2.2. The Fluid-dynamic Limit

In the sequel, we outline the passage from the velocity discrete model to the fluid-dynamic limit equation of our model problem.

The focus of our interest lies on the approximation of the density r that is modeled by the heat equation by solutions of the velocity discrete system in the fluid-dynamic limit. In our model problem, the latter one is a linear advection system, namely the Goldstein-Taylor model. While the heat equation is the prototype of a parabolic equation, the Goldstein-Taylor model turns out to be of hyperbolic type. Hence, we are faced with a singularly perturbed limit problem.

The equations of the Goldstein-Taylor model are the starting point for our numerical discretizations. The aim is to obtain lattice Boltzmann discretizations in order to take advantage of all its nice properties that are pointed out in this work.

The heat equation with viscosity ν reads

$$\partial_t r - \nu \partial_x^2 r = f \quad \text{in } \Omega_T.$$

On a bounded interval, the heat equation has to be equipped with suited boundary conditions. We choose, for example, Dirichlet boundary conditions

$$r(\cdot, x_L) = r_L^{\mathrm{Di}}(\cdot) \qquad \text{in } (0, T),$$
$$r(\cdot, x_R) = r_R^{\mathrm{Di}}(\cdot) \qquad \text{in } (0, T).$$

Furthermore, an initial condition

$$r(0, \cdot) = r_0(\cdot) \quad \text{in } \Omega$$

has to be supplied for the time-dependent problem.

The given data f, r_L^{Di}, r_R^{Di} and r_0 are assumed to be smooth. In addition, the data have to fulfill compatibility conditions to ensure sufficient regularity of the solution. The data for the heat equation are then used as data for the Goldstein-Taylor model. The equations of the Goldstein-Taylor model in the diffusive scaling depend on the additional small parameter $\epsilon > 0$. We have to examine a hyperbolic system of the form

$$\partial_t u^\epsilon + \frac{1}{\epsilon} \partial_x u^\epsilon + \frac{1}{2\nu\epsilon^2}(u^\epsilon - v^\epsilon) = \frac{1}{2}f - \frac{\nu\epsilon}{2}\partial_x f \qquad \text{in } \Omega_T,$$
$$\partial_t v^\epsilon - \frac{1}{\epsilon} \partial_x v^\epsilon - \frac{1}{2\nu\epsilon^2}(u^\epsilon - v^\epsilon) = \frac{1}{2}f + \frac{\nu\epsilon}{2}\partial_x f \qquad \text{in } \Omega_T,$$

together with the boundary conditions

$$u^\epsilon(\cdot, x_L) + v^\epsilon(\cdot, x_L) = r_L^{\mathrm{Di}}(\cdot) \qquad \text{in } (0, T),$$
$$u^\epsilon(\cdot, x_R) + v^\epsilon(\cdot, x_R) = r_R^{\mathrm{Di}}(\cdot) \qquad \text{in } (0, T),$$

and the initial conditions

$$u^\epsilon(0, \cdot) = \frac{1}{2}r_0(\cdot) - \frac{\nu\epsilon}{2}\partial_x r_0(\cdot) \quad \text{in } \Omega,$$
$$v^\epsilon(0, \cdot) = \frac{1}{2}r_0(\cdot) + \frac{\nu\epsilon}{2}\partial_x r_0(\cdot) \quad \text{in } \Omega.$$

The density r, modeled by the heat equation, is approximated by

$$r^\epsilon := u^\epsilon + v^\epsilon,$$

when ϵ approaches zero.

By defining the flux $j^\epsilon := u^\epsilon - v^\epsilon$, the equations of the Goldstein-Taylor model can be written as

$$\partial_t r^\epsilon + \frac{1}{\epsilon}\partial_x j^\epsilon = f \qquad \text{in } \Omega_T,$$

$$\partial_t j^\epsilon + \frac{1}{\epsilon}\partial_x r^\epsilon + \frac{1}{\nu\epsilon^2}j^\epsilon = -\nu\epsilon\partial_x f \qquad \text{in } \Omega_T.$$

These equations, together with the given initial and boundary conditions, can be decoupled and rewritten as two independent telegraph equations for the density $r^\epsilon := u^\epsilon + v^\epsilon$

$$\nu\epsilon^2\partial_t^2 r^\epsilon + \partial_t r^\epsilon - \nu\partial_x^2 r^\epsilon = f + \nu\epsilon^2\partial_t f + \nu^2\epsilon^2\partial_x^2 f \quad \text{in } \Omega_T,$$

$$r^\epsilon(\cdot, x_L) = r_L^{\text{Di}}(\cdot) \qquad\qquad\qquad \text{in } (0,T),$$

$$r^\epsilon(\cdot, x_R) = r_R^{\text{Di}}(\cdot) \qquad\qquad\qquad \text{in } (0,T),$$

$$r^\epsilon(0, \cdot) = r_0(\cdot) \qquad\qquad\qquad\quad \text{in } \Omega,$$

$$\partial_t r^\epsilon(0, \cdot) = \nu\partial_x^2 r_0(\cdot) + f(0, \cdot) \qquad\quad \text{in } \Omega,$$

and for the scaled flux $k^\epsilon := j^\epsilon/\epsilon$

$$\nu\epsilon^2\partial_t^2 k^\epsilon + \partial_t k^\epsilon - \nu\partial_x^2 k^\epsilon = -\nu\partial_x f - \nu^2\epsilon^2\partial_t\partial_x f \quad \text{in } \Omega_T,$$

$$\partial_x k^\epsilon(\cdot, x_L) = -\partial_t r_L^{\text{Di}}(\cdot) + f(\cdot, x_L) \qquad \text{in } (0,T),$$

$$\partial_x k^\epsilon(\cdot, x_R) = -\partial_t r_R^{\text{Di}}(\cdot) + f(\cdot, x_R) \qquad \text{in } (0,T),$$

$$k^\epsilon(0, \cdot) = -\nu\partial_x r_0(\cdot) \qquad\qquad\qquad \text{in } \Omega,$$

$$\partial_t k^\epsilon(0, \cdot) = -\nu\partial_x f(0, \cdot) \qquad\qquad\quad \text{in } \Omega.$$

Initial conditions for $\partial_t r^\epsilon$ and $\partial_t \kappa^\epsilon$ have to be supplied due to the hyperbolic type of the telegraph equations. The additional boundary conditions of Neumann type for the scaled flux are gained from the equations of the advection system that are extended onto the boundary. If the solutions of the advection system shall be approximations of the solutions of the heat equation, we have to postulate $\kappa^\epsilon \approx -\nu\partial_x r^\epsilon$. With this knowledge we can complete the missing initial conditions.

In the form of the telegraph equations, it is easy to discover the singularly perturbed nature of the Goldstein-Taylor model. In the limit ϵ to zero, we get two heat equations for the density and the scaled flux, endowed with Dirichlet and Neumann boundary conditions, respectively. The hyperbolic type of the differential equations changes in the limit ϵ to zero to parabolic type. At the same time, we are faced with too many initial conditions.

The telegraph equations are completely solvable in terms of Fourier series, as well as the heat equation. The solutions for the advection system are then given by the backward transformations

$$u^\epsilon = \frac{1}{2}(r^\epsilon + \epsilon k^\epsilon), \quad v^\epsilon = \frac{1}{2}(r^\epsilon - \epsilon k^\epsilon).$$

In the limit $\epsilon \to 0$, we prove

$$r^\epsilon \to r, \qquad \partial_x r^\epsilon \to \partial_x r \qquad \text{with order } \epsilon^2,$$

$$k^\epsilon \to -\nu\partial_x r, \quad \partial_x k^\epsilon \to -\nu\partial_x^2 r \quad \text{with order } \epsilon$$

in $L^\infty((0,T); L^2(\Omega))$ and

$$k^\epsilon \to -\nu\partial_x r, \quad \partial_x k^\epsilon \to -\nu\partial_x^2 r \quad \text{with order } \epsilon^2$$

in $L^2((0,T); L^2(\Omega))$

The advection system in the variables u^ϵ and v^ϵ is the starting point for the numerical discretizations. The discretizations of the differential operators in advection form

$$\partial_t \pm a\partial_x \quad \text{with } a := \frac{1}{\epsilon}$$

lead to lattice Boltzmann schemes. In the lattice Boltzmann discretizations the grid size h is linked to the parameter ϵ. This fact expresses the parabolic limit of our discretizations.

2.3. The Heat Equation

In this section we examine the solutions of the heat equation on a bounded interval subject to Dirichlet, Neumann, Robin or periodic boundary conditions. The results on the solution properties are necessary for the ongoing convergence proofs.

The heat equation with viscosity ν and a source term f is given by

$$Lr := \partial_t r - \nu\partial_x^2 r = f \quad \text{in } \Omega_T. \tag{2.1}$$

The initial condition is

$$r(0, \cdot) = r_0(\cdot) \qquad \text{in } \Omega. \tag{2.2}$$

Four different types of boundary conditions are possible in our context. We can provide Dirichlet values

$$\begin{aligned}
r(\cdot, x_L) &= r_L^{\mathrm{Di}}(\cdot) &&\text{in } (0, T), \\
r(\cdot, x_R) &= r_R^{\mathrm{Di}}(\cdot) &&\text{in } (0, T),
\end{aligned} \tag{2.3}$$

Neumann values

$$\begin{aligned}
\partial_x r(\cdot, x_L) &= r_L^{\mathrm{Neu}}(\cdot) &&\text{in } (0, T), \\
\partial_x r(\cdot, x_R) &= r_R^{\mathrm{Neu}}(\cdot) &&\text{in } (0, T),
\end{aligned} \tag{2.4}$$

or a specific choice of Robin boundary values

$$\begin{aligned}
\frac{1}{2}r(\cdot, x_L) - \frac{\vartheta}{2}\partial_x r(\cdot, x_L) &= r_L^{\mathrm{Rob}}(\cdot) &&\text{in } (0, T), \\
\frac{1}{2}r(\cdot, x_R) + \frac{\vartheta}{2}\partial_x r(\cdot, x_R) &= r_R^{\mathrm{Rob}}(\cdot) &&\text{in } (0, T),
\end{aligned} \tag{2.5}$$

where $\vartheta > 0$ is chosen as an additional parameter that has to be determined later. For the continuous approximation problem, ϑ is linked to the parameter ϵ of the Goldstein-Taylor model and to the viscosity of the heat equation. For the approximation of the discrete lattice Boltzmann solutions, ϑ depends on the grid size. In both cases, ϑ approaches zero in the limiting case. For the stability considerations we have to be aware of this property.

In the periodic case we assume periodic initial data r_0 as well as a periodic source term f. Periodic boundary conditions are then chosen by

$$r(\cdot, x_L) = r(\cdot, x_R) \quad \text{in } (0, T). \tag{2.6}$$

Globally continuous classical solutions of the heat equation (2.1) subject to Dirichlet boundary conditions (2.3) can only exist if the compatibility conditions

$$r_L^{\mathrm{Di}}(0) = r_0(x_L), \quad r_R^{\mathrm{Di}}(0) = r_0(x_R) \tag{2.7}$$

for the data are fulfilled. In the Neumann case,

$$r_L^{\mathrm{Neu}}(0) = \partial_x r_0(x_L), \quad r_R^{\mathrm{Neu}}(0) = \partial_x r_0(x_R) \tag{2.8}$$

has to be postulated for the regularity up to the boundary. The necessary compatibility conditions in the Robin case are

$$r_L^{\text{Rob}}(0) = \frac{1}{2}r_0(x_L) - \frac{\vartheta}{2}\partial_x r_0(x_L), \quad r_R^{\text{Rob}}(0) = \frac{1}{2}r_0(x_R) + \frac{\vartheta}{2}\partial_x r_0(x_R). \qquad (2.9)$$

Furthermore, in the Dirichlet case the validity of the heat equation in the corner of the parabolic boundary may be required, that is,

$$\partial_t r_L^{\text{Di}}(0) - \nu\partial_x^2 r_0(x_L) = f(0, x_L), \quad \partial_t r_R^{\text{Di}}(0) - \nu\partial_x^2 r_0(x_R) = f(0, x_R), \qquad (2.10)$$

if the differentiability of the solution at the boundary is a prerequisite. For higher regularity of the classical solutions at the boundary,

$$\partial_t^k r_L^{\text{Di}}(0) - \nu^k\partial_x^{2k} r_0(x_L) = \sum_{l=0}^{k-1} \nu^{k-l-1}\partial_t^l\partial_x^{2(k-l-1)} f(0, x_L), \quad k = 2, 3, \ldots,$$

$$\partial_t^k r_R^{\text{Di}}(0) - \nu^k\partial_x^{2k} r_0(x_R) = \sum_{l=0}^{k-1} \nu^{k-l-1}\partial_t^l\partial_x^{2(k-l-1)} f(0, x_R), \quad k = 2, 3, \ldots, \tag{2.11}$$

have to hold. In the Neumann case

$$\partial_t r_L^{\text{Neu}}(0) - \nu\partial_x^3 r_0(x_L) = \partial_x f(0, x_L), \quad \partial_t r_R^{\text{Neu}}(0) - \nu\partial_x^3 r_0(x_R) = \partial_x f(0, x_R) \tag{2.12}$$

and the corresponding extensions to higher orders (as in (2.11)) might be necessary. In the Robin case

$$\partial_t r_L^{\text{Rob}}(0) - \frac{\nu}{2}\left(\partial_x^2 r_0(x_L) - \vartheta\partial_x^3 r_0(x_L)\right) = \frac{1}{2}\left(f(0, x_L) - \vartheta\partial_x f(0, x_L)\right),$$

$$\partial_t r_R^{\text{Rob}}(0) - \frac{\nu}{2}\left(\partial_x^2 r_0(x_R) + \vartheta\partial_x^3 r_0(x_R)\right) = \frac{1}{2}\left(f(0, x_R) + \vartheta\partial_x f(0, x_R)\right)$$

and extensions for higher orders may be imposed.

If some of the preceding compatibility conditions are not fulfilled, classical solutions of the heat equation exist in the interior of the space-time domain due to the smoothing property of the parabolic operator. But then, singularities in the space-time corners may occur.

2.4. Existence and Regularity of the Solutions

In a first step we consider classical solutions of the initial-boundary value problem for the heat equation. We call $r \in C^{2,1}(\Omega_T) \cap C^0\left(\overline{\Omega_T}\right)$ a classical solution to the Dirichlet problem for the heat equation, if it solves (2.1), (2.2) and (2.3).

For the parabolic operator L, we have the weak maximum principle for classical solutions:

Theorem 2.1. *If $Lr \leq 0$ for $r \in C^{2,1}(\Omega_T) \cap C^0\left(\overline{\Omega_T}\right)$ holds in Ω_T, then the maximum of r is achieved at the parabolic boundary, that is,*

$$\max_{(t,x)\in\overline{\Omega_T}} r(t, x) = \max_{(t,x)\in\Gamma_T} r(t, x).$$

PROOF. See Ref. [**15**, Theorem 2.2.6]. $\square$

The assertion of Theorem 2.1 can be strengthened. If we have $Lr < 0$ in Ω_T, then the maximum is achieved only at the parabolic boundary.

The weak maximum principle ensures uniqueness of the solution. It also allows to give statements on the positivity of the solutions. For non-negative initial data r_0, non-negative boundary data r_L^{Di} and r_R^{Di} and for a non-negative source term f, non-negativity of the solution follows by the weak maximum principle. In this case, the non-negative minimum is achieved on the parabolic boundary. The same result follows in the Neumann case, if we assume $r_L^{\mathrm{Neu}} \geq 0$ and $r_R^{\mathrm{Neu}} \leq 0$ (confer Ref. [**33**, Problem 8.14]).

In the Dirichlet case we define

$$r_\Gamma^{\mathrm{Di}}(t,x) := r_0(x) + \frac{x-x_L}{|\Omega|}(r_R^{\mathrm{Di}}(t) - r_R^{\mathrm{Di}}(0)) + \frac{x_R-x}{|\Omega|}(r_L^{\mathrm{Di}}(t) - r_L^{\mathrm{Di}}(0)),$$

which interpolates to the initial-boundary conditions, if the compatibility conditions (2.7) are fulfilled. Then we have the following existence result:

Theorem 2.2. *Let $f \in C^\alpha\left(\overline{\Omega_T}\right)$ and $r_\Gamma^{\mathrm{Di}} \in C^{2+\alpha,1+\alpha}\left(\overline{\Omega_T}\right)$ and assume the compatibility conditions (2.10) to hold true. Then the Dirichlet problem (2.1), (2.2) and (2.3) for the heat equation has a unique solution $r \in C^{2+\alpha,1+\alpha}\left(\overline{\Omega_T}\right)$. Further on, we have the a priori estimate*

$$\|r\|_{2+\alpha,1+\alpha} \leq C(\|r_\Gamma^{\mathrm{Di}}\|_{2+\alpha,1+\alpha} + \|f\|_\alpha).$$

PROOF. See Ref. [**15**, Theorems 3.3.6 and 3.3.7]. $\square$

The regularity assumption on r_Γ^{Di} in Theorem 2.2 imposes the compatibility conditions (2.7). With respect to the assumptions of Theorem 2.2, the heat equation (2.1) is valid up to the boundary. Additional smoothness of the initial-boundary data directly carries over to additional smoothness of the solution with respect to the validity of the compatibility conditions (2.11); see Ref. [**15**, Theorems 3.4.10 - 3.4.12].

If the regularity of the initial-boundary data is omitted or the compatibility conditions are not fulfilled, classical solutions still exist in the interior of the domain. However, the boundedness of the derivatives at the boundaries does not hold any more in general; see also Ref. [**15**, Theorems 3.3.5 and 3.4.9].

Weak solutions for the Dirichlet problem are sought for by considering the variational formulation. We define the symmetric bilinear form

$$a(u,v) := \nu \langle u,v \rangle_1 \quad \text{for all } u,v \in H^1(\Omega). \tag{2.13}$$

In the Dirichlet case the boundary conditions for the density r can be reduced by subtracting

$$r_{LR}^{\mathrm{Di}}(t,x) := \frac{x-x_L}{|\Omega|} r_R^{\mathrm{Di}}(t) + \frac{x_R-x}{|\Omega|} r_L^{\mathrm{Di}}(t). \tag{2.14}$$

If the compatibility conditions (2.7) are fulfilled, we get vanishing boundary conditions for $r_0^{\mathrm{Di}} := r_0 - r_{LR}^{\mathrm{Di}}(0,\cdot)$. We are then considering the solution $r^{\mathrm{Di}} := r - r_{LR}^{\mathrm{Di}}$ of the heat equation with vanishing Dirichlet values and right hand side $f_{LR}^{\mathrm{Di}} := f - Lr_{LR}^{\mathrm{Di}}$. Hence, we seek a solution $r^{\mathrm{Di}} \in L^2((0,T); H_0^1(\Omega)) \cap C^0([0,T]; L^2(\Omega))$ of the heat equation such that

$$\lim_{t \to 0} \|r^{\mathrm{Di}}(t,\cdot) - r_0^{\mathrm{Di}}\|_0 = 0$$

and

$$\frac{d}{dt}(r^{\mathrm{Di}}(t,\cdot),v) + a(r^{\mathrm{Di}}(t,\cdot),v) = (f_{LR}^{\mathrm{Di}}(t,\cdot),v) \quad \text{for all } v \in H_0^1(\Omega). \tag{2.15}$$

The derivative with respect to the time is meant in the weak sense. Now we have the following weak existence result:

Theorem 2.3. *For $r_0^{\mathrm{Di}} \in H_0^1(\Omega)$, $f \in L^2(\Omega_T)$ and $r_L^{\mathrm{Di}}, r_R^{\mathrm{Di}} \in H^1([0,T])$ there is a unique solution $r = r^{\mathrm{Di}} + r_{LR}^{\mathrm{Di}}$ to the Dirichlet problem (2.1), (2.2), (2.3) for the heat equation, where r_{LR}^{Di} is given by (2.14) and $r^{\mathrm{Di}} \in L^2((0,T); H_0^1(\Omega)) \cap C^0([0,T]; L^2(\Omega))$ is a weak solution in the sense of (2.15).*

PROOF. See Ref. [**41**, Theorem 7.2.1]. $\square$

The condition $r_0^{\mathrm{Di}} \in H_0^1(\Omega)$ requires the compatibility condition (2.7). Apart from Theorem 2.2, the compatibility assumption (2.10) is not required in Theorem 2.3. The regularity of the right hand side f can even be relaxed to $f \in L^2((0,T), H^{-1}(\Omega))$, where $H^{-1}(\Omega)$ is the dual space of $H_0^1(\Omega)$. Since we have the embedding $H^k(\Omega) \subseteq C^{k-1,\alpha}\left(\overline{\Omega}\right)$ for $0 < \alpha \leq 1/2$ (see Ref. [1, Theorem 8.13]), the solution in Theorem 2.3 is continuous. Especially, we have

$$|f(x) - f(x')| = \left| \int_{x'}^{x} f' \right| \leq |x - x'|^{1/2}|f|_1 \quad \text{for } f \in H^1(\Omega).$$

The proof of Theorem 2.3 uses the fact that there exists an orthogonal Hilbert space basis $\left\{\phi_k^{\mathrm{Di}}\right\}_{k \geq 1}$ of $H_0^1(\Omega)$, consisting of eigenfunctions of the associated stationary eigenvalue problem

$$a(\phi_k^{\mathrm{Di}}, v) = \lambda_k(\phi_k^{\mathrm{Di}}, v) \quad \text{for all } v \in H_0^1(\Omega).$$

The normalized eigenfunctions are given by

$$\phi_k^{\mathrm{Di}}(x) := \sqrt{\frac{2}{|\Omega|}} \sin\left(\frac{k\pi}{|\Omega|}(x - x_L)\right) \quad \text{for } k = 1, 2, \ldots \tag{2.16}$$

and the corresponding eigenvalues are

$$\lambda_k := \nu \frac{k^2 \pi^2}{|\Omega|^2} \quad \text{for } k = 1, 2, \ldots . \tag{2.17}$$

We have orthogonality of the eigenfunctions in $L^2(\Omega)$ as well as in $H_0^1(\Omega)$, that is,

$$\langle \phi_l^{\mathrm{Di}}, \phi_k^{\mathrm{Di}} \rangle_1 = (\phi_l^{\mathrm{Di}}, \phi_k^{\mathrm{Di}}) = 0 \quad \text{for } l \neq k.$$

The solution r^{Di} in the sense of Theorem 2.3 can be written as a Fourier series

$$r^{\mathrm{Di}}(t,x) = \sum_{k=1}^{\infty} \left((r_0^{\mathrm{Di}}, \phi_k^{\mathrm{Di}})e^{-\lambda_k t} + \int_0^t (f_{LR}^{\mathrm{Di}}(s,\cdot), \phi_k^{\mathrm{Di}})e^{-\lambda_k(t-s)}\, ds \right) \phi_k^{\mathrm{Di}}(x). \tag{2.18}$$

For the derivation of further a priori estimates for the solution of the heat equation subject to Dirichlet boundary conditions, we need vanishing boundary values for higher order derivatives. The procedure is as follows. The boundary conditions for r and $\partial_x^2 r$ can be reduced at the same time by subtracting

$$r_{LR}^{\mathrm{Di}}(t,x) := \frac{x - x_L}{|\Omega|}r_R^{\mathrm{Di}}(t) + \frac{x_R - x}{|\Omega|}r_L^{\mathrm{Di}}(t)$$

$$- \frac{(x - x_L)(x_R - x)}{6\nu|\Omega|}(x - x_L + |\Omega|)(\partial_t r_R^{\mathrm{Di}}(t) - f(t, x_R)) \tag{2.19}$$

$$- \frac{(x - x_L)(x_R - x)}{6\nu|\Omega|}(x_R - x + |\Omega|)(\partial_t r_L^{\mathrm{Di}}(t) - f(t, x_L))$$

for smooth right hand side f and smooth r_L^{Di}, r_R^{Di}. We obtain vanishing boundary values for r, $\partial_x^2 r$ and for the modified right hand side $f_{LR}^{\mathrm{Di}} := f - L r_{LR}^{\mathrm{Di}}$. Vanishing boundary values for $\partial_x^2 r_0^{\mathrm{Di}}$ require (2.10). By evaluating the Fourier series (2.18), we get the following a priori estimates:

Corollary 2.4. *With respect to the assumptions of Theorem 2.3, with r_{LR}^{Di} given in (2.19), we gain the a priori estimates*

$$\|r^{\mathrm{Di}}(t,\cdot)\|_0^2 \le 2\|r_0^{\mathrm{Di}}\|_0^2 e^{-2\lambda_1 t} + \frac{1}{\lambda_1}\int_0^t \|f_{LR}^{\mathrm{Di}}(s,\cdot)\|_0^2\,ds, \tag{2.20}$$

$$\nu|r^{\mathrm{Di}}(t,\cdot)|_1^2 \le 2\nu|r_0^{\mathrm{Di}}|_1^2 e^{-2\lambda_1 t} + \int_0^t \|f_{LR}^{\mathrm{Di}}(s,\cdot)\|_0^2\,ds, \tag{2.21}$$

$$\nu^2|r^{\mathrm{Di}}(t,\cdot)|_2^2 \le 2\nu^2|r_0^{\mathrm{Di}}|_2^2 e^{-2\lambda_1 t} + \nu\int_0^t |f_{LR}^{\mathrm{Di}}(s,\cdot)|_1^2\,ds, \tag{2.22}$$

where the last estimate requires $r_0^{\mathrm{Di}} \in H_0^1(\Omega) \cap H^2(\Omega)$ and $f_{LR}^{\mathrm{Di}} \in L^2((0,T); H_0^1(\Omega))$. If we have in addition the validity of (2.10), we get

$$\nu^3|r^{\mathrm{Di}}(t,\cdot)|_3^2 \le 2\nu^3|r_0^{\mathrm{Di}}|_3^2 e^{-2\lambda_1 t} + \nu^2\int_0^t |f_{LR}^{\mathrm{Di}}(s,\cdot)|_2^2\,ds \tag{2.23}$$

for $\partial_x^2 r_0^{\mathrm{Di}} \in H_0^1(\Omega)$ and $f_{LR}^{\mathrm{Di}} \in L^2((0,T); H_0^1(\Omega) \cap H^2(\Omega))$. Furthermore, we have

$$\|\partial_t r^{\mathrm{Di}}(t,\cdot)\|_0 \le \nu|r^{\mathrm{Di}}(t,\cdot)|_2 + \|f_{LR}^{\mathrm{Di}}(t,\cdot)\|_0. \tag{2.24}$$

If the initial data r_0^{Di} is less regular, that is, $r_0^{\mathrm{Di}} \in L^2(\Omega)$ only, then there is a H^1-solution for $t > 0$ only. For $t \to 0$ there may be an initial layer. In particular, we find

$$\nu|r^{\mathrm{Di}}(t,\cdot)|_1^2 \le \frac{1}{te}\|r_0^{\mathrm{Di}}\|_0^2 + \int_0^t \|f_{LR}^{\mathrm{Di}}(s,\cdot)\|_0^2\,ds \qquad \text{for } t > 0,$$

$$\nu^2|r^{\mathrm{Di}}(t,\cdot)|_2^2 \le \frac{2}{(te)^2}\|r_0^{\mathrm{Di}}\|_0^2 + \nu\int_0^t |f_{LR}^{\mathrm{Di}}(s,\cdot)|_1^2\,ds \quad \text{for } t > 0. \tag{2.25}$$

PROOF. Use

$$\|r^{\mathrm{Di}}(t,\cdot)\|_0^2 \le 2\sum_{k=1}^{\infty}(r_0^{\mathrm{Di}},\phi_k^{\mathrm{Di}})^2 e^{-2\lambda_k t}$$

$$+ 2\sum_{k=1}^{\infty}\left(\int_0^t (f_{LR}^{\mathrm{Di}}(s,\cdot),\phi_k^{\mathrm{Di}})^2\,ds\right)\left(\int_0^t e^{-2\lambda_k(t-s)}\,ds\right)$$

$$\le 2e^{-2\lambda_1 t}\sum_{k=1}^{\infty}(r_0^{\mathrm{Di}},\phi_k^{\mathrm{Di}})^2$$

$$+ \frac{1}{\lambda_1}\int_0^t \sum_{k=1}^{\infty}(f_{LR}^{\mathrm{Di}}(s,\cdot),\phi_k^{\mathrm{Di}})^2\,ds,$$

$$\nu^l|r^{\mathrm{Di}}(t,\cdot)|_l^2 \le 2\sum_{k=1}^{\infty}(r_0^{\mathrm{Di}},\phi_k^{\mathrm{Di}})^2 \lambda_k^l e^{-2\lambda_k t}$$

$$+ 2\sum_{k=1}^{\infty}\left(\int_0^t \lambda_k^{l-1}(f_{LR}^{\mathrm{Di}}(s,\cdot),\phi_k^{\mathrm{Di}})^2\,ds\right)\left(\int_0^t \lambda_k e^{-2\lambda_k(t-s)}\,ds\right),$$

and $x^l e^{-x} \le (l/e)^l$ for $x \ge 0$ and $l \in \mathbb{N}$. $\qquad\square$

We made use of

Lemma 2.5. *For $r \in L^2(\Omega)$ we have*

$$\|r\|_0^2 = \sum_{k=1}^{\infty}(r,\phi_k^{\mathrm{Di}})^2.$$

We have $r \in H_0^1(\Omega)$, if and only if

$$\nu |r|_1^2 = \sum_{k=1}^{\infty} \lambda_k (r, \phi_k^{\mathrm{Di}})^2 < \infty.$$

Furthermore, we have $r \in H_0^1(\Omega) \cap H^2(\Omega)$ if and only if

$$\nu^2 |r|_2^2 = \sum_{k=1}^{\infty} \lambda_k^2 (r, \phi_k^{\mathrm{Di}})^2 < \infty.$$

PROOF. See Ref. [**33**, Theorem 6.4]. $\qquad\qquad\qquad\qquad\qquad\qquad$ $\square$

Estimates for higher order derivatives can be achieved in the following way. Define

$$r_{2k}^{\mathrm{Di}} := \partial_x^{2k} r - r_{LR,2k}^{\mathrm{Di}} \quad \text{for } k \geq 1,$$

where $r_{LR,2k}^{\mathrm{Di}}$ interpolates the boundary conditions of $\partial_x^{2k} r$ and $\partial_x^{2k+2} r$ instead of r and $\partial_x^2 r$ like in (2.19). The data are obtained from the differentiated heat equations $\partial_t^{k-1} Lr = \partial_t^{k-1} f$ and $\partial_t^k Lr = \partial_t^k f$. Now r_{2k}^{Di} has vanishing boundary values and it solves the heat equation with right hand side $f_{LR,2k}^{\mathrm{Di}} := \partial_x^{2k} f - Lr_{LR,2k}^{\mathrm{Di}}$, which also has vanishing boundary values. Hence, we can apply the estimates (2.20) - (2.23) in Corollary 2.4 to r_{2k}^{Di} and $f_{LR,2k}^{\mathrm{Di}}$ in connection with the assumption of the compatibility conditions in (2.11).

Next, we take a look at the Neumann and the Robin problem for the heat equation. We call $r \in C^{2,1}(\Omega_T) \cap C^1 (\overline{\Omega_T})$ a classical solution to the Neumann problem or the Robin problem for the heat equation, if it solves (2.1), (2.2) and (2.4) or (2.5), respectively.

The boundary conditions are reduced by subtracting

$$r_{LR}^{\mathrm{Neu}}(t, x) := \frac{(x - x_L)^2}{2|\Omega|} r_R^{\mathrm{Neu}}(t) - \frac{(x_R - x)^2}{2|\Omega|} r_L^{\mathrm{Neu}}(t) \qquad (2.26)$$

in the Neumann case and

$$r_{LR}^{\mathrm{Rob}}(t, x) := \frac{x - x_L + \vartheta}{|\Omega| + 2\vartheta} r_R^{\mathrm{Rob}}(t) + \frac{x_R - x + \vartheta}{|\Omega| + 2\vartheta} r_L^{\mathrm{Rob}}(t) \qquad (2.27)$$

in the Robin case. The variational formulation now reads

$$\frac{d}{dt}(r(t, \cdot), v) + a(r(t, \cdot), v) = (f_{LR}(t, \cdot), v) \quad \text{for all } v \in H^1(\Omega). \qquad (2.28)$$

In the Neumann case we take $a(\cdot, \cdot)$ as in (2.13). In the Robin case we define

$$a(u, v) := \nu \langle u, v \rangle_1 + \frac{\nu}{\vartheta} \left(u(x_L) v(x_L) + u(x_R) v(x_R) \right) \quad \text{for all } u, v \in H^1(\Omega). \qquad (2.29)$$

The right hand side is defined as $f_{LR} := f_{LR}^{\mathrm{Neu}} := f - Lr_{LR}^{\mathrm{Neu}}$ or $f_{LR} := f_{LR}^{\mathrm{Rob}} := f - Lr_{LR}^{\mathrm{Rob}}$, respectively. The initial condition translates to

$$\lim_{t \to 0} \| r(t, \cdot) - r_0^{\mathrm{BC}} \|_0 = 0$$

with $r_0^{\mathrm{BC}} := r_0^{\mathrm{Neu}} := r_0 - r_{LR}^{\mathrm{Neu}}(0, \cdot)$ in the Neumann case and $r_0^{\mathrm{BC}} := r_0^{\mathrm{Rob}} := r_0 - r_{LR}^{\mathrm{Rob}}(0, \cdot)$ in the Robin case.

In the Neumann case the orthogonal eigenfunctions $\left\{ \phi_k^{\mathrm{Neu}} \right\}_{k \geq 0}$, which now form a Hilbert basis in $H^1(\Omega)$, are given by

$$\phi_k^{\mathrm{Neu}}(x) := \sqrt{\frac{2}{|\Omega|}} \cos \left(\frac{k\pi}{|\Omega|}(x - x_L) \right) \quad \text{for } k = 0, 1, \ldots \qquad (2.30)$$

with the same eigenvalues as in (2.17). In the Robin case, the eigenfunctions $\{\phi_k^{\mathrm{Rob}}\}_{k\geq 1}$, which also form a Hilbert basis in $H^1(\Omega)$, are given by

$$\phi_k^{\mathrm{Rob}}(x) := \sqrt{c_k}\,\vartheta\mu_k \cos\left(\mu_k(x-x_L)\right) + \sqrt{c_k}\sin\left(\mu_k(x-x_L)\right) \quad \text{for } k = 1, 2, \ldots, \tag{2.31}$$

where μ_k are the positive solutions of the transcendental equation

$$\tan(\mu_k|\Omega|) = \frac{2\vartheta\mu_k}{\vartheta^2\mu_k^2 - 1}$$

and c_k is chosen by

$$c_k := \frac{2}{|\Omega| + \vartheta + \vartheta^2\mu_k^2|\Omega|}.$$

We find

$$(k-1)\frac{\pi}{|\Omega|} < \mu_k < k\frac{\pi}{|\Omega|} \quad \text{for } k = 1, 2, \ldots,$$

$$\lim_{\vartheta\to 0}\mu_k = k\frac{\pi}{|\Omega|} \quad \text{for a fixed } k,$$

$$\lim_{k\to\infty}\left(\mu_k - (k-1)\frac{\pi}{|\Omega|}\right) = 0 \quad \text{for fixed } \vartheta.$$

See also Ref. [47, Chapter 4.3]. The eigenvalues of the corresponding stationary problem are given by $\nu\mu_k^2$. In the Neumann and in the Robin case we have the orthogonality of the eigenfunctions in $L^2(\Omega)$ and in $H^1(\Omega)$. In the H^1-case, orthogonality has to be considered with respect to the scalar products $(\cdot,\cdot)_1$ for the Neumann case and $a(\cdot,\cdot)$ as in (2.29) for the Robin case, respectively. Now we have the following existence results:

Theorem 2.6. *For $r_0^{\mathrm{Neu}} \in H^1(\Omega)$, $f \in L^2(\Omega_T)$ and $r_L^{\mathrm{Neu}}, r_R^{\mathrm{Neu}} \in H^1([0,T])$ there is a unique solution $r = r^{\mathrm{Neu}} + r_{LR}^{\mathrm{Neu}}$ to the Neumann problem (2.1), (2.2), (2.4) for the heat equation, where r_{LR}^{Neu} is given by (2.26) and $r^{\mathrm{Neu}} \in L^2((0,T); H^1(\Omega)) \cap C^0([0,T]; L^2(\Omega))$ is a weak solution in the sense of (2.28) with $a(\cdot,\cdot)$ given by (2.13).*

PROOF. See Ref. [41, Theorem 7.2.1]. $\qquad\square$

Furthermore, we have

Theorem 2.7. *For $r_0^{\mathrm{Rob}} \in H^1(\Omega)$, $f \in L^2(\Omega_T)$ and $r_L^{\mathrm{Rob}}, r_R^{\mathrm{Rob}} \in H^1([0,T])$ there is a unique solution $r = r^{\mathrm{Rob}} + r_{LR}^{\mathrm{Rob}}$ to the Robin problem (2.1), (2.2), (2.5) for the heat equation, where r_{LR}^{Rob} is given by (2.27) and $r^{\mathrm{Rob}} \in L^2((0,T); H^1(\Omega)) \cap C^0([0,T]; L^2(\Omega))$ is a weak solution in the sense of (2.28) with $a(\cdot,\cdot)$ given by (2.29).*

PROOF. See Ref. [41, Theorem 7.2.1]. $\qquad\square$

The solutions r^{Neu} and r^{Rob} in the sense of the Theorems 2.6 and 2.7 can be written as Fourier series

$$r^{\mathrm{Neu}}(t,x) = \frac{1}{2}\left((r_0^{\mathrm{Neu}}, \phi_0^{\mathrm{Neu}}) + \int_0^t (f_{LR}^{\mathrm{Neu}}(s,\cdot), \phi_0^{\mathrm{Neu}})\,ds\right)\phi_0^{\mathrm{Neu}}(x)$$

$$+ \sum_{k=1}^{\infty}\left((r_0^{\mathrm{Neu}}, \phi_k^{\mathrm{Neu}})e^{-\lambda_k t} + \int_0^t (f_{LR}^{\mathrm{Neu}}(s,\cdot), \phi_k^{\mathrm{Neu}})e^{-\lambda_k(t-s)}\,ds\right)\phi_k^{\mathrm{Neu}}(x),$$

$$r^{\mathrm{Rob}}(t,x) = \sum_{k=1}^{\infty}\left((r_0^{\mathrm{Rob}}, \phi_k^{\mathrm{Rob}})e^{-\nu\mu_k^2 t} + \int_0^t (f_{LR}^{\mathrm{Rob}}(s,\cdot), \phi_k^{\mathrm{Rob}})e^{-\nu\mu_k^2(t-s)}\,ds\right)\phi_k^{\mathrm{Rob}}(x).$$

From the Fourier representations of the solutions for the Neumann problem we find

Corollary 2.8. *With respect to the assumptions of Theorem 2.6, we have*

$$\|r^{\mathrm{Neu}}(t,\cdot)\|_0^2 \le 2\|r_0^{\mathrm{Neu}}\|_0^2 + \max\left\{2t, \frac{1}{\lambda_1}\right\} \int_0^t \|f_{LR}^{\mathrm{Neu}}(s,\cdot)\|_0^2\, ds, \tag{2.32}$$

$$\nu|r^{\mathrm{Neu}}(t,\cdot)|_1^2 \le 2\nu|r_0^{\mathrm{Neu}}|_1^2 e^{-2\lambda_1 t} + \int_0^t \|f_{LR}^{\mathrm{Neu}}(s,\cdot)\|_0^2\, ds. \tag{2.33}$$

If we have the compatibility condition (2.8), *then we get* $\partial_x r_0^{\mathrm{Neu}} \in H_0^1(\Omega)$. *If the right hand side is in* $L^2(0,T;H^1(\Omega))$, *then the solution is in* $H^2(\Omega)$, *provided* $r_0^{\mathrm{Neu}} \in H^2(\Omega)$. *We have*

$$\nu^2|r^{\mathrm{Neu}}(t,\cdot)|_2^2 \le 2\nu^2|r_0^{\mathrm{Neu}}|_2^2 + \nu\int_0^t |f_{LR}^{\mathrm{Neu}}(s,\cdot)|_1^2\, ds. \tag{2.34}$$

Furthermore, the estimates (2.24) *and* (2.25) *are valid.*

Note that there is no exponential decay in $L^2(\Omega)$ with respect to the initial data as in the Dirichlet case. In Corollary 2.8 we employed

Lemma 2.9. *For* $r \in L^2(\Omega)$ *we have*

$$\|r\|_0^2 = \frac{1}{2}(r, \phi_0^{\mathrm{Neu}})^2 + \sum_{k=1}^{\infty}(r, \phi_k^{\mathrm{Neu}})^2.$$

Furthermore, we have $r \in H^1(\Omega)$, *if and only if*

$$\nu|r|_1^2 = \sum_{k=1}^{\infty} \lambda_k(r, \phi_k^{\mathrm{Neu}})^2 < \infty.$$

We have $r \in H^2(\Omega)$, $\partial_x r \in H_0^1(\Omega)$, *if and only if*

$$\nu^2|r|_2^2 = \sum_{k=1}^{\infty} \lambda_k^2(r, \phi_k^{\mathrm{Neu}})^2 < \infty.$$

Estimates for higher order derivatives can be achieved as follows. Define

$$r_{2k}^{\mathrm{Neu}} := \partial_x^{2k} r - r_{LR,2k}^{\mathrm{Neu}} \quad \text{for } k \ge 1,$$

where $r_{LR,2k}^{\mathrm{Neu}}$ interpolates the boundary conditions of $\partial_x^{2k+1} r$ like in (2.26) by using the known data of the differentiated heat equation $\partial_x \partial_t^{k-1} Lr = \partial_x \partial_t^{k-1} f$. Especially, we use

$$r_{LR,2}^{\mathrm{Neu}}(t,x) = \frac{(x-x_L)^2}{2|\Omega|}\frac{1}{\nu}\left(\partial_x f(t,x_R) - \partial_t r_R^{\mathrm{Neu}}(t)\right)$$

$$- \frac{(x_R-x)^2}{2|\Omega|}\frac{1}{\nu}\left(\partial_x f(t,x_L) - \partial_t r_L^{\mathrm{Neu}}(t)\right).$$

Since now r_{2k}^{Neu} solves the heat equation with right hand side $\partial_x^{2k} f - Lr_{LR,2k}^{\mathrm{Neu}}$, we can apply (2.32) (with exponential decay with respect to the initial data and constant λ_1^{-1} instead of $\max\{2t, \lambda_1^{-1}\}$, confer (2.20)) and (2.33) of Corollary 2.8. With respect to the compatibility conditions (2.12) for $k = 1$ and its higher order extensions for $k \ge 2$, we have vanishing boundary conditions for $\partial_x r_{2k}^{\mathrm{Neu}}(0,\cdot)$ and hence we can apply (2.34).

In the Robin case we do not have the orthogonality of the eigenfunctions with respect to $\langle\cdot,\cdot\rangle_1$ (which only induces a semi-norm here), but we do have orthogonality with respect to $(\cdot,\cdot)$, $a(\cdot,\cdot)$ and $\langle\cdot,\cdot\rangle_2$. More precisely, we get

Lemma 2.10. *For $r \in L^2(\Omega)$ we have*

$$\|r\|_0^2 = \sum_{k=1}^{\infty} (r, \phi_k^{\mathrm{Rob}})^2.$$

For $r \in H^1(\Omega)$ we get

$$\nu|r|_1^2 \leq a(r, r) = \sum_{k=1}^{\infty} \nu\mu_k^2 (r, \phi_k^{\mathrm{Rob}})^2 < \infty,$$

where the equal sign holds for $r \in H_0^1(\Omega)$. Furthermore, we have $r \in H^2(\Omega)$ with $(r - \vartheta\partial_x r)(x_L) = 0$ and $(r + \vartheta\partial_x r)(x_R) = 0$, if and only if

$$\nu^2|r|_2^2 = \sum_{k=1}^{\infty} (\nu\mu_k^2)^2 (r, \phi_k^{\mathrm{Rob}})^2 < \infty.$$

From the Fourier representations of the solutions for the Robin problem we find

Corollary 2.11. *With respect to the assumptions of Theorem 2.7, we have*

$$\|r^{\mathrm{Rob}}(t, \cdot)\|_0^2 \leq 2\|r_0^{\mathrm{Rob}}\|_0^2 e^{-2\nu\mu_1^2 t} + \frac{1}{\nu\mu_1^2} \int_0^t \|f_{LR}^{\mathrm{Rob}}(s, \cdot)\|_0^2 \, ds,$$

$$\nu|r^{\mathrm{Rob}}(t, \cdot)|_1^2 \leq 2a(r_0^{\mathrm{Rob}}, r_0^{\mathrm{Rob}})e^{-2\nu\mu_1^2 t} + \int_0^t \|f_{LR}^{\mathrm{Rob}}(s, \cdot)\|_0^2 \, ds.$$

In the case of the compatibility conditions (2.9), we have $(r_0^{\mathrm{Rob}} - \vartheta\partial_x r_0^{\mathrm{Rob}})(x_L) = 0$, $(r_0^{\mathrm{Rob}} + \vartheta\partial_x r_0^{\mathrm{Rob}})(x_R) = 0$. For $r_0^{\mathrm{Rob}} \in H^2(\Omega)$ and $f_{LR}^{\mathrm{Rob}} \in L^2((0, T); H^1(\Omega))$ we get

$$\nu^2|r^{\mathrm{Rob}}(t, \cdot)|_2^2 \leq 2|r_0^{\mathrm{Rob}}|_2^2 e^{-2\nu\mu_1^2 t} + \int_0^t a(f_{LR}^{\mathrm{Rob}}, f_{LR}^{\mathrm{Rob}})(s, \cdot) \, ds.$$

Furthermore, the estimates (2.24) and (2.25) are valid.

Estimates for higher order derivatives can be derived by applying similar procedures as described before.

In the periodic case we seek for solutions in

$$H_{\mathrm{P}}^1(\Omega) := \{u \in H^1(\Omega) : u(x_L) = u(x_R)\}.$$

The variational formulation is given by

$$\frac{d}{dt}(r^{\mathrm{P}}(t, \cdot), v) + a(r^{\mathrm{P}}(t, \cdot), v) = (f(t, \cdot), v) \quad \text{for all } v \in H_{\mathrm{P}}^1(\Omega), \tag{2.35}$$

with $a(\cdot, \cdot)$ as in (2.13). A Hilbert basis in $H_{\mathrm{P}}^1(\Omega)$, consisting of orthogonal eigenfunctions, is obtained by $\{\phi_k^{\mathrm{P}}\}_{k \geq 0} \cup \{\psi_k^{\mathrm{P}}\}_{k \geq 1}$ with

$$\phi_k^{\mathrm{P}}(x) := \sqrt{\frac{2}{|\Omega|}} \cos\left(\frac{2k\pi}{|\Omega|}(x - x_L)\right) \quad \text{for } k = 0, 1, \ldots,$$

$$\psi_k^{\mathrm{P}}(x) := \sqrt{\frac{2}{|\Omega|}} \sin\left(\frac{2k\pi}{|\Omega|}(x - x_L)\right) \quad \text{for } k = 1, 2, \ldots.$$

The corresponding eigenvalues are given by $4\lambda_k$ with λ_k as in (2.17).

Theorem 2.12. *For $r_0 \in H_{\mathrm{P}}^1(\Omega)$ and $f \in L^2(\Omega_T)$ there is a unique solution $r^{\mathrm{P}} \in L^2((0, T); H_{\mathrm{P}}^1(\Omega)) \cap C^0([0, T]; L^2(\Omega))$ to the periodic problem (2.1), (2.2) and (2.6) for the heat equation in the sense of (2.35) with $a(\cdot, \cdot)$ given by (2.13).*

PROOF. Since we have $H_0^1(\Omega) \subseteq H_{\mathrm{P}}^1(\Omega) \subseteq H^1(\Omega)$ we can use Ref. [**41**, Theorem 7.2.1]. $\square$

The Fourier series of the periodic solution reads

$$r^{\mathrm{P}}(t,x) = \frac{1}{2}\left((r_0,\phi_0^{\mathrm{P}}) + \int_0^t (f(s,\cdot),\phi_0^{\mathrm{P}})\,ds\right)\phi_0^{\mathrm{P}}(x)$$

$$+ \sum_{k=1}^{\infty}\left((r_0,\phi_k^{\mathrm{P}})e^{-4\lambda_k t} + \int_0^t (f(s,\cdot),\phi_k^{\mathrm{P}})e^{-4\lambda_k(t-s)}\,ds\right)\phi_k^{\mathrm{P}}(x)$$

$$+ \sum_{k=1}^{\infty}\left((r_0,\psi_k^{\mathrm{P}})e^{-4\lambda_k t} + \int_0^t (f(s,\cdot),\psi_k^{\mathrm{P}})e^{-4\lambda_k(t-s)}\,ds\right)\psi_k^{\mathrm{P}}(x).$$

No we use

Lemma 2.13. *For $r \in L^2(\Omega)$ we have*

$$\|r\|_0^2 = \frac{1}{2}(r,\phi_0^{\mathrm{P}})^2 + \sum_{k=1}^{\infty}\left((r,\phi_k^{\mathrm{P}})^2 + (r,\psi_k^{\mathrm{P}})^2\right).$$

Furthermore, we define for $l \geq 2$

$$H_{\mathrm{P}}^l(\Omega) := \left\{ r \in L^2(\Omega) \,:\, \nu^l|r|_l^2 = \sum_{k=1}^{\infty}(4\lambda_k)^l\left((r,\phi_k^{\mathrm{P}})^2 + (r,\psi_k^{\mathrm{P}})^2\right) < \infty \right\}.$$

Due to the uniform convergence, we have $\partial_x^{l-1}r(\cdot,x_L) = \partial_x^{l-1}r(\cdot,x_R)$ for $r \in H_{\mathrm{P}}^l(\Omega)$ and $l \geq 1$. In the periodic case there are no compatibility conditions for higher order regularity. We find

Corollary 2.14. *With respect to the assumptions of Theorem 2.12, we have*

$$\|r^{\mathrm{P}}(t,\cdot)\|_0^2 \leq 2\|r_0\|_0^2 + \max\left\{2t,\frac{1}{4\lambda_1}\right\}\int_0^t \|f(s,\cdot)\|_0^2\,ds,$$

$$\nu|r^{\mathrm{P}}(t,\cdot)|_1^2 \leq 2\nu|r_0|_1^2 e^{-2\lambda_1 t} + \int_0^t \|f(s,\cdot)\|_0^2\,ds.$$

If for $l \geq 2$ the right hand side is in $L^2(0,T;H_{\mathrm{P}}^{l-1}(\Omega))$, then the solution is in $H_{\mathrm{P}}^l(\Omega)$, provided $r_0 \in H_{\mathrm{P}}^l(\Omega)$. We have

$$\nu^l|r^{\mathrm{P}}(t,\cdot)|_l^2 \leq 2\nu^l|r_0|_l^2 + \nu^{l-1}\int_0^t |f(s,\cdot)|_{l-1}^2\,ds.$$

The total mass

$$m(t) := \int_{\Omega} r(t,\cdot)$$

in the domain Ω changes in time only due to a flux through the boundaries or the influence of the source term. We have

$$\frac{d}{dt}m(t) = \nu\big(\partial_x r(t,x_R) - \partial_x r(t,x_L)\big) + \int_{\Omega} f(t,\cdot). \qquad (2.36)$$

For vanishing right hand side, the total mass is constant in the periodic case or for vanishing Neumann boundary conditions. In the case of vanishing Robin boundary conditions, the total mass decreases provided that r is positive at the boundary.

2.5. Energy Estimates

The energy of a solution r of the heat equation is defined by

$$E(t) := \|r(t,\cdot)\|_0^2 + 2\nu\int_0^t |r(s,\cdot)|_1^2\,ds.$$

For vanishing Dirichlet boundary conditions and vanishing right hand side, we find the typical decay of energy, that is,

$$E(t) \leq E(0) \quad \text{for } t \geq 0.$$

More detailed, we prove the following results.

Theorem 2.15. *Let* $r = r^{\mathrm{Di}}$, $r = r^{\mathrm{Neu}}$ *or* $r = r^{\mathrm{P}}$ *be the solutions as in Theorem 2.3, 2.6 or 2.12. Then we have the a priori estimate*

$$\|r(T, \cdot)\|_0^2 + \frac{1}{T} \int_0^T \|r(s, \cdot)\|_0^2 \, ds + 2\nu \int_0^T |r(s, \cdot)|_1^2 \leq 4\|r_0\|_0^2 + 13T \int_0^T \|f_{LR}\|_0^2 \tag{2.37}$$

with $r_0 = r_0^{\mathrm{Di}}$, $r_0 = r_0^{\mathrm{Neu}}$ *or* $r_0 = r_0$ *and* $f_{LR} = f_{LR}^{\mathrm{Di}}$, $f_{LR} = f_{LR}^{\mathrm{Neu}}$ *or* $f_{LR} = f$. *If* $r = r^{\mathrm{Rob}}$ *as in Theorem 2.7, then the term*

$$\frac{2\nu}{\vartheta} \int_0^T \left(r^2(s, x_L) + r^2(s, x_R) \right) ds$$

can be added on the left hand side of (2.37). *Here,* $\vartheta > 0$ *is the small parameter provided by the Robin boundary conditions* (2.5). *Choose* $r_0 = r_0^{\mathrm{Rob}}$ *and* $f_{LR} = f_{LR}^{\mathrm{Rob}}$ *in this case. In the Dirichlet case with* $r = r^{\mathrm{Di}}$, *we get in addition*

$$\|r^{\mathrm{Di}}(T, \cdot)\|_0^2 + \frac{\nu \pi^2}{2|\Omega|^2} \int_0^T \|r^{\mathrm{Di}}(s, \cdot)\|_0^2 \, ds + \frac{\nu}{2} \int_0^T |r^{\mathrm{Di}}(s, \cdot)|_1^2$$

$$\leq \|r_0^{\mathrm{Di}}\|_0^2 + \frac{|\Omega|^2}{\nu \pi^2} \int_0^T \|f_{LR}^{\mathrm{Di}}\|_0^2. \tag{2.38}$$

PROOF. Multiplying the heat equation by r yields

$$\frac{1}{2} \partial_t r^2 - \nu r \partial_x^2 r = r f_{LR}.$$

Integration along Ω gives

$$\frac{d}{dt} \|r\|_0^2 + 2\nu \|\partial_x r\|_0^2 - 2\nu \left[r \partial_x r \right]_{x_L}^{x_R} \leq 2\|r\|_0 \|f_{LR}\|_0.$$

The boundary term cancels in the Dirichlet, Neumann and periodic case. For Robin boundary conditions, we get

$$-2\nu \left[r \partial_x r \right]_{x_L}^{x_R} = \frac{2\nu}{\vartheta} \left(r^2(\cdot, x_L) + r^2(\cdot, x_R) \right).$$

Integration over $[0, T]$ yields

$$\|r(T)\|_0^2 + 2\nu \int_0^T |r|_1^2 \leq \|r_0\|_0^2 + 2 \int_0^T \|r\|_0 \|f_{LR}\|_0. \tag{2.39}$$

From

$$\int_0^T \|r\|_0^2 \leq T\|r_0\|_0^2 + 2T \left(\int_0^T \|r\|_0^2 \right)^{1/2} \left(\int_0^T \|f_{LR}\|_0^2 \right)^{1/2}$$

we get

$$\left(\int_0^T \|r\|_0^2 \right)^{1/2} \leq \sqrt{T}\|r_0\|_0 + 2T \left(\int_0^T \|f_{LR}\|_0^2 \right)^{1/2}.$$

Hence, we end up with

$$\|r(T)\|_0^2 + 2\nu \int_0^T |r|_1^2 \leq 2\|r_0\|_0^2 + 5T \int_0^T \|f_{LR}\|_0^2.$$

By the combination of the obtained results, the assertion (2.37) follows. Using Poincaré's inequality in the Dirichlet case on the right hand side of (2.39), we obtain

$$2 \int_0^T \|r\|_0 \|f_{LR}\|_0 \le 2 \frac{|\Omega|}{\pi} \int_0^T |r|_1 \|f_{LR}\|_0 \le \nu \int_0^T |r|_1^2 + \frac{|\Omega|^2}{\nu \pi^2} \int_0^T \|f_{LR}\|_0^2.$$

Absorbing yields

$$\|r(T)\|_0^2 + \nu \int_0^T |r|_1^2 \le \|r_0\|_0^2 + \frac{|\Omega|^2}{\nu \pi^2} \int_0^T \|f_{LR}\|_0^2.$$

If we use Poincaré's inequality on the left hand side of (2.39), we find

$$\|r(T)\|_0^2 + 2 \frac{\nu \pi^2}{|\Omega|^2} \int_0^T \|r\|_0^2 \le \|r_0\|_0^2 + \frac{|\Omega|^2}{\nu \pi^2} \int_0^T \|f_{LR}\|_0^2 + \frac{\nu \pi^2}{|\Omega|^2} \int_0^T \|r\|_0^2.$$

By collecting the previous two results, the assertion (2.38) follows. $\square$

In the Robin case the boundary conditions contribute to the decay of the energy by the additional term on the left hand side of the above estimate. For this reason, Robin boundary conditions are called dissipative.

Theorem 2.16. *Let $r = r^{\mathrm{Neu}}$ or $r = r^{\mathrm{P}}$ be the solutions as in Theorem 2.6 or 2.12. Then we have the a priori estimate*

$$|r(T, \cdot)|_1^2 + \frac{1}{T} \int_0^T |r(s, \cdot)|_1^2 \, ds + 2\nu \int_0^T |r(s, \cdot)|_2^2$$

$$\le 4|r_0|_1^2 + 13T \int_0^T |f_{LR}|_1^2.$$

The initial data r_0 and the right hand side f_{LR} has to be chosen according to the underlying problem. In the Neumann case with $r = r^{\mathrm{Neu}}$ we get in addition

$$|r^{\mathrm{Neu}}(T, \cdot)|_1^2 + \frac{\nu \pi^2}{2|\Omega|^2} \int_0^T |r^{\mathrm{Neu}}(s, \cdot)|_1^2 \, ds + \frac{\nu}{2} \int_0^T |r^{\mathrm{Neu}}(s, \cdot)|_2^2$$

$$\le |r_0^{\mathrm{Neu}}|_1^2 + \frac{|\Omega|^2}{\nu \pi^2} \int_0^T |f_{LR}^{\mathrm{Neu}}|_1^2.$$

PROOF. Differentiate the heat equation with respect to x and multiply by $\partial_x r$. Then proceed as in the proof of Theorem 2.15. $\square$

Furthermore, we get

Theorem 2.17. *Let $r = r^{\mathrm{Di}}$, $r = r^{\mathrm{Neu}}$ or $r = r^{\mathrm{P}}$ be the solutions as in Theorem 2.3, 2.6 or 2.12. Then we have the a priori estimate*

$$\nu |r(T, \cdot)|_1^2 + \int_0^T \|\partial_t r(s, \cdot)\|_0^2 \, ds + \nu^2 \int_0^T |r(s, \cdot)|_2^2 \, ds \le 3\nu |r_0|_1^2 + 5 \int_0^T \|f_{LR}\|_0^2.$$

If $r = r^{\mathrm{Rob}}$ as in Theorem 2.7, then the terms

$$\frac{\nu}{\vartheta} \left(r^2(T, x_L) + r^2(T, x_R) \right), \quad \frac{\nu}{\vartheta} \left(r_0^2(x_L) + r_0^2(x_R) \right)$$

have to be added to the left and right hand side of the estimate. The initial data r_0 and the right hand side f_{LR} has to be chosen according to the underlying problem.

PROOF. Multiplying the heat equation by $\partial_t r$ yields

$$|\partial_t r|^2 - \nu \partial_t r \partial_x^2 r = \partial_t r f_{LR}.$$

Integration along Ω gives

$$2\|\partial_t r\|_0^2 + \nu \partial_t \|\partial_x r\|_0^2 - 2\nu \left[\partial_t r \partial_x r \right]_{x_L}^{x_R} \le 2\|\partial_t r\|_0 \|f_{LR}\|_0.$$

The boundary term cancels in the Dirichlet, Neumann and periodic case. For Robin boundary conditions we get

$$-2\nu\left[\partial_t r\partial_x r\right]_{x_L}^{x_R} = \frac{\nu}{\vartheta}\partial_t\left(r^2(\cdot,x_L) + r^2(\cdot,x_R)\right).$$

Integration over $[0,T]$ yields

$$\nu|r(T,\cdot)|_1^2 + 2\int_0^T \|\partial_t r(t,\cdot)\|_0^2\,dt \le \nu|r_0|_1^2 + 2\int_0^T \|\partial_t r\|_0\|f_{LR}\|_0.$$

Now use

$$\nu^2|r|_2^2 \le 2\|\partial_t r\|_0^2 + 2\|f_{LR}\|_0^2.$$

$\square$

Theorem 2.18. *Let* $r = r^{\mathrm{Di}}$, $r = r^{\mathrm{Neu}}$ *or* $r = r^{\mathrm{P}}$ *be the solutions as in Theorem 2.3, 2.6 or 2.12. Then we have the estimates*

$$T\|r(T,\cdot)\|_0^2 + 2\nu\int_0^T t|r|_1^2 \le 2\int_0^T \|r\|_0^2 + \int_0^T t^2\|f_{LR}\|_0^2, \tag{2.40}$$

$$\nu T|r(T)|_1^2 + \int_0^T t\|\partial_t r\|_0^2 \le \nu\int_0^T |r|_1^2 + \int_0^T t\|f_{LR}\|_0^2. \tag{2.41}$$

In the Dirichlet case we find in addition

$$T\|r^{\mathrm{Di}}(T,\cdot)\|_0^2 + \nu\int_0^T t|r^{\mathrm{Di}}|_1^2 \le \int_0^T \|r^{\mathrm{Di}}\|_0^2 + \frac{|\Omega|^2}{\nu\pi^2}\int_0^T t\|f_{LR}^{\mathrm{Di}}\|_0^2.$$

If $r = r^{\mathrm{Rob}}$ *as in Theorem 2.7, then the term*

$$\frac{2\nu}{\vartheta}\int_0^T s\left(r^2(s,x_L) + r^2(s,x_R)\right)\,ds$$

can be added on the left hand side of (2.40) *and*

$$\frac{\nu}{\vartheta}T\left(r^2(T,x_L) + r^2(T,x_R)\right), \quad \frac{\nu}{\vartheta}\int_0^T \left(r^2(s,x_L) + r^2(s,x_R)\right)\,ds$$

have to be added on the left and right hand side of (2.41), *respectively.*

PROOF. Upon multiplying the heat equation (2.1) by tr and integrating, we get

$$t\frac{d}{dt}\|r\|_0^2 + 2\nu t\|\partial_x r\|_0^2 - 2\nu t\left[r\partial_x r\right]_{x_L}^{x_R} \le 2t\|r\|_0\|f_{LR}\|_0.$$

Upon multiplying the heat equation by $t\partial_t r$ and integrating, the ensuing expression yields

$$\nu t\frac{d}{dt}\|\partial_x r\|_0^2 + 2t\|\partial_t r\|_0^2 - 2\nu t\left[\partial_t r\partial_x r\right]_{x_L}^{x_R} \le 2t\|\partial_t r\|_0\|f_{LR}\|_0.$$

$\square$

CHAPTER 3

The Goldstein-Taylor Model

In this chapter, we study the velocity discrete system for our linear model problem. This a linear advection system, known as the Goldstein-Taylor model. The purpose of our investigation is to approximate the solutions of the heat equation by solutions of the advection system under consideration. Variables in the advection system are the distribution functions u^ϵ and v^ϵ, which replace the notation f_1 and f_2 from Section 1.3 in order to avoid the lower indices. The upper index expresses the dependence on the singular perturbation parameter in the diffusive scaling. The distribution functions refer to particles traveling into the positive and the negative x-direction as depicted in Figure 1.3. The density r, modeled by the heat equation, is approximated by the macroscopic quantity $r^\epsilon := u^\epsilon + v^\epsilon$, while the flux $\partial_x r$ is approximated by $-(u^\epsilon - v^\epsilon)/(\nu\epsilon)$, where ν is the viscosity of the heat equation.

The advection system of the Goldstein-Taylor model is the starting point for the discretizations performed in Chapter 5. The objective is to achieve lattice Boltzmann discretizations by exploiting the derivatives with respect to the characteristic directions.

In the following sections we describe how the boundary conditions for the heat equation and the corresponding initial data and the source terms have to be interpreted for the advection system in order to ensure the convergence of the solutions. For the a priori estimates of the solutions, the boundary data of the advection system have to be reduced. By employing energy principles, we prove a priori estimates for the solutions of the advection system. These estimates lead to a rigorous convergence result in terms of the perturbation parameter ϵ.

3.1. The Advection System in the Diffusion Scaling

For a given viscosity ν and the small singular perturbation parameter $\epsilon > 0$ we consider the Goldstein-Taylor model in the diffusive scaling with source terms f^ϵ and g^ϵ. This is a hyperbolic advection system with right hand side in directions $[1; 1]$ and $[1; -1]$, which reads

$$\partial_t u^\epsilon + \frac{1}{\epsilon}\partial_x u^\epsilon + \frac{1}{2\nu\epsilon^2}(u^\epsilon - v^\epsilon) = \frac{1}{2}(f^\epsilon + g^\epsilon) \qquad \text{in } \Omega_T,$$

$$\partial_t v^\epsilon - \frac{1}{\epsilon}\partial_x v^\epsilon - \frac{1}{2\nu\epsilon^2}(u^\epsilon - v^\epsilon) = \frac{1}{2}(f^\epsilon - g^\epsilon) \qquad \text{in } \Omega_T.$$

$$(3.1)$$

The characteristics of these equations are

$$X^+(t, x) := x + at, \quad X^-(t, x) := x - at, \quad \text{with } a := \frac{1}{\epsilon}.$$

By defining the density $r^\epsilon := u^\epsilon + v^\epsilon$ and the flux $j^\epsilon := u^\epsilon - v^\epsilon$, the advection system (3.1) transforms to

$$\partial_t r^\epsilon + \frac{1}{\epsilon}\partial_x j^\epsilon = f^\epsilon \qquad \text{in } \Omega_T,$$

$$\partial_t j^\epsilon + \frac{1}{\epsilon}\partial_x r^\epsilon + \frac{1}{\nu\epsilon^2}j^\epsilon = g^\epsilon \qquad \text{in } \Omega_T.$$

$$(3.2)$$

29

Initial conditions for the advection system (3.1) are supplied by

$$u^\epsilon(0,\cdot) = u_0^\epsilon(\cdot) \quad \text{in } \Omega,$$
$$v^\epsilon(0,\cdot) = v_0^\epsilon(\cdot) \quad \text{in } \Omega.$$

Boundary conditions for the advection system (3.1) are provided at the inflow boundaries, that is,

$$u^\epsilon(\cdot, x_L) = u_L^\epsilon(\cdot) \quad \text{in } (0,T),$$
$$v^\epsilon(\cdot, x_R) = v_R^\epsilon(\cdot) \quad \text{in } (0,T).$$

In the sequel, we are interested in approximations of solutions of the heat equation. For this reason we use the data of the heat equation for the advection system. Let r_L^{Di}, r_R^{Di} r_L^{Neu}, r_R^{Neu}, r_L^{Rob} and r_R^{Rob} be the boundary data for the heat equation belonging to the Dirichlet, Neumann or Robin problem. The following choices for the boundary conditions are possible for the advection system. In the Dirichlet case (2.3) we choose density boundary conditions

$$u_L^\epsilon(\cdot) := r_L^{\mathrm{Di}}(\cdot) - v^\epsilon(\cdot, x_L) \quad \text{in } (0,T),$$
$$v_R^\epsilon(\cdot) := r_R^{\mathrm{Di}}(\cdot) - u^\epsilon(\cdot, x_R) \quad \text{in } (0,T). \tag{3.3}$$

In the Neumann case (2.4) we supply flux boundary conditions

$$u_L^\epsilon(\cdot) := v^\epsilon(\cdot, x_L) - \nu\epsilon r_L^{\mathrm{Neu}}(\cdot) \quad \text{in } (0,T),$$
$$v_R^\epsilon(\cdot) := u^\epsilon(\cdot, x_R) + \nu\epsilon r_R^{\mathrm{Neu}}(\cdot) \quad \text{in } (0,T). \tag{3.4}$$

In the Robin case (2.5) we can define inflow boundary conditions

$$u_L^\epsilon(\cdot) := r_L^{\mathrm{Rob}}(\cdot) \quad \text{in } (0,T),$$
$$v_R^\epsilon(\cdot) := r_R^{\mathrm{Rob}}(\cdot) \quad \text{in } (0,T). \tag{3.5}$$

In the periodic case (2.6) we use

$$u_L^\epsilon(\cdot) := u^\epsilon(\cdot, x_R) \quad \text{in } (0,T),$$
$$v_R^\epsilon(\cdot) := v^\epsilon(\cdot, x_L) \quad \text{in } (0,T). \tag{3.6}$$

Let r_0 be the initial data for the heat equation supplied in (2.2). Initial values for u^ϵ and v^ϵ are chosen by

$$u_0^\epsilon(\cdot) := \frac{1}{2}r_0(\cdot) - \frac{\nu\epsilon}{2}\partial_x r_0(\cdot) \quad \text{in } \Omega,$$
$$v_0^\epsilon(\cdot) := \frac{1}{2}r_0(\cdot) + \frac{\nu\epsilon}{2}\partial_x r_0(\cdot) \quad \text{in } \Omega, \tag{3.7}$$

if the approximation of solutions for the heat equation is the goal. This choice corresponds to

$$r^\epsilon(0,\cdot) = r_0^\epsilon(\cdot) := r_0(\cdot) \quad \text{in } \Omega,$$
$$j^\epsilon(0,\cdot) = j_0^\epsilon(\cdot) := -\nu\epsilon\partial_x r_0(\cdot) \quad \text{in } \Omega.$$

3.2. A Priori Estimates

Stability for solutions of the advection system (3.1) in the L^2-Norm is given by the following estimates. In a first step, the boundary conditions are reduced. In the case of density boundary conditions (3.3), we use the decompositions $u^\epsilon = u^{\mathrm{Den}} + u_{LR}^{\mathrm{Den}}$ and $v^\epsilon = v^{\mathrm{Den}} + v_{LR}^{\mathrm{Den}}$, and accordingly $r^\epsilon = r^{\mathrm{Den}} + r_{LR}^{\mathrm{Den}}$ and $j^\epsilon = j^{\mathrm{Den}} + j_{LR}^{\mathrm{Den}}$, where we put $u^{\text{Den}} := (r^{\text{Den}} + j^{\text{Den}})/2$ and $v^{\text{Den}} := (r^{\text{Den}} - j^{\text{Den}})/2$. For smooth data, we simultaneously subtract the boundary values for r^ϵ, $\partial_x j^\epsilon$ and $\partial_x^2 r^\epsilon$ by using

$$r_{LR}^{\text{Den}}(t,x) := \frac{x - x_L}{|\Omega|} r_R^{\text{Di}}(t) + \frac{x_R - x}{|\Omega|} r_L^{\text{Di}}(t)$$
$$- \frac{(x - x_L)(x_R - x)}{6\nu|\Omega|}(x - x_L + |\Omega|)D^2 r_R^{\text{Di}}(t)$$
$$- \frac{(x - x_L)(x_R - x)}{6\nu|\Omega|}(x_R - x + |\Omega|)D^2 r_L^{\text{Di}}(t),$$

$$j_{LR}^{\text{Den}}(t,x) := \frac{(x - x_L)^2}{2|\Omega|}\epsilon(f^\epsilon(t,x_R) - \partial_t r_R^{\text{Di}}(t)) - \frac{(x_R - x)^2}{2|\Omega|}\epsilon(f^\epsilon(t,x_L) - \partial_t r_L^{\text{Di}}(t)),$$

$$(3.8)$$

where we employ the definitions

$$D^2 r_R^{\text{Di}}(t) := \nu\epsilon^2 \partial_t^2 r_R^{\text{Di}}(t) + \partial_t r_R^{\text{Di}}(t) - f^\epsilon(t,x_R) - \nu\epsilon^2 \partial_t f^\epsilon(t,x_R) + \nu\epsilon \partial_x g^\epsilon(t,x_R),$$
$$D^2 r_L^{\text{Di}}(t) := \nu\epsilon^2 \partial_t^2 r_L^{\text{Di}}(t) + \partial_t r_L^{\text{Di}}(t) - f^\epsilon(t,x_L) - \nu\epsilon^2 \partial_t f^\epsilon(t,x_L) + \nu\epsilon \partial_x g^\epsilon(t,x_L).$$

We get

$$r^{\text{Den}}(\cdot, x_L) = r^{\text{Den}}(\cdot, x_R) = 0 \qquad \text{in } (0,T),$$
$$\partial_x j^{\text{Den}}(\cdot, x_L) = \partial_x j^{\text{Den}}(\cdot, x_R) = 0 \quad \text{in } (0,T).$$

$$(3.9)$$

After reducing the boundary conditions, we have to consider the modified right hand sides

$$f_{LR}^{\text{Den}} := f^\epsilon - \partial_t r_{LR}^{\text{Den}} - \frac{1}{\epsilon}\partial_x j_{LR}^{\text{Den}},$$
$$g_{LR}^{\text{Den}} := g^\epsilon - \partial_t j_{LR}^{\text{Den}} - \frac{1}{\epsilon}\partial_x r_{LR}^{\text{Den}} - \frac{1}{\nu\epsilon^2} j_{LR}^{\text{Den}},$$

obeying

$$f_{LR}^{\text{Den}}(\cdot, x_L) = f_{LR}^{\text{Den}}(\cdot, x_R) = 0 \qquad \text{in } (0,T),$$
$$\partial_x g_{LR}^{\text{Den}}(\cdot, x_L) = \partial_x g_{LR}^{\text{Den}}(\cdot, x_R) = 0 \quad \text{in } (0,T).$$

The initial conditions have to be adapted by $r_0^{\text{Den}} := r_0^\epsilon - r_{LR}^{\text{Den}}(0,\cdot)$ and $j_0^{\text{Den}} := j_0^\epsilon - j_{LR}^{\text{Den}}(0,\cdot)$. Compatibility conditions are required to assure $r_0^{\text{Den}}(x_L) = r_0^{\text{Den}}(x_R) = 0$, $\partial_x j_0^{\text{Den}}(x_L) = \partial_x j_0^{\text{Den}}(x_R) = 0$ and $\partial_x^2 r_0^{\text{Den}}(x_L) = \partial_x^2 r_0^{\text{Den}}(x_R) = 0$.

In the case of flux boundary conditions (3.4), we use the decomposition $u^\epsilon = u^{\text{Flu}} + u_{LR}^{\text{Flu}}$ and $v^\epsilon = v^{\text{Flu}} + v_{LR}^{\text{Flu}}$, and accordingly $r^\epsilon = r^{\text{Flu}} + r_{LR}^{\text{Flu}}$ and $j^\epsilon = j^{\text{Flu}} + j_{LR}^{\text{Flu}}$. We subtract the boundary values for j^ϵ, $\partial_x r^\epsilon$ and $\partial_x^2 j^\epsilon$ by setting

$$r_{LR}^{\text{Flu}}(t,x) := \frac{(x - x_L)^2}{2|\Omega|}\left(\epsilon g^\epsilon(t,x_R) + \nu\epsilon^2 \partial_t r_R^{\text{Neu}}(t) + r_R^{\text{Neu}}(t)\right)$$
$$- \frac{(x_R - x)^2}{2|\Omega|}\left(\epsilon g^\epsilon(t,x_L) + \nu\epsilon^2 \partial_t r_L^{\text{Neu}}(t) + r_L^{\text{Neu}}(t)\right),$$

$$j_{LR}^{\text{Flu}}(t,x) := -\frac{x - x_L}{|\Omega|}\nu\epsilon r_R^{\text{Neu}}(t) - \frac{x_R - x}{|\Omega|}\nu\epsilon r_L^{\text{Neu}}(t)$$
$$+ \frac{(x - x_L)(x_R - x)}{6|\Omega|}(x - x_L + |\Omega|)\epsilon D^2 j_R^{\text{Neu}}(t)$$
$$+ \frac{(x - x_L)(x_R - x)}{6|\Omega|}(x_R - x + |\Omega|)\epsilon D^2 j_L^{\text{Neu}}(t),$$

$$(3.10)$$

where we use the definitions

$$D^2 j_R^{\text{Neu}}(t) := \nu\epsilon^2 \partial_t^2 r_R^{\text{Neu}}(t) + \partial_t r_R^{\text{Neu}}(t) + \epsilon\partial_t g^\epsilon(t,x_R) - \partial_x f^\epsilon(t,x_R),$$
$$D^2 j_L^{\text{Neu}}(t) := \nu\epsilon^2 \partial_t^2 r_L^{\text{Neu}}(t) + \partial_t r_L^{\text{Neu}}(t) + \epsilon\partial_t g^\epsilon(t,x_L) - \partial_x f^\epsilon(t,x_L).$$

We get

$$\partial_x r^{\mathrm{Flu}}(\cdot, x_L) = \partial_x r^{\mathrm{Flu}}(\cdot, x_R) = 0 \quad \text{in } (0, T),$$
$$j^{\mathrm{Flu}}(\cdot, x_L) = j^{\mathrm{Flu}}(\cdot, x_R) = 0 \qquad \text{in } (0, T). \tag{3.11}$$

The modified right hand sides

$$f_{LR}^{\mathrm{Flu}} := f^\epsilon - \partial_t r_{LR}^{\mathrm{Flu}} - \frac{1}{\epsilon}\partial_x j_{LR}^{\mathrm{Flu}},$$
$$g_{LR}^{\mathrm{Flu}} := g^\epsilon - \partial_t j_{LR}^{\mathrm{Flu}} - \frac{1}{\epsilon}\partial_x r_{LR}^{\mathrm{Flu}} - \frac{1}{\nu\epsilon^2} j_{LR}^{\mathrm{Flu}}$$

fulfill

$$\partial_x f_{LR}^{\mathrm{Flu}}(\cdot, x_L) = \partial_x f_{LR}^{\mathrm{Flu}}(\cdot, x_R) = 0 \quad \text{in } (0, T),$$
$$g_{LR}^{\mathrm{Flu}}(\cdot, x_L) = g_{LR}^{\mathrm{Flu}}(\cdot, x_R) = 0 \qquad \text{in } (0, T).$$

The initial conditions have to be adapted by $r_0^{\mathrm{Flu}} := r_0^\epsilon - r_{LR}^{\mathrm{Flu}}(0, \cdot)$ and $j_0^{\mathrm{Flu}} := j_0^\epsilon - j_{LR}^{\mathrm{Flu}}(0, \cdot)$. Compatibility conditions are required to assure $j_0^{\mathrm{Flu}}(x_L) = j_0^{\mathrm{Flu}}(x_R) = 0$, $\partial_x r_0^{\mathrm{Flu}}(x_L) = \partial_x r_0^{\mathrm{Flu}}(x_R) = 0$ and $\partial_x^2 j_0^{\mathrm{Flu}}(x_L) = \partial_x^2 j_0^{\mathrm{Flu}}(x_R) = 0$.

For inflow boundary conditions (3.5) we use the decompositions $u^\epsilon = u^{\mathrm{In}} + u_{LR}^{\mathrm{In}}$ and $v^\epsilon = v^{\mathrm{In}} + v_{LR}^{\mathrm{In}}$. We only subtract the boundary values for u^ϵ and v^ϵ at the inflow boundaries by setting

$$u_{LR}^{\mathrm{In}}(t, x) := r_L^{\mathrm{Rob}}(t) + \frac{x - x_L}{|\Omega| + 2\nu\epsilon}\left(r_R^{\mathrm{Rob}}(t) - r_L^{\mathrm{Rob}}(t)\right),$$
$$v_{LR}^{\mathrm{In}}(t, x) := r_R^{\mathrm{Rob}}(t) - \frac{x_R - x}{|\Omega| + 2\nu\epsilon}\left(r_R^{\mathrm{Rob}}(t) - r_L^{\mathrm{Rob}}(t)\right).$$

Here, we use $r_{LR}^{\mathrm{In}} = u_{LR}^{\mathrm{In}} + v_{LR}^{\mathrm{In}}$ and $j_{LR}^{\mathrm{In}} = u_{LR}^{\mathrm{In}} - v_{LR}^{\mathrm{In}}$ with

$$-\nu\epsilon\partial_x r_{LR}^{\mathrm{In}} = j_{LR}^{\mathrm{In}} \quad \text{in } \Omega_T.$$

We get

$$u^{\mathrm{In}}(\cdot, x_L) = v^{\mathrm{In}}(\cdot, x_R) = 0 \quad \text{in } (0, T). \tag{3.12}$$

For the solutions r^ϵ and j^ϵ of the transformed advection system (3.2) we define the energy

$$E^\epsilon(t) := \|r^\epsilon(t, \cdot)\|_0^2 + \|j^\epsilon(t, \cdot)\|_0^2 + \frac{2}{\nu\epsilon^2}\int_0^t \|j^\epsilon(s, \cdot)\|_0^2\, ds.$$

We find a decay of the energy, that is,

$$E^\epsilon(t) \le E^\epsilon(0) \quad \text{for } t \ge 0,$$

for vanishing boundary conditions and vanishing right hand sides. Estimates for the distribution functions u^ϵ and v^ϵ can be gained by observing $r^2 + j^2 = 2u^2 + 2v^2$. More detailed, we prove

Lemma 3.1. *Let $r = r^{\mathrm{Den}}$, $j = j^{\mathrm{Den}}$ or $r = r^{\mathrm{Flu}}$, $j = j^{\mathrm{Flu}}$ or $r = r^{\mathrm{P}}$, $j = j^{\mathrm{P}}$ be the solutions of the advection system (3.2) with reduced density boundary conditions (3.9), reduced flux boundary conditions (3.11) or periodic boundary conditions implied by (3.6). Then we have*

$$\|r(T, \cdot)\|_0^2 + \|j(T, \cdot)\|_0^2 + \frac{1}{\nu\epsilon^2}\int_0^T \|j(s, \cdot)\|_0^2\, ds$$

$$\le 2\|r(0, \cdot)\|_0^2 + 2\|j(0, \cdot)\|_0^2 + 2\nu\epsilon^2\int_0^T \|g_{LR}(s, \cdot)\|_0^2\, ds + 5T\int_0^T \|f_{LR}(s, \cdot)\|_0^2\, ds.$$

If we take $r = r^{\mathrm{In}}$, $j = j^{\mathrm{In}}$ in the reduced inflow case (3.12), then the outflow values

$$\frac{2}{\epsilon} \int_0^T \left(u^2(x_R, s) + v^2(x_L, s) \right) ds$$

can be added to the left hand side of the estimate.

PROOF. Test the equations in (3.1) with u and v, respectively. Adding the equations leads to

$$\partial_t(u^2 + v^2) + \frac{1}{\epsilon}\partial_x(u^2 - v^2) + \frac{1}{\nu\epsilon^2}(u - v)^2 = f_{LR}(u + v) + g_{LR}(u - v).$$

Integration over Ω_T yields

$$\|u(T, \cdot)\|_0^2 + \|v(T, \cdot)\|_0^2 + \frac{1}{2\nu\epsilon^2} \int_0^T \|u - v\|_0^2$$
$$\leq \|u_0\|_0^2 + \|v_0\|_0^2 + \int_0^T \|f_{LR}\|_0 \|u + v\|_0 + \frac{\nu\epsilon^2}{2} \int_0^T \|g_{LR}\|_0^2. \tag{3.13}$$

From this we find

$$\left(\int_0^T \left(\|u\|_0^2 + \|v\|_0^2 \right) \right)^{1/2} \leq \sqrt{T} \left(\|u_0\|_0^2 + \|v_0\|_0^2 + \frac{\nu\epsilon^2}{2} \int_0^T \|g_{LR}\|_0^2 \right)^{1/2}$$
$$+ \sqrt{2}T \left(\int_0^T \|f_{LR}\|_0^2 \right)^{1/2}.$$

$\square$

Estimates for the first order derivatives in space are given in the following way. In the case of reduced density boundary conditions (3.9), we get $\partial_x j^{\mathrm{Den}} = 0$ at the boundary. In the reduced flux case (3.11), we have $\partial_x r^{\mathrm{Flu}} = 0$ at the boundary. Then we have

Lemma 3.2. *Let $r = r^{\mathrm{Den}}$, $j = j^{\mathrm{Den}}$ or $r = r^{\mathrm{Flu}}$, $j = j^{\mathrm{Flu}}$ be the solutions of the transformed advection system (3.2) with reduced density boundary conditions (3.9) or reduced flux boundary conditions (3.11). Then we have*

$$|r(T, \cdot)|_1^2 + |j(T, \cdot)|_1^2 + \frac{1}{\nu\epsilon^2} \int_0^T |j(s, \cdot)|_1^2 \, ds$$
$$\leq 2|r(0, \cdot)|_1^2 + 2|j(0, \cdot)|_1^2 + 2\nu\epsilon^2 \int_0^T |g_{LR}(s, \cdot)|_1^2 \, ds + 5T \int_0^T |f_{LR}(s, \cdot)|_1^2 \, ds.$$

PROOF. Take the derivative with respect to x in the advection system (3.1) and test with $\partial_x u$ and $\partial_x v$, respectively. Proceed as in the proof of Lemma 3.1. $\square$

In the periodic case we have

Lemma 3.3. *For the solutions r^{P} and j^{P} of the transformed advection system (3.2) with periodic boundary conditions implied by (3.6), we get for $k \in \mathbb{N}$*

$$|r^{\mathrm{P}}(T, \cdot)|_k^2 + |j^{\mathrm{P}}(T, \cdot)|_k^2 + \frac{1}{\nu\epsilon^2} \int_0^T |j^{\mathrm{P}}(s, \cdot)|_k^2 \, ds$$
$$\leq 2|r^{\mathrm{P}}(0, \cdot)|_k^2 + 2|j^{\mathrm{P}}(0, \cdot)|_k^2 + 2\nu\epsilon^2 \int_0^T |g^\epsilon(s, \cdot)|_k^2 \, ds + 5T \int_0^T |f^\epsilon(s, \cdot)|_k^2 \, ds,$$

with the assumption that the data are sufficiently smooth.

For positive solutions, stability estimates of another kind can be achieved by applying the *H-Theorem*. Let therefore be the entropy density defined by

$$q^\epsilon := u^\epsilon \log u^\epsilon + v^\epsilon \log v^\epsilon \quad \text{for } u^\epsilon, v^\epsilon > 0.$$

We find

$$\partial_t q^\epsilon = \partial_t u^\epsilon (\log u^\epsilon + 1) + \partial_t v^\epsilon (\log v^\epsilon + 1)$$

$$= \left(\frac{1}{2} g^\epsilon + \frac{v^\epsilon}{2\nu\epsilon^2} \left(1 - \frac{u^\epsilon}{v^\epsilon} \right) \right) \log \frac{u^\epsilon}{v^\epsilon} + \frac{1}{2} f^\epsilon (\log u^\epsilon + \log v^\epsilon + 2)$$

$$- \frac{1}{\epsilon} \partial_x \left(u^\epsilon \log u^\epsilon - v^\epsilon \log v^\epsilon \right).$$

For $f^\epsilon = g^\epsilon = 0$ we have due to $(1 - x) \log x \leq 0$

$$\int_{x_L}^{x_R} \partial_t q^\epsilon \leq \frac{1}{\epsilon} (u_L^\epsilon(t) \log u_L^\epsilon(t) + v_R^\epsilon(t) \log v_R^\epsilon(t))$$

$$- \frac{1}{\epsilon} (u^\epsilon(t, x_R) \log u^\epsilon(t, x_R) + v^\epsilon(t, x_L) \log v^\epsilon(t, x_L)).$$

For the total entropy

$$H^\epsilon(t) := \int_{x_L}^{x_R} q^\epsilon(t, \cdot)$$

we find for positive u^ϵ and v^ϵ and for periodic or vanishing flux boundary conditions and for vanishing source terms

$$\frac{d}{dt} H^\epsilon(t) \leq 0.$$

Hence, we have $H^\epsilon(t) \leq H^\epsilon(0)$ for $t \geq 0$. Since $H^\epsilon(t)$ is bounded from below by $-2|\Omega|/e$, we have

$$\lim_{t \to \infty} \frac{d}{dt} H^\epsilon(t) = 0$$

for decreasing H^ϵ. This implies that $\lim_{t \to \infty} u^\epsilon = \lim_{t \to \infty} v^\epsilon = r^\epsilon/2$ holds true, since we have $(1 - x) \log x \leq 0$ for $x \geq 0$ and $(1 - x) \log x = 0$ if and only if $x = 1$. Hence, we get $\lim_{t \to \infty} j^\epsilon = 0$.

The total mass for solutions of the advection system is defined by

$$m^\epsilon(t) := \int_\Omega r^\epsilon(t, \cdot).$$

The change of mass is given by

$$\frac{d}{dt} m^\epsilon(t) = \frac{1}{\epsilon} \left(j^\epsilon(t, x_L) - j^\epsilon(t, x_R) \right) + \int_\Omega f^\epsilon(t, \cdot).$$

For vanishing right hand side we have conservation of mass in the vanishing flux case and in the periodic case.

3.3. Convergence of the Solutions of the Advection System

By a reformulation of the heat equation as an advection system, we get a convergence result for the solutions of the advection system (3.2) towards the density modeled by the heat equation and its derivative.

Let r be a solution of the heat equation (2.1) with source term f. Define $j := -\nu\epsilon\partial_x r$. Then the heat equation can be reformulated as an advection system in the diffusive scaling

$$\partial_t r + \frac{1}{\epsilon}\partial_x j \qquad\quad = f,$$
$$\partial_t j + \frac{1}{\epsilon}\partial_x r + \frac{1}{\nu\epsilon^2}j = \partial_t j.$$

The transformation to the variables

$$u := \frac{1}{2}(r + j), \quad v := \frac{1}{2}(r - j)$$

yields

$$\partial_t u + \frac{1}{\epsilon}\partial_x u + \frac{1}{2\nu\epsilon^2}(u - v) = \frac{1}{2}(f + \partial_t j),$$
$$\partial_t v - \frac{1}{\epsilon}\partial_x v - \frac{1}{2\nu\epsilon^2}(u - v) = \frac{1}{2}(f - \partial_t j). \tag{3.14}$$

Now we have the following result.

Theorem 3.4. *Let r be the solution of the heat equation (2.1) with viscosity ν, right hand side f and*

    i) *Dirichlet boundary conditions (2.3),*
    ii) *Neumann boundary conditions (2.4),*
    iii) *Robin boundary conditions (2.5) with $\vartheta := \nu\epsilon$,*
    iv) *periodic boundary conditions (2.6).*

We choose ϵ as the parameter of the advection system (3.1). Define $j := -\nu\epsilon\partial_x r$. Take $u := (r + j)/2$ and $v := (r - j)/2$.

Let u^ϵ and v^ϵ be the solutions of the advection system (3.1) with right hand side $f^\epsilon := f$ and $g^\epsilon := -\nu\epsilon\partial_x f$, viscosity ν, and the parameter $\epsilon > 0$. Boundary conditions are chosen as

    i) *density boundary conditions (3.3),*
    ii) *flux boundary conditions (3.4),*
    iii) *inflow boundary conditions (3.5),*
    iv) *periodic boundary conditions (3.6).*

Define $r^\epsilon := u^\epsilon + v^\epsilon$, $j^\epsilon := u^\epsilon - v^\epsilon$. Then the errors $e_r := r - r^\epsilon$ and $e_j := j - j^\epsilon$ can be estimated by

$$\|e_r(T, \cdot)\|_0^2 + \|e_j(T, \cdot)\|_0^2 + \frac{1}{\nu\epsilon^2}\int_0^T \|e_j(s, \cdot)\|_0^2\,ds$$
$$\leq \|e_r(0, \cdot)\|_0^2 + \|e_j(0, \cdot)\|_0^2 + \nu^5\epsilon^4 \int_0^T \|\partial_x^3 r(s, \cdot)\|_0^2\,ds.$$

PROOF. The errors $e_u := u - u^\epsilon$ and $e_v := v - v^\epsilon$ fulfill the advection system (3.1) with right hand side $f_e := 0$ and

$$g_e := \partial_t j - g^\epsilon = -\nu\epsilon\partial_t\partial_x r + \nu\epsilon\partial_x f = -\nu^2\epsilon\partial_x^3 r.$$

In the Dirichlet/density case i), we have $e_r(\cdot, x_L) = e_r(\cdot, x_R) = 0$. In the Neumann/flux case ii), $e_j(\cdot, x_L) = e_j(\cdot, x_R) = 0$ holds. For the Robin/inflow case iii), we have $e_u(\cdot, x_L) = e_v(\cdot, x_R) = 0$. In the periodic case, there are $e_r(\cdot, x_L) = e_r(\cdot, x_R)$ and $e_j(\cdot, x_L) = e_j(\cdot, x_R)$. The stability estimate (3.13) gives

$$\|e_u(T, \cdot)\|_0^2 + \|e_v(T, \cdot)\|_0^2 + \frac{1}{2\nu\epsilon^2}\int_0^T \|e_u - e_v\|_0^2$$
$$\leq \|e_u(0, \cdot)\|_0^2 + \|e_v(0, \cdot)\|_0^2 + \frac{\nu\epsilon^2}{2}\int_0^T \|\nu^2\epsilon\partial_x^3 r\|_0^2.$$

In the case of Robin/inflow boundary conditions, the outflow errors

$$\frac{1}{\epsilon} \int_0^T e_u^2(\cdot, x_R) + \frac{1}{\epsilon} \int_0^T e_v^2(\cdot, x_L)$$

additionally appear on the left hand side of the estimate. $\qquad\square$

If the initial values for the advection system are chosen as in (3.7), then the initial errors vanish, that is, $e_r(0, \cdot) = e_j(0, \cdot) = 0$ in Ω. This yields convergence of order two for the density and the ϵ-scaled flux in the $L^\infty((0, T); L^2(\Omega))$-norm and third order convergence for the ϵ-scaled flux in the $L^2((0, T); L^2(\Omega))$-norm.

Corollary 3.5. *Let the assumptions of Theorem 3.4 hold true. Choose the initial conditions for r^ϵ and j^ϵ by (3.7). Then we have*

$$\|e_r(T, \cdot)\|_0^2 + \|e_j(T, \cdot)\|_0^2 + \frac{1}{\nu \epsilon^2} \int_0^T \|e_j\|_0^2 \leq C\epsilon^4,$$

where the constant C only depends on the viscosity ν, the end time T and the boundary and the initial data for the heat equation.

PROOF. By the a priori estimates for the solutions of the heat equation, the integral expression on the right hand side is bounded by the initial and the boundary data. $\qquad\square$

If the initial data of the advection system are chosen by $r^\epsilon(0, \cdot) = r_0$ and $j^\epsilon(0, \cdot) = 0$ in Ω, then the convergence is only of order ϵ. A convergence result of order $\sqrt{\epsilon}$ is obtained, if the data of a Dirichlet problem for the heat equation is chosen to initialize the boundary values for the inflow problem, that is, one chooses

$$u_L^\epsilon(\cdot) = r_L^{\mathrm{Di}}(\cdot) \qquad \text{in } (0, T),$$
$$v_R^\epsilon(\cdot) = r_R^{\mathrm{Di}}(\cdot) \qquad \text{in } (0, T).$$

Furthermore, we find

Corollary 3.6. *Let the assumptions of Theorem 3.4 hold true. Choose the initial conditions by (3.7). Then we have in the cases i) and iv)*

$$|e_r(T, \cdot)|_1^2 + |e_j(T, \cdot)|_1^2 + \frac{1}{\nu \epsilon^2} \int_0^T |e_j|_1^2 \leq C\epsilon^4.$$

PROOF. In the Dirichlet/density case i) we find $\partial_x e_j(x_L) = \partial_x e_j(x_R) = 0$ by the evaluation of $\partial_t e_r + \partial_x e_j / \epsilon = 0$. Apply Lemma 3.2 and Lemma 3.3. $\qquad\square$

In the velocity non-discrete case, the convergence of the solutions of a Boltzmann-like equation to the solutions of the heat equation is only of order $\mathcal{O}(\epsilon)$. In addition, boundary layer correctors are required (Ref. [**38**]).

CHAPTER 4

The Telegraph Equations

For a deeper understanding of the solution behavior, we continue the discussion of the advection system. We demonstrate that the advection system can be transformed into two independent telegraph equations. A solution formula for the telegraph equation on unbounded domains is presented. By an extension of the data, this formula turns out to be applicable even in boundary cases.

The telegraph equations form a singularly perturbed problem for the heat equation. In the limit ϵ to zero, the hyperbolic type of the telegraph equations and the advection system changes to parabolic type. This transition is articulated in the behavior of the Fourier solutions for the telegraph equation that are presented in this chapter. By back-transformations we gain Fourier representations for the solutions of the advection system.

Energy principles for the telegraph equation render additional a priori estimates for higher order derivatives of the solutions of the advection system.

4.1. Transformation to Telegraph Equations

The advection system in the diffusive scaling in the variables r^ϵ and j^ϵ is given by

$$\partial_t r^\epsilon + \frac{1}{\epsilon}\partial_x j^\epsilon = f^\epsilon \qquad \text{in } \Omega_T, \tag{4.1}$$

$$\partial_t j^\epsilon + \frac{1}{\epsilon}\partial_x r^\epsilon + \frac{1}{\nu\epsilon^2}j^\epsilon = g^\epsilon \qquad \text{in } \Omega_T. \tag{4.2}$$

A telegraph equation for r^ϵ is gained in the following way. We derive (4.1) with respect to t and (4.2) with respect to x. Then we eliminate $\partial_x\partial_t j^\epsilon$. Furthermore, we eliminate $\partial_x j^\epsilon$ by using (4.1). A telegraph equation for j^ϵ is achieved by deriving (4.1) with respect to x and (4.2) with respect to t. Then we eliminate $\partial_x\partial_t r^\epsilon$. We get

$$\nu\epsilon^2\partial_t^2 r^\epsilon + \partial_t r^\epsilon - \nu\partial_x^2 r^\epsilon = F^\epsilon := f^\epsilon + \nu\epsilon^2\partial_t f^\epsilon - \nu\epsilon\partial_x g^\epsilon \quad \text{in } \Omega_T, \tag{4.3}$$

$$\nu\epsilon^2\partial_t^2 j^\epsilon + \partial_t j^\epsilon - \nu\partial_x^2 j^\epsilon = G^\epsilon := \nu\epsilon^2\partial_t g^\epsilon - \nu\epsilon\partial_x f^\epsilon \qquad \text{in } \Omega_T. \tag{4.4}$$

Telegraph equations for u^ϵ and v^ϵ are achieved by taking the sum and the difference of (4.3) and (4.4).

These telegraph equations have to be equipped with initial conditions

$$\begin{aligned}
r^\epsilon(0,\cdot) &= r_0^\epsilon(\cdot) \quad \text{in } \Omega, \\
j^\epsilon(0,\cdot) &= j_0^\epsilon(\cdot) \quad \text{in } \Omega.
\end{aligned} \tag{4.5}$$

Since we now have partial differential equations of order two with respect to time, we need to supply initial data for $\partial_t r^\epsilon$ and $\partial_t j^\epsilon$. This is done by using (4.1) and (4.2). We define

$$\partial_t r^\epsilon(0,\cdot) = r_1^\epsilon(\cdot) := f^\epsilon(0,\cdot) - \frac{1}{\epsilon}\partial_x j_0^\epsilon(\cdot) \qquad \text{in } \Omega,$$

$$\partial_t j^\epsilon(0,\cdot) = j_1^\epsilon(\cdot) := g^\epsilon(0,\cdot) - \frac{1}{\epsilon}\partial_x r_0^\epsilon(\cdot) - \frac{1}{\nu\epsilon^2}j_0^\epsilon(\cdot) \quad \text{in } \Omega. \tag{4.6}$$

37

In order to ensure convergence of order two in ϵ towards the solutions of the heat equation, we use the data

$$r_0^\epsilon := r_0, \quad j_0^\epsilon := -\nu\epsilon\partial_x r_0,$$
$$f^\epsilon := f, \quad g^\epsilon := -\nu\epsilon\partial_x f, \tag{4.7}$$

as in Theorem 3.4 and Corollary 3.5. Hence, we get

$$r_1^\epsilon(\cdot) = f(0,\cdot) + \nu\partial_x^2 r_0(\cdot) \quad \text{in } \Omega,$$
$$j_1^\epsilon(\cdot) = -\nu\epsilon\partial_x f(0,\cdot) \qquad \text{in } \Omega.$$

Periodic boundary conditions for the advection system translate to periodic boundary conditions for the telegraph equations. Hence, the equations (4.3) and (4.4) have to be solved using

$$r^\epsilon(\cdot, x_L) = r^\epsilon(\cdot, x_R) \quad \text{in } (0,T),$$
$$j^\epsilon(\cdot, x_L) = j^\epsilon(\cdot, x_R) \quad \text{in } (0,T).$$

If we want to approximate the solutions of the Dirichlet problem for the heat equation, we only know Dirichlet data for r^ϵ. Hence, we have to solve (4.3) subject to Dirichlet values

$$r^\epsilon(\cdot, x_L) = r_L^{\mathrm{Di}}(\cdot) \quad \text{in } (0,T),$$
$$r^\epsilon(\cdot, x_R) = r_R^{\mathrm{Di}}(\cdot) \quad \text{in } (0,T).$$

In order to obtain a solution of the advection system subject to the density boundary conditions (3.3), we use (4.1) to determine

$$\partial_x j^\epsilon(\cdot, x_L) = \epsilon f^\epsilon(\cdot, x_L) - \epsilon\partial_t r_L^{\mathrm{Di}}(\cdot) \quad \text{in } (0,T),$$
$$\partial_x j^\epsilon(\cdot, x_R) = \epsilon f^\epsilon(\cdot, x_R) - \epsilon\partial_t r_R^{\mathrm{Di}}(\cdot) \quad \text{in } (0,T).$$

Hence, equation (4.4) has to be solved in presence of Neumann boundary values.

If we want to approximate the solutions of the Neumann problem for the heat equation, we can only provide Dirichlet data for j^ϵ. Hence, we have to solve (4.4) subject to the Dirichlet values

$$j^\epsilon(\cdot, x_L) = -\nu\epsilon r_L^{\mathrm{Neu}}(\cdot) \quad \text{in } (0,T),$$
$$j^\epsilon(\cdot, x_R) = -\nu\epsilon r_R^{\mathrm{Neu}}(\cdot) \quad \text{in } (0,T).$$

In order to obtain a solution of the advection system subject to the flux boundary conditions (3.4), we use (4.2) to determine

$$\partial_x r^\epsilon(\cdot, x_L) := \epsilon g^\epsilon(\cdot, x_L) + \nu\epsilon^2\partial_t r_L^{\mathrm{Neu}}(\cdot) + r_L^{\mathrm{Neu}}(\cdot) \quad \text{in } (0,T),$$
$$\partial_x r^\epsilon(\cdot, x_R) := \epsilon g^\epsilon(\cdot, x_R) + \nu\epsilon^2\partial_t r_R^{\mathrm{Neu}}(\cdot) + r_R^{\mathrm{Neu}}(\cdot) \quad \text{in } (0,T).$$

Hence, we have to solve a Neumann problem for (4.3).

For the inflow problem there is no information on r^ϵ or j^ϵ at the boundaries. For u^ϵ and v^ϵ only the inflow values (and their derivatives) are known. Hence, we cannot determine a solution of this problem by using this approach.

4.2. The Telegraph Equation

We have to consider solutions of the telegraph equation

$$L^\epsilon s^\epsilon := \nu\epsilon^2\partial_t^2 s^\epsilon + \partial_t s^\epsilon - \nu\partial_x^2 s^\epsilon = h^\epsilon \quad \text{in } \Omega_T. \tag{4.8}$$

This equation is singularly perturbed by the small parameter ϵ. In the limit ϵ to zero, we formally gain the heat equation. Initial conditions for (4.8) are given by

$$s^\epsilon(0,\cdot) = s_0^\epsilon(\cdot) \quad \text{in } \Omega,$$
$$\partial_t s^\epsilon(0,\cdot) = s_1^\epsilon(\cdot) \quad \text{in } \Omega. \tag{4.9}$$

We impose either Dirichlet boundary conditions

$$s^\epsilon(\cdot, x_L) = s_L^{\mathrm{Di}}(\cdot) \quad \text{in } (0, T),$$
$$s^\epsilon(\cdot, x_R) = s_R^{\mathrm{Di}}(\cdot) \quad \text{in } (0, T),$$

or Neumann boundary conditions

$$\partial_x s^\epsilon(\cdot, x_L) = s_L^{\mathrm{Neu}}(\cdot) \quad \text{in } (0, T),$$
$$\partial_x s^\epsilon(\cdot, x_R) = s_R^{\mathrm{Neu}}(\cdot) \quad \text{in } (0, T).$$

In the Dirichlet case, compatibility conditions for the supplied data are determined by

$$s_0^\epsilon(x_L) = s_L^{\mathrm{Di}}(0), \qquad s_0^\epsilon(x_R) = s_R^{\mathrm{Di}}(0),$$
$$s_1^\epsilon(x_L) = \partial_t s_L^{\mathrm{Di}}(0), \quad s_1^\epsilon(x_R) = \partial_t s_R^{\mathrm{Di}}(0)$$

and

$$\nu\epsilon^2 \partial_t^2 s_L^{\mathrm{Di}}(0) + \partial_t s_L^{\mathrm{Di}}(0) - \nu\partial_x^2 s_0^\epsilon(x_L) = h^\epsilon(0, x_L),$$
$$\nu\epsilon^2 \partial_t^2 s_R^{\mathrm{Di}}(0) + \partial_t s_R^{\mathrm{Di}}(0) - \nu\partial_x^2 s_0^\epsilon(x_R) = h^\epsilon(0, x_R).$$

Extensions are needed for the higher order derivatives. In the Neumann case we have to impose

$$\partial_x s_0^\epsilon(x_L) = s_L^{\mathrm{Neu}}(0), \qquad \partial_x s_0^\epsilon(x_R) = s_R^{\mathrm{Neu}}(0),$$
$$\partial_x s_1^\epsilon(x_L) = \partial_t s_L^{\mathrm{Neu}}(0), \quad \partial_x s_1^\epsilon(x_R) = \partial_t s_R^{\mathrm{Neu}}(0)$$

and

$$\nu\epsilon^2 \partial_t^2 s_L^{\mathrm{Neu}}(0) + \partial_t s_L^{\mathrm{Neu}}(0) - \nu\partial_x^3 s_0^\epsilon(x_L) = \partial_x h^\epsilon(0, x_L),$$
$$\nu\epsilon^2 \partial_t^2 s_R^{\mathrm{Neu}}(0) + \partial_t s_R^{\mathrm{Neu}}(0) - \nu\partial_x^3 s_0^\epsilon(x_R) = \partial_x h^\epsilon(0, x_R).$$

For the derivation of a solution formula for the telegraph equation on unbounded domains, we follow Ref. [**31**, 8.36, 9.42-9.43]. We switch to characteristic variables by taking

$$\xi := \frac{1}{2\nu\epsilon^2}(t + \epsilon x),$$

$$\eta := \frac{1}{2\nu\epsilon^2}(t - \epsilon x).$$

We define

$$S^\epsilon(\xi, \eta) := s^\epsilon(\nu\epsilon^2(\xi + \eta), \nu\epsilon(\xi - \eta))e^{(\xi+\eta)/2}$$

and find

$$\left(\partial_{\xi\eta} S^\epsilon - \frac{1}{4}S^\epsilon\right)(\xi, \eta) = H^\epsilon(\xi, \eta) := \nu\epsilon^2 h^\epsilon(\nu\epsilon^2(\xi + \eta), \nu\epsilon(\xi - \eta))e^{(\xi+\eta)/2}.$$

For fixed $(\xi_0, \eta_0) \in \Omega_T$, we define the Riemann's function

$$\chi(\xi, \eta) := \sum_{n=0}^{\infty} \frac{(\xi_0 - \xi)^n(\eta_0 - \eta)^n}{4^n(n!)^2}.$$

This Riemann's function obeys $\partial_{\xi\eta}\chi - \chi/4 = 0$ and $\chi \equiv 1$ along the characteristics $\xi = \xi_0$ and $\eta = \eta_0$. Now we apply Green's formula

$$\iint_{D_0} (vu_{\xi\eta} - uv_{\xi\eta})\, d(\xi, \eta) = \int_{\partial D_0} (vu_\eta\, d\eta + uv_\xi\, d\xi)$$

to $u = \chi$, $v = S^\epsilon$ and $u = S^\epsilon$, $v = \chi$. We integrate over the triangular domain D_0 with vertex in (ξ_0, η_0) and the edges $\xi = \xi_0$, $\eta = \eta_0$ and $t = 0$ and its positively

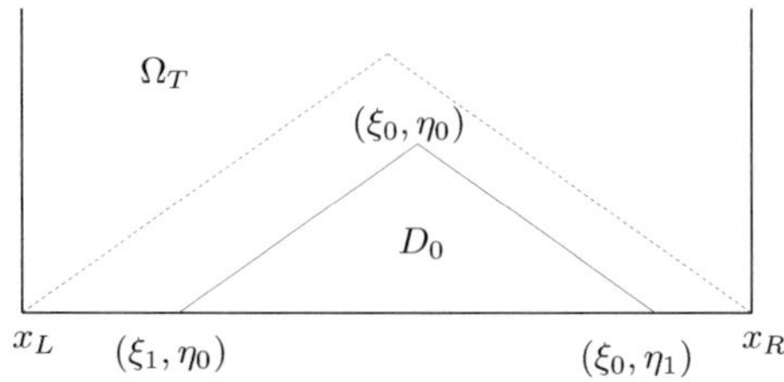

FIGURE 4.1. Integration in D_0.

oriented boundary curve ∂D_0; see Figure 4.1. Initially, we assume that the characteristics $\xi = \xi_0$ and $\eta = \eta_0$ do not intersect with the boundaries $x = x_L$ or $x = x_R$. The solution in (ξ_0, η_0) is given by

$$S^\epsilon(\xi_0, \eta_0) = \frac{1}{2} S^\epsilon(\xi_1, \eta_0) + \frac{1}{2} S^\epsilon(\xi_0, \eta_1) + \iint_{D_0} \chi H^\epsilon$$

$$+ \frac{1}{4} \int_{(\xi_1, \eta_0)}^{(\xi_0, \eta_1)} \big(\chi(S_\xi^\epsilon + S_\eta^\epsilon) - S^\epsilon(\chi_\xi + \chi_\eta) \big)(d\xi - d\eta).$$

By using the modified Bessel function

$$I_0(r) := \sum_{n=0}^{\infty} \frac{(r/2)^{2n}}{(n!)^2}, \quad J_0(r) := \frac{I_0'(r)}{r} = \frac{1}{2} \sum_{n=0}^{\infty} \frac{(r/2)^{2n}}{(n+1)! n!}$$

and

$$r = r(s, y) := \frac{1}{2\nu\epsilon^2} \sqrt{(t-s)^2 - \epsilon^2 (x-y)^2},$$

we get

Theorem 4.1. *In the periodic case or for $x_L + t/\epsilon \le x \le x_R - t/\epsilon$, the solution of the telegraph equation (4.8) in (t, x), $t \ge 0$, can be represented by*

$$s^\epsilon(t, x) = \frac{1}{2} \big(s_0^\epsilon(x - t/\epsilon) + s_0^\epsilon(x + t/\epsilon) \big) e^{-t/(2\nu\epsilon^2)}$$

$$+ \frac{1}{2\nu\epsilon} \int_0^t \int_{x-(t-s)/\epsilon}^{x+(t-s)/\epsilon} e^{-(t-s)/(2\nu\epsilon^2)} h^\epsilon(s, y) I_0(r(s, y)) \, dy \, ds$$

$$+ \frac{t}{8\nu^2\epsilon^3} e^{-t/(2\nu\epsilon^2)} \int_{x-t/\epsilon}^{x+t/\epsilon} s_0^\epsilon(y) J_0(r(0, y)) \, dy \tag{4.10}$$

$$+ \frac{\epsilon}{2} e^{-t/(2\nu\epsilon^2)} \int_{x-t/\epsilon}^{x+t/\epsilon} \left(s_1^\epsilon(y) + \frac{1}{2\nu\epsilon^2} s_0^\epsilon(y) \right) I_0(r(0, y)) \, dy.$$

If we have $h^\epsilon \ge 0$ in Ω_T, $s_0^\epsilon \ge 0$ in Ω and $s_1^\epsilon + s_0^\epsilon/(2\nu\epsilon^2) \ge 0$ in Ω, then the solution is non-negative.

In contrast to the heat equation, we have finite speed of information transport. The modulus of the speed is $1/\epsilon$ and turns to infinity in the limit $\epsilon \to 0$.

In Ref. [31, 9.44], the author intends to construct solutions for the initial-boundary value problems by following the characteristics and computing surface elements. The unknown derivatives at the boundary shall be computed by using the derivatives given at $t = 0$. The formula 9.44.(3) provided is obviously wrong. It only holds in the case of the wave equation $\partial_{\xi\eta} S = H$, where the Riemann's function does not depend on ξ and η. Hence, the suggested construction of solutions for the boundary value problems fails.

Apart from this fact, in the homogeneous boundary case, solutions can be gained by extensions of the data, which are described in the following. For vanishing Dirichlet data, we use odd extensions for h^ϵ, s_0^ϵ and s_1^ϵ. We define

$$s_0^\epsilon(x) := -s_0^\epsilon(2x_L - x) \quad \text{for } x_L - |\Omega| < x < x_L,$$
$$s_0^\epsilon(x) := s_0^\epsilon(x - 2k|\Omega|) \quad \text{for } x_R + 2(k-1)|\Omega| < x \leq x_R + 2k|\Omega|, \ k \in \mathbb{Z},$$

and apply the same procedure to s_1^ϵ and $h^\epsilon(t, \cdot)$. In the homogeneous Neumann case we use even extensions, that is,

$$s_0^\epsilon(x) := s_0^\epsilon(2x_L - x) \quad \text{for } x_L - |\Omega| < x < x_L,$$
$$s_0^\epsilon(x) := s_0^\epsilon(x - 2k|\Omega|) \quad \text{for } x_R + 2(k-1)|\Omega| < x \leq x_R + 2k|\Omega|, \ k \in \mathbb{Z},$$

and analogous extensions for s_1^ϵ and $h^\epsilon(t, \cdot)$. For vanishing Dirichlet boundary conditions with $s_0^\epsilon(x_L) = s_0^\epsilon(x_R) = s_1^\epsilon(x_L) = s_1^\epsilon(x_R) = 0$ or vanishing Neumann boundary conditions, (4.10) applies for all $(t, x) \in \Omega_T$. In the homogeneous Neumann case, non-negativity of the initial data implies non-negativity of the solution in Ω_T. In the Dirichlet case, the non-negativity of the data does not carry over to the odd extensions. For the closely related wave equation $\nu\epsilon^2\partial_t^2 s - \nu\partial_x^2 s = 0$, Dirichlet boundary conditions lead to a phase reversal. Hence, positive solutions can turn into negative ones. See also Ref. [**39**, Chapter 4].

4.3. Fourier Solutions for the Telegraph Equations

Another possibility to present solution formulas for the homogeneous boundary value problems is the use of Fourier series in terms of the orthogonal eigenfunctions introduced in Section 2.4 for the solution of the heat equation.

We define $p^\epsilon := s^\epsilon e^{t/(2\nu\epsilon^2)}$ and consider the transformed telegraph equation

$$M^\epsilon p^\epsilon := \nu\epsilon^2 \partial_t^2 p^\epsilon - \nu\partial_x^2 p^\epsilon - \frac{1}{4\nu\epsilon^2}p^\epsilon = h^\epsilon e^{t/(2\nu\epsilon^2)},$$

which is known as the *Klein-Gordon equation with imaginary mass*. Let

$$t^\epsilon := \frac{t}{2\nu\epsilon^2}.$$

By using the orthogonal eigenfunctions defined in (2.16) and (2.30), we define the Green's functions

$$G^{\text{Di}}(t, x, z) := \sum_{k=1}^\infty \frac{2\nu\epsilon^2}{\alpha_k^\epsilon} \sinh(\alpha_k^\epsilon t^\epsilon)\phi_k^{\text{Di}}(x)\phi_k^{\text{Di}}(z),$$

$$G^{\text{Neu}}(t, x, z) := \frac{2\nu\epsilon^2}{|\Omega|} \sinh(t^\epsilon) + \sum_{k=1}^\infty \frac{2\nu\epsilon^2}{\alpha_k^\epsilon} \sinh(\alpha_k^\epsilon t^\epsilon)\phi_k^{\text{Neu}}(x)\phi_k^{\text{Neu}}(z)$$

for the Dirichlet problem and the Neumann problem. Here, we use the definition

$$\alpha_k^\epsilon := \sqrt{1 - 4\nu\epsilon^2\lambda_k} \text{ for } k \geq 0,$$

where $\lambda_k = \nu k^2\pi^2/|\Omega|^2$ are the eigenvalues of the stationary problem for the heat equation. For small values of ϵ and small k, the quantity α_k^ϵ takes real values. For large k, the quantity α_k^ϵ becomes complex. We remark that the Green's functions given in Ref. [**37**, 7.33.B.2.1 and 7.33.B.2.2] are not written down correctly.

We apply Green's formula to the domain Ω_t

$$\iint_{\Omega_t} (vM^\epsilon u - uM^\epsilon v)d\tau dz = \nu\epsilon^2 \int_{\partial\Omega_t} (u\partial_\tau v - v\partial_\tau u)dz + \nu \int_{\partial\Omega_t} (u\partial_z v - v\partial_z u)d\tau$$

and take $u(\tau, z) := p^\epsilon(\tau, z)$ with $M^\epsilon p^\epsilon = h^\epsilon e^{t^\epsilon}$ and $v(\tau, z) = G^{\text{Di}}(t - \tau, x, z)$ or $v(\tau, z) = G^{\text{Neu}}(t - \tau, x, z)$. The evaluation of the integrals and the transformation to $s^\epsilon = p^\epsilon e^{-t^\epsilon}$ leads to the following solution formulas. For abbreviation, we define for $k \in \mathbb{N}$

$$a_0^\epsilon(t) \equiv 1,$$

$$b_0^\epsilon(t) := 1 - e^{-2t^\epsilon},$$

$$a_k^\epsilon(t) := \frac{\alpha_k^\epsilon + 1}{2\alpha_k^\epsilon} e^{(\alpha_k^\epsilon - 1)t^\epsilon} + \frac{\alpha_k^\epsilon - 1}{2\alpha_k^\epsilon} e^{-(\alpha_k^\epsilon + 1)t^\epsilon},$$

$$b_k^\epsilon(t) := \frac{1}{\alpha_k^\epsilon} \left(e^{(\alpha_k^\epsilon - 1)t^\epsilon} - e^{-(\alpha_k^\epsilon + 1)t^\epsilon} \right).$$

Theorem 4.2. *Let s_ϵ^{Di} be the solution of the telegraph equation (4.8) subject to the initial conditions (4.9) and vanishing Dirichlet boundary conditions $s_\epsilon^{\mathrm{Di}}(\cdot, x_L) = s_\epsilon^{\mathrm{Di}}(\cdot, x_R) = 0$. Then s_ϵ^{Di} is given by*

$$
\begin{aligned}
s_\epsilon^{\mathrm{Di}}(t, x) = &\sum_{k=1}^{\infty} \int_0^t b_k^\epsilon(t - \tau)(h^\epsilon(\tau, \cdot), \phi_k^{\mathrm{Di}}) \, d\tau \, \phi_k^{\mathrm{Di}}(x) \\
&+ \sum_{k=1}^{\infty} a_k^\epsilon(t)(s_0^\epsilon, \phi_k^{\mathrm{Di}})\phi_k^{\mathrm{Di}}(x) + \nu\epsilon^2 \sum_{k=1}^{\infty} b_k^\epsilon(t)(s_1^\epsilon, \phi_k^{\mathrm{Di}})\phi_k^{\mathrm{Di}}(x),
\end{aligned}
\tag{4.11}
$$

provided that s_0^ϵ and s_1^ϵ vanish at the boundary.

The condition that s_0^ϵ and s_1^ϵ vanish at the boundary is achieved by the fulfillment of the corresponding compatibility conditions.

Theorem 4.3. *A solution $s_\epsilon^{\mathrm{Neu}}$ of the telegraph equation (4.8) subject to the initial conditions (4.9) and vanishing Neumann boundary conditions $s_\epsilon^{\mathrm{Neu}}(\cdot, x_L) = s_\epsilon^{\mathrm{Neu}}(\cdot, x_R) = 0$ is given by*

$$
\begin{aligned}
s_\epsilon^{\mathrm{Neu}}(t, x) = &\frac{1}{2} \int_0^t b_0^\epsilon(t - \tau)(h^\epsilon(\tau, \cdot), \phi_0^{\mathrm{Neu}}) \, d\tau \, \phi_0^{\mathrm{Neu}}(x) \\
&+ \frac{1}{2}(s_0^\epsilon, \phi_0^{\mathrm{Neu}})\phi_0^{\mathrm{Neu}}(x) + \frac{\nu\epsilon^2}{2}b_0^\epsilon(t)(s_1^\epsilon, \phi_0^{\mathrm{Neu}})\phi_0^{\mathrm{Neu}}(x) \\
&+ \sum_{k=1}^{\infty} \int_0^t b_k^\epsilon(t - \tau)(h^\epsilon(\tau, \cdot), \phi_k^{\mathrm{Neu}}) \, d\tau \, \phi_k^{\mathrm{Neu}}(x) \\
&+ \sum_{k=1}^{\infty} a_k^\epsilon(t)(s_0^\epsilon, \phi_k^{\mathrm{Neu}})\phi_k^{\mathrm{Neu}}(x) + \nu\epsilon^2 \sum_{k=1}^{\infty} b_k^\epsilon(t)(s_1^\epsilon, \phi_k^{\mathrm{Neu}})\phi_k^{\mathrm{Neu}}(x).
\end{aligned}
\tag{4.12}
$$

PROOF. The time kernels a_k^ϵ and b_k^ϵ fulfill

$$\nu\epsilon^2\partial_t^2 a_k^\epsilon + \partial_t a_k^\epsilon + \lambda_k a_k^\epsilon = 0 \quad \text{in } (0, T),$$

$$\nu\epsilon^2\partial_t^2 b_k^\epsilon + \partial_t b_k^\epsilon + \lambda_k b_k^\epsilon = 0 \quad \text{in } (0, T)$$

for $k \in \mathbb{N}_0$, subject to the initial conditions

$$a_k^\epsilon(0) = 1, \quad \nu\epsilon^2\partial_t a_k^\epsilon(0) = 0,$$

$$b_k^\epsilon(0) = 0, \quad \nu\epsilon^2\partial_t b_k^\epsilon(0) = 1.$$

The assertion of the Theorem 4.2 and 4.3 follows by direct verification. $\square$

Furthermore, we observe

$$\partial_t a_k^\epsilon + \lambda_k b_k^\epsilon = 0 \quad \text{in } (0, T),$$

$$\nu\epsilon^2\partial_t b_k^\epsilon + b_k^\epsilon = a_k^\epsilon \quad \text{in } (0, T).$$

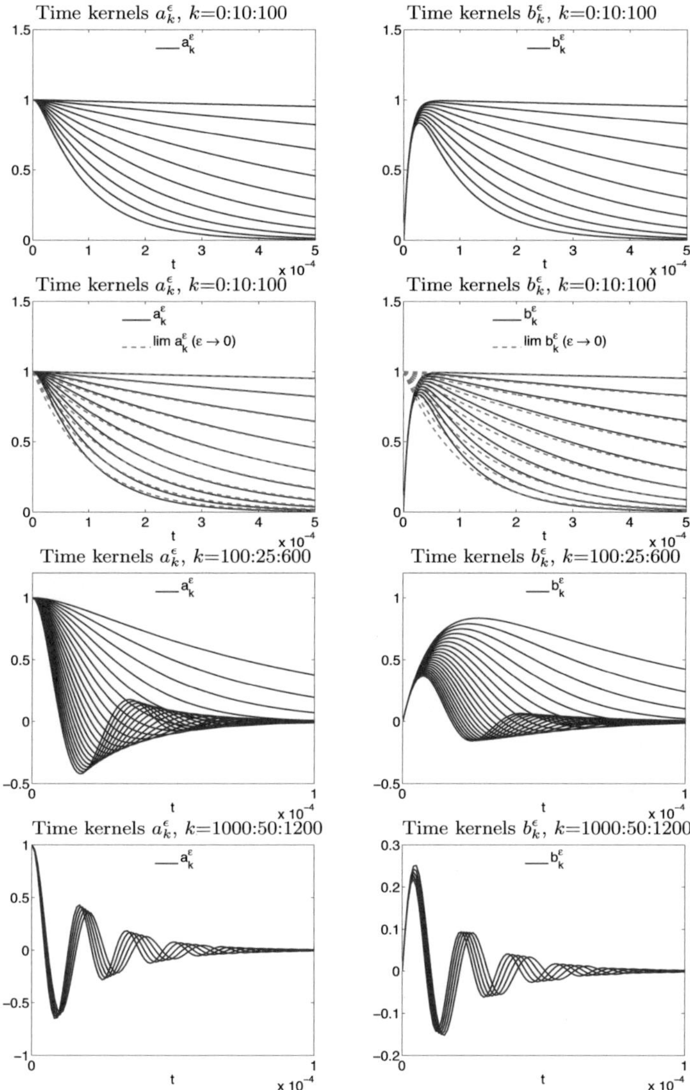

FIGURE 4.2. Time kernels of the Fourier coefficients for data $\epsilon = 0.01$, $\nu = 0.1$, $|\Omega| = 1$ and $n_0 = 159$.

The exponential terms with argument $-(\alpha_k^\epsilon + 1)t^\epsilon$ decay very fast, whereas the exponential terms with argument $(\alpha_k^\epsilon - 1)t^\epsilon$ approximate the time kernels of the heat equation for small k.

Since we have

$$\lim_{\epsilon \to 0} \alpha_k^\epsilon = 1,$$

$$\lim_{\epsilon \to 0} \frac{\alpha_k^\epsilon - 1}{2\nu\epsilon^2} = -\lambda_k$$

for fixed k, we get

$$\lim_{\epsilon \to 0} a_k^\epsilon(t) = e^{-\lambda_k t} \quad \text{for } t \geq 0,$$

$$\lim_{\epsilon \to 0} b_k^\epsilon(t) = e^{-\lambda_k t} \quad \text{for } t > 0.$$

In the latter case, convergence cannot be uniform due to $b_k^\epsilon(0) = 0$. Furthermore, we find

$$\lim_{\epsilon \to 0} \partial_t a_k^\epsilon(t) = -\lambda_k e^{-\lambda_k t} \quad \text{for } t > 0,$$

$$\lim_{\epsilon \to 0} \partial_t b_k^\epsilon(t) = -\lambda_k e^{-\lambda_k t} \quad \text{for } t > 0.$$

Hence, we formally have

$$\lim_{\epsilon \to 0} s_\epsilon^{\mathrm{Di}} = s^{\mathrm{Di}}, \quad \lim_{\epsilon \to 0} s_\epsilon^{\mathrm{Neu}} = s^{\mathrm{Neu}},$$

where s^{Di} and s^{Neu} are solutions of the heat equation subject to vanishing Dirichlet or vanishing Neumann boundary conditions.

Let

$$n_0 := n_0(\epsilon) := \left\lfloor \frac{|\Omega|}{2\nu\epsilon\pi} \right\rfloor \in \mathbb{N}_0,$$

where $\lfloor \gamma \rfloor$ denotes the largest integer lower or equal the real value γ. Then, for $k \leq n_0$, the time kernels a_k^ϵ and b_k^ϵ are positive and show an exponential decay as it is expected for the solutions of parabolic problems. For $k > n_0$, we have oscillating time kernels a_k^ϵ and b_k^ϵ. They show a fast exponential decay, but there are frequent changes in the sign; see Figure 4.2. For $k > n_0$, we have

$$a_k^\epsilon(t) = \left(\frac{1}{\beta_k^\epsilon} \sin\left(\beta_k^\epsilon t^\epsilon\right) + \cos\left(\beta_k^\epsilon t^\epsilon\right) \right) e^{-t^\epsilon},$$

$$b_k^\epsilon(t) = \frac{2}{\beta_k^\epsilon} \sin\left(\beta_k^\epsilon t^\epsilon\right) e^{-t^\epsilon}$$

with $\beta_k^\epsilon := i\alpha_k^\epsilon = \sqrt{4\nu\epsilon^2\lambda_k - 1} \in \mathbb{R}$.

Now we consider the function

$$A_k^\epsilon(t, x) := \frac{1}{2} e^{-\pi/\beta_k^\epsilon} + a_k^\epsilon(t) \sin\left(k\pi(x - x_L)/|\Omega|\right)$$

for $n > n_0$. On the interval $I_k := (x_L, x_L + |\Omega|/k)$, the initial value $A_k^\epsilon(0, \cdot)$ is positive. Furthermore, A_k^ϵ is a solution of the telegraph equation $L^\epsilon A_k^\epsilon = 0$ with $\partial_t A_k^\epsilon(0, \cdot) = 0$ and $A_k^\epsilon > 0$ on ∂I_k. But for the time $T_k := 2\nu\epsilon^2\pi/\beta_k^\epsilon > 0$, we have $A_k^\epsilon(T_k, x_L + |\Omega|/(2k)) < 0$, since $a_k^\epsilon(T_k) = -e^{-\pi/\beta_k^\epsilon}$ takes its negative minimum there. Hence, non-negativity of the data does not guarantee non-negativity of the solutions.

The initial layers of the time kernels b_ϵ^k (see Figure 4.2) express the fact that we are examining a singularly perturbed problem. The presence of the second order derivative in time in (4.8) requires initial conditions for $\partial_t s^\epsilon$.

4.4. Fourier Solutions for the Advection System

With the results of the previous section we can give Fourier solutions for the advection system of the Goldstein-Taylor model.

Theorem 4.4. *Let r^ϵ and j^ϵ be solutions of the advection system* (4.1) *and* (4.2) *subject to the boundary conditions* $r^\epsilon(\cdot, x_L) = r^\epsilon(\cdot, x_R) = 0$ *and* $\partial_x j^\epsilon(\cdot, x_L) = \partial_x j^\epsilon(\cdot, x_R) = 0$ *in* $(0, T)$. *Let the initial conditions be given by* (4.5) *and* (4.6) *with* $r_0^\epsilon(x_L) = r_0^\epsilon(x_R) = 0$ *and* $\partial_x j_0^\epsilon(x_L) = \partial_x j_0^\epsilon(x_R) = 0$. *Let* F^ϵ *and* G^ϵ *be defined in* (4.3) *and* (4.4) *with* f^ϵ *and* g^ϵ *fulfilling* $f^\epsilon(\cdot, x_L) = f^\epsilon(\cdot, x_R) = 0$ *and* $\partial_x g^\epsilon(\cdot, x_L) = \partial_x g^\epsilon(\cdot, x_R) = 0$ *in* $(0, T)$. *Then we have*

$$r^\epsilon(t, x) = \sum_{k=1}^\infty \int_0^t b_k^\epsilon(t - \tau)(F^\epsilon(\tau, \cdot), \phi_k^{\mathrm{Di}}) \, d\tau \, \phi_k^{\mathrm{Di}}(x)$$

$$+ \sum_{k=1}^\infty a_k^\epsilon(t)(r_0^\epsilon, \phi_k^{\mathrm{Di}})\phi_k^{\mathrm{Di}}(x) + \nu\epsilon^2 \sum_{k=1}^\infty b_k^\epsilon(t)(r_1^\epsilon, \phi_k^{\mathrm{Di}})\phi_k^{\mathrm{Di}}(x),$$

$$j^\epsilon(t, x) = \frac{1}{|\Omega|} \int_0^t b_0^\epsilon(t - \tau)(G^\epsilon(\tau, \cdot), 1) \, d\tau + \frac{1}{|\Omega|} a_0^\epsilon(t)(j_0^\epsilon, 1) + \frac{\nu\epsilon^2}{|\Omega|} b_0^\epsilon(t)(j_1^\epsilon, 1)$$

$$+ \sum_{k=1}^\infty \int_0^t b_k^\epsilon(t - \tau)(G^\epsilon(\tau, \cdot), \phi_k^{\mathrm{Neu}}) \, d\tau \, \phi_k^{\mathrm{Neu}}(x)$$

$$+ \sum_{k=1}^\infty a_k^\epsilon(t)(j_0^\epsilon, \phi_k^{\mathrm{Neu}})\phi_k^{\mathrm{Neu}}(x) + \nu\epsilon^2 \sum_{k=1}^\infty b_k^\epsilon(t)(j_1^\epsilon, \phi_k^{\mathrm{Neu}})\phi_k^{\mathrm{Neu}}(x).$$

PROOF. Direct computation. $\qquad\square$

In the case of density boundary conditions (3.3), the assumptions on the boundary conditions of the data are achieved by subtracting r_{LR}^{Den} and j_{LR}^{Den} given in (3.8), where the data have to fulfill adequate compatibility conditions.

Theorem 4.5. *Let r^ϵ and j^ϵ be solutions of the advection system* (4.1) *and* (4.2) *subject to the boundary conditions* $\partial_x r^\epsilon(\cdot, x_L) = \partial_x r^\epsilon(\cdot, x_R) = 0$ *and* $j^\epsilon(\cdot, x_L) = j^\epsilon(\cdot, x_R) = 0$ *in* $(0, T)$. *Let the initial conditions be given by* (4.5) *and* (4.6) *with* $\partial_x r_0^\epsilon(x_L) = \partial_x r_0^\epsilon(x_R) = 0$ *and* $j_0^\epsilon(x_L) = j_0^\epsilon(x_R) = 0$. *Let* F^ϵ *and* G^ϵ *be defined in* (4.3) *and* (4.4) *with* f^ϵ *and* g^ϵ *fulfilling* $\partial_x f^\epsilon(\cdot, x_L) = \partial_x f^\epsilon(\cdot, x_R) = 0$ *and* $g^\epsilon(\cdot, x_L) = g^\epsilon(\cdot, x_R) = 0$ *in* $(0, T)$. *Then we have*

$$r^\epsilon(t, x) = \frac{1}{|\Omega|} \int_0^t b_0^\epsilon(t - \tau)(F^\epsilon(\tau, \cdot), 1) \, d\tau + \frac{1}{|\Omega|} a_0^\epsilon(t)(r_0^\epsilon, 1) + \frac{\nu\epsilon^2}{|\Omega|} b_0^\epsilon(t)(r_1^\epsilon, 1)$$

$$+ \sum_{k=1}^\infty \int_0^t b_k^\epsilon(t - \tau)(F^\epsilon(\tau, \cdot), \phi_k^{\mathrm{Neu}}) \, d\tau \, \phi_k^{\mathrm{Neu}}(x)$$

$$+ \sum_{k=1}^\infty a_k^\epsilon(t)(r_0^\epsilon, \phi_k^{\mathrm{Neu}})\phi_k^{\mathrm{Neu}}(x) + \nu\epsilon^2 \sum_{k=1}^\infty b_k^\epsilon(t)(r_1^\epsilon, \phi_k^{\mathrm{Neu}})\phi_k^{\mathrm{Neu}}(x),$$

$$j^\epsilon(t, x) = \sum_{k=1}^\infty \int_0^t b_k^\epsilon(t - \tau)(G^\epsilon(\tau, \cdot), \phi_k^{\mathrm{Di}}) \, d\tau \, \phi_k^{\mathrm{Di}}(x)$$

$$+ \sum_{k=1}^\infty a_k^\epsilon(t)(j_0^\epsilon, \phi_k^{\mathrm{Di}})\phi_k^{\mathrm{Di}}(x) + \nu\epsilon^2 \sum_{k=1}^\infty b_k^\epsilon(t)(j_1^\epsilon, \phi_k^{\mathrm{Di}})\phi_k^{\mathrm{Di}}(x).$$

PROOF. Direct computation. $\qquad\square$

In the case of flux boundary conditions (3.4), the assumptions on the boundary conditions of the data are achieved by subtracting r_{LR}^{Flu} and j_{LR}^{Flu} given in (3.10), where the data have to fulfill adequate compatibility conditions.

A similar representation of the solutions in terms of Fourier series can be obtained for periodic boundary conditions (3.6).

4.5. Energy estimates

For the derivation of a priori estimates we employ the energy functional. Let the total energy associated to solutions of the telegraph equation (4.8) be defined by

$$F^\epsilon(t) := \nu\epsilon^2 \|\partial_t s^\epsilon(t,\cdot)\|_0^2 + \nu\|\partial_x s^\epsilon(t,\cdot)\|_0^2.$$

For vanishing right hand side h^ϵ and constant Dirichlet boundary values, vanishing Neumann boundary values or periodic boundary values, we find decay of the total energy. More detailed, we prove

Lemma 4.6. *Let s^ϵ be a solution of the telegraph equation (4.8) subject to constant Dirichlet boundary values, vanishing Neumann boundary values or periodic boundary values. Then we have*

$$F^\epsilon(t) + \int_0^t \|\partial_t s^\epsilon(s,\cdot)\|_0^2\, ds \le F^\epsilon(0) + \int_0^t \|h^\epsilon(s,\cdot)\|_0^2\, ds \quad \text{for } t \ge 0.$$

Proof. Due to

$$\partial_t\left(\nu\epsilon^2(\partial_t s^\epsilon)^2 + \nu(\partial_x s^\epsilon)^2\right) = -2(\partial_t s^\epsilon)^2 + 2\nu\partial_x(\partial_x s^\epsilon \partial_t s^\epsilon) + 2\partial_t s^\epsilon h^\epsilon$$

we find

$$\frac{d}{dt}F^\epsilon(t) + 2\|\partial_t s^\epsilon(t,\cdot)\|_0^2 \le \|\partial_t s^\epsilon(t,\cdot)\|_0^2 + \|h^\epsilon(t,\cdot)\|_0^2.$$

$\square$

By employing the energy principle of the telegraph equation, we find a priori estimates for the solutions of the advection system.

Corollary 4.7. *Let the assumptions of Theorem 4.4 or Theorem 4.5 hold true or consider periodic boundary conditions (3.6). Use the data given in (4.7) in order to ensure convergence of the solutions (confer Theorem 3.4 and Corollary 3.5). Then we get the following a priori estimates for the solutions r^ϵ and j^ϵ of the advection system (4.1) and (4.2):*

$$\nu\|\partial_x r^\epsilon(T,\cdot)\|_0^2 + \int_0^T \|\partial_t r^\epsilon(s,\cdot)\|_0^2\, ds + \nu\epsilon^2\|\partial_t r^\epsilon(T,\cdot)\|_0^2$$

$$\le \nu\|\partial_x r_0(t,\cdot)\|_0^2 + 2\nu^3\epsilon^2\|\partial_x^2 r_0\|_0^2 + 2\nu\epsilon^2\|f(0,\cdot)\|_0^2$$

$$+ 3\int_0^T \|f(s,\cdot)\|_0^2\, ds + 3\nu^2\epsilon^4\int_0^T \|\partial_t f(s,\cdot)\|_0^2\, ds + 3\nu^4\epsilon^4\int_0^T \|\partial_x^2 f(s,\cdot)\|_0^2\, ds,$$

$$\nu\|\partial_x j^\epsilon(T,\cdot)\|_0^2 + \int_0^T \|\partial_t j^\epsilon(s,\cdot)\|_0^2\, ds + \nu\epsilon^2\|\partial_t j^\epsilon(T,\cdot)\|_0^2$$

$$\le \nu^3\epsilon^2\|\partial_x^2 r_0\|_0^2 + \nu^3\epsilon^4\|\partial_x f(0,\cdot)\|_0^2$$

$$+ 2\nu^2\epsilon^2\int_0^T \|\partial_x f(s,\cdot)\|_0^2\, ds + 2\nu^4\epsilon^6\int_0^T \|\partial_t\partial_x f(s,\cdot)\|_0^2\, ds.$$

Proof. Apply Lemma 4.6. $\square$

For higher order regularity, further compatibility conditions for the data are required. In the case of desired convergence with data chosen as in (4.7), these compatibility conditions have to be seen as extensions of the compatibility conditions for the heat equation given in Section 2.3. In the case of density boundary conditions, compatibility for the heat equation leads to zero boundary conditions for

r_0^{Den} and $\partial_x j_0^{\text{Den}}$. Zero boundary conditions for $\partial_x^2 r_0^{\text{Den}}$ require in addition $\partial_x^4 r_0 = 0$ at the boundary. For flux boundary conditions, compatibility for the heat equation leads to zero boundary conditions for j_0^{Flu}. Zero boundary conditions for $\partial_x r_0^{\text{Flu}}$ and $\partial_x^2 j_0^{\text{Flu}}$ require $\partial_x^3 r_0 = 0$ and $\nu \partial_x^5 r_0 + \partial_x^3 f = 0$ at the boundaries. In the case of violations, the perturbations are of orders ϵ^2.

In the lattice Boltzmann community it is a widespread custom to use Hilbert expansions or Chapman-Enskog expansions (Ref. [24]) of a function f in terms of the small parameter ϵ, with the aim to gain insight to the limiting behavior of the discrete velocity models or the lattice Boltzmann schemes. One uses representations of the form

$$f = \sum_{l=0}^{\infty} \epsilon^l f^l.$$

However, it is a well-known fact that there is no expansion for e^{-1/ϵ^2} about $\epsilon = 0$. If we consider initial data, where the eigenfunctions ϕ_k^{Di} or ϕ_k^{Neu} are incorporated, then the solutions of the advection system are equipped with the time kernels a_k^ϵ and b_k^ϵ, which have parts of the form $e^{-t/(2\nu\epsilon^2)}$. Hence, not all terms appearing in the solution are matched by the asymptotic expansions in ϵ.

CHAPTER 5

Lattice Boltzmann Schemes

The goal of our study is the computation of discrete solutions of the heat equation by using explicit lattice Boltzmann (LB) discretizations of the form

$$U(t + \tau, x + h) = U(t, x) + Q(U, V)(t, x) + F_+(t, x),$$
$$V(t + \tau, x - h) = V(t, x) - Q(U, V)(t, x) + F_-(t, x),$$

where Q is the collision term and F_+ and F_- are the source terms. This is done upon discretizing the advection system (3.1) under consideration of the characteristic directions. The fluid-dynamic limit, which corresponds to the limit ϵ to zero, is attained by choosing the fixed space-time coupling

$$h = \gamma \epsilon, \quad \tau = \gamma \epsilon^2.$$

The parameter γ has to be chosen as a function of the viscosity of the limiting equation and the relaxation parameter of the collision term. The respective dependence may be different for different discretizations.

In this chapter, the lattice Boltzmann discretizations are performed in a finite difference and in a finite volume context on vertex centered grids and on cell centered grids. We are treating four different kinds of lattice Boltzmann schemes. Finite element methods are disregarded, since their application is equipped with additional technical difficulties. But, as a consequence, the abandonment of variational formulations is a considerable deficiency.

We present diverse boundary conditions that are chosen in accordance with the type of the boundary conditions for the heat equation and the underlying grid situation. The scaling of the lattice Boltzmann schemes is discussed. We identify the Knudsen number and the Mach number, and we determine the relationship to the numerical viscosity.

In Section 5.7, we switch to matrix formulations in terms of the macroscopic quantities. The simultaneous reduction of the boundary values for the density and for the flux is described. We perform a detailed eigenvalue analysis for the time evolution matrices and we present discrete Fourier solutions for the lattice Boltzmann schemes that reveal the coupling of the flux and the density. The distribution of the eigenvalues gives rise to the conjecture that the lattice Boltzmann schemes work badly in the case of nonsmooth data.

A discrete stability estimate is proved in the case of the finite difference lattice Boltzmann schemes. This stability estimate is the basis for the convergence proofs in Chapter 6. Stability for the finite volume lattice Boltzmann schemes is derived from the Fourier representations.

5.1. Vertex Centered and Cell Centered Grids

For the discretization of the equations we introduce the following grid on the interval $\Omega := (x_L, x_R)$ for $N \in \mathbb{N}$. We define the grid size

$$h := \frac{|\Omega|}{N}$$

49

and the grid points

$$x_l := x_L + lh \quad \text{for } l = 0, \dots, N.$$

The intermediate grid points $x_{l-1/2}$ are defined by $x_{l-1/2} := (x_l + x_{l-1})/2$ for $l = 1, \dots, N$. Furthermore, we define the intervals $K_l := (x_{l-1/2}, x_{l+1/2})$ for $l = 0, \dots, N$ and $K_{l-1/2} := (x_{l-1}, x_l)$ for $l = 1, \dots, N$, where we employ the straightforward definition of grid points outside the interval Ω.

By using the constant γ that has to be determined later, we define the uniform time step

$$\tau := \frac{h^2}{\gamma}.$$

The time interval $(0, T)$ is split by the intermediate times $t_k := k\tau$ for $k = 0, \dots, M$ with $t_{M-1} < T \le t_M$. We define $I_k := (t_k, t_{k+1})$ for $k = 0, \dots, M - 1$ and $\Omega_M := (0, t_M) \times \Omega$.

The discretizations are done in a *finite difference* (FD) or in a *finite volume* (FV) context. In the FD context the discrete values U_l^k and V_l^k are interpreted as point values in (t_k, x_l), that is, we use

$$U_l^k \approx u(t_k, x_l), \quad V_l^k \approx v(t_k, x_l).$$

In the FV context we consider integral mean values, that is, we make use of the approximation properties

$$U_l^k \approx \frac{1}{h} \int_{x_{l-1/2}}^{x_{l+1/2}} u(t_k, \cdot), \quad V_l^k \approx \frac{1}{h} \int_{x_{l-1/2}}^{x_{l+1/2}} v(t_k, \cdot).$$

We introduce the rectangular cells $X_l^k := I_k \times K_l$ and the corresponding discrete FV space with piecewise constant functions

$$V_h^V := \left\{ F \in L^2(\Omega_M) \,:\, F(t, x) = F_l^k \text{ for } (t, x) \in X_l^k, \begin{array}{l} l = 0, \dots, N, \\ k = 0, \dots, M - 1 \end{array} \right\}.$$

This grid is referred to as the *vertex centered finite volume grid* (VCFV grid). The domain of V_h^V is $(0, t_M) \times (x_L - h/2, x_R + h/2)$. Hence, for the discretization the data has to be extended on the boundary cells X_0^k and X_N^k. Characteristic cells Y_l^k and Z_l^k are defined by having the basis K_l at time t_k and the tops K_{l+1} and K_{l-1}, respectively, at time t_{k+1}. These cells are bounded by the characteristics $(x_i \pm sh, t_k \pm s\tau)$ for $s \in (0, 1)$ and $i = l - 1/2, l + 1/2$; see Figure 5.1.

In the same manner, we introduce the rectangular cells $X_{l-1/2}^k := I_k \times K_{l-1/2}$ for $l = 1, \dots, N$ and $k = 0, \dots, M - 1$. Then we define the corresponding discrete FV space with piecewise constant functions

$$V_h^C := \left\{ F \in L^2(\Omega_M) \,:\, F(t, x) = F_{l-1/2}^k \text{ for } (t, x) \in X_{l-1/2}^k, \begin{array}{l} l = 1, \dots, N, \\ k = 0, \dots, M - 1 \end{array} \right\}.$$

This construction is called *cell centered finite volume grid* (CCFV grid). Characteristic cells $Y_{l-1/2}^k$ and $Z_{l-1/2}^k$ are defined by having the basis $K_{l-1/2}$ at time t_k and the tops $K_{l+1/2}$ and $K_{l-3/2}$, respectively, at time t_{k+1}. These cells are bounded by the characteristics $(x_i \pm sh, t_k \pm s\tau)$ for $s \in (0, 1)$ and $i = l - 1, l$. The characteristic cells are cut off at the boundary; see Figure 5.2.

For $F \in V_h^V$ or $F \in V_h^C$ we use

$$F^k := F_{|\cup_l X_l^k} \quad \text{or} \quad F^k := F_{|\cup_l X_{l-1/2}^k} \quad \text{for } k = 0, \dots, M - 1.$$

In accordance with the above definitions, we call a grid designed for the application of FD schemes a *vertex centered finite difference grid* (VCFD grid), if we evaluate the functions at the grid points x_l for $l = 0, \dots, N$. In the case of function

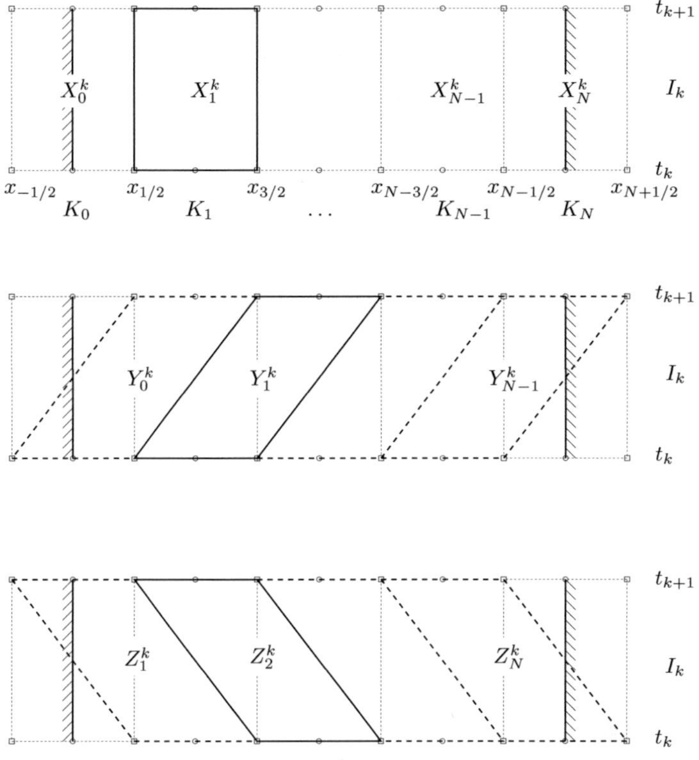

FIGURE 5.1. Vertex centered finite volume grid (VCFV).

evaluations at the interval midpoints $x_{l-1/2}$ for $l = 1, \ldots, N$, we refer to *cell centered finite difference grids* (CCFD grids). Since the resulting discrete equations are of a similar structure for finite difference methods and for finite volume methods, we sometimes refer to vertex centered grids (VC grids) or cell centered grids (CC grids).

5.2. Lattice Boltzmann Discretizations

In the finite difference context we discretize the advection system (3.1) by replacing the derivatives by difference quotients, where we take care of the characteristic directions. We use $U_l^k \approx u(t_k, x_l)$ and $V_l^k \approx v(t_k, x_l)$. Here, and in the following, the upper index ϵ for the solutions u^ϵ and v^ϵ of the advection system is omitted. For the time derivatives we take

$$\partial_t u(t_k, x_{l+1}) \approx \frac{U_{l+1}^{k+1} - U_{l+1}^k}{\tau}, \quad \partial_t v(t_k, x_{l-1}) \approx \frac{V_{l-1}^{k+1} - V_{l-1}^k}{\tau}.$$

The space derivatives are replaced by using the one-sided differences

$$\partial_x u(t_k, x_{l+1}) \approx \frac{U_{l+1}^k - U_l^k}{h}, \quad \partial_x v(t_k, x_{l-1}) \approx \frac{V_l^k - V_{l-1}^k}{h}.$$

The collision terms $u - v$ are evaluated in the points (t_k, x_l), where the characteristics start that meet the points (t_{k+1}, x_{l+1}) for the u-equation and (t_{k+1}, x_{l-1}) for the

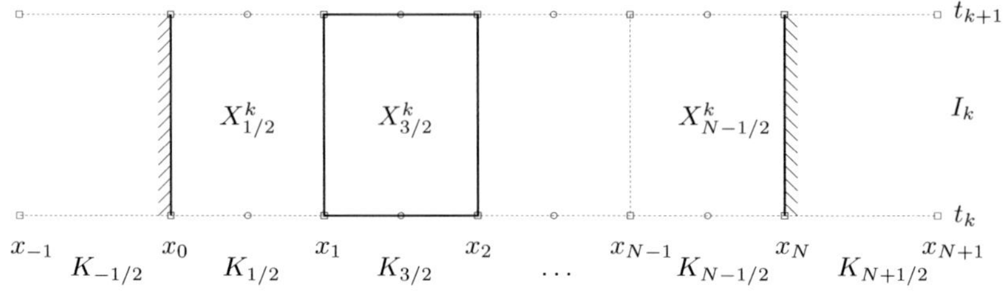

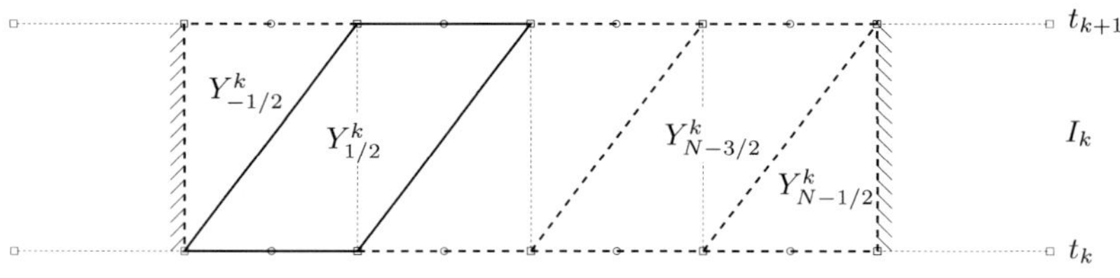

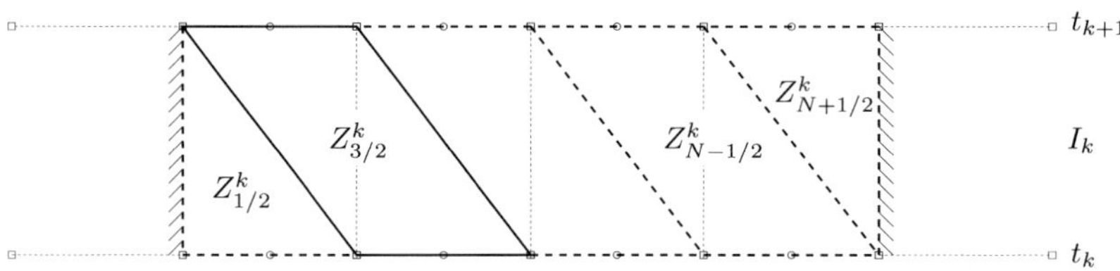

FIGURE 5.2. Cell centered finite volume grid (CCFV).

v-equation. The source terms are evaluated along the corresponding characteristics. We take

$$F_{l,+}^k := f(t_k + s\tau, x_l + sh), \quad F_{l,-}^k := f(t_k + s\tau, x_l - sh)$$

for a fixed $s \in [0,1]$. A discretization of the source term g in (3.1) is disregarded. By caring of $1/\epsilon = \gamma/h$ and replacing $1/(2\nu\epsilon^2)$ by ω/τ with a relaxation parameter $\omega \in (0,1)$, we find the VCFD lattice Boltzmann equations

$$U_{l+1}^{k+1} = U_l^k - \omega(U_l^k - V_l^k) + \frac{\tau}{2}F_{l,+}^k \quad \text{for } l = 0,\ldots,N-1,$$

$$V_{l-1}^{k+1} = V_l^k + \omega(U_l^k - V_l^k) + \frac{\tau}{2}F_{l,-}^k \quad \text{for } l = 1,\ldots,N. \tag{5.1}$$

The missing inflow values U_0^{k+1} and V_N^{k+1} have to be determined by the boundary conditions. Initial values have to be supplied for U_l^0 and V_l^0 for $l = 0,\ldots,N$. Then the lattice Boltzmann (LB) equations have to be solved for $k = 0,\ldots,M-1$.

On CCFD grids we get the CCFD lattice Boltzmann equations

$$U_{l+1/2}^{k+1} = U_{l-1/2}^k - \omega(U_{l-1/2}^k - V_{l-1/2}^k) + \frac{\tau}{2}F_{l-1/2,+}^k \quad \text{for } l = 1,\ldots,N-1,$$

$$V_{l-1/2}^{k+1} = V_{l+1/2}^k + \omega(U_{l+1/2}^k - V_{l+1/2}^k) + \frac{\tau}{2}F_{l+1/2,-}^k \quad \text{for } l = 1,\ldots,N-1. \tag{5.2}$$

Here, we have to determine the missing values $U_{1/2}^{k+1}$ and $V_{N-1/2}^{k+1}$ by an evaluation of the given boundary conditions.

For FV discretizations we write the advection system in the form

$$\frac{d}{dt}u(t, x + (t - t_k)/\epsilon) + \frac{\omega}{\tau}(u - v)(t, x + (t - t_k)/\epsilon) = \frac{1}{2}f(t, x + (t - t_k)/\epsilon),$$

$$\frac{d}{dt}v(t, x - (t - t_k)/\epsilon) - \frac{\omega}{\tau}(u - v)(t, x - (t - t_k)/\epsilon) = \frac{1}{2}f(t, x - (t - t_k)/\epsilon).$$

Integration over I_k yields

$$u(t_{k+1}, x + h) = u(t_k, x) - \frac{\omega}{\tau}\int_{I_k}(u - v)(s, X^{k,+})\,ds + \frac{1}{2}\int_{I_k}f(s, X^{k,+})\,ds,$$

$$v(t_{k+1}, x - h) = v(t_k, x) + \frac{\omega}{\tau}\int_{I_k}(u - v)(s, X^{k,-})\,ds + \frac{1}{2}\int_{I_k}f(s, X^{k,-})\,ds,$$

where we use $X^{k,\pm} := X^{k,\pm}(s, x) := x \pm (s - t_k)/\epsilon$. On VCFV grids we integrate along K_l and get

$$\frac{1}{h}\int_{K_{l+1}}u(t_{k+1}, \cdot) = \frac{1}{h}\int_{K_l}u(t_k, \cdot) - \frac{\omega}{\tau h}\iint_{Y_l^k}(u - v) + \frac{1}{2h}\iint_{Y_l^k}f,$$

$$\frac{1}{h}\int_{K_{l-1}}v(t_{k+1}, \cdot) = \frac{1}{h}\int_{K_l}v(t_k, \cdot) + \frac{\omega}{\tau h}\iint_{Z_l^k}(u - v) + \frac{1}{2h}\iint_{Z_l^k}f.$$

Now we replace u and v by $U \in V_h^V$ and $V \in V_h^V$. We end up with the VCFV lattice Boltzmann equations

$$U_{l+1}^{k+1} = U_l^k - \frac{\omega}{2}(U_{l+1}^k + U_l^k - V_{l+1}^k - V_l^k) + \frac{\tau}{2}F_{l,+}^k \quad \text{for } l = 0, \ldots, N - 1,$$

$$V_{l-1}^{k+1} = V_l^k + \frac{\omega}{2}(U_l^k + U_{l-1}^k - V_l^k - V_{l-1}^k) + \frac{\tau}{2}F_{l,-}^k \quad \text{for } l = 1, \ldots, N,$$

$$(5.3)$$

where we define

$$F_{l,+}^k := \frac{1}{|Y_l^k|}\iint_{Y_l^k}f, \quad F_{l,-}^k := \frac{1}{|Z_l^k|}\iint_{Z_l^k}f.$$

The values U_0^{k+1} and V_N^{k+1} have to be computed from the boundary conditions.

On CCFV grids we integrate along $K_{l-1/2}$ and $K_{l+1/2}$ and plug in $U \in V_h^C$ and $V \in V_h^C$. Hence, we gain the CCFV lattice Boltzmann equations

$$U_{l+1/2}^{k+1} = U_{l-1/2}^k - \frac{\omega}{2}(U_{l+1/2}^k + U_{l-1/2}^k - V_{l+1/2}^k - V_{l-1/2}^k) + \frac{\tau}{2}F_{l-1/2,+}^k,$$

$$V_{l-1/2}^{k+1} = V_{l+1/2}^k + \frac{\omega}{2}(U_{l+1/2}^k + U_{l-1/2}^k - V_{l+1/2}^k - V_{l-1/2}^k) + \frac{\tau}{2}F_{l+1/2,-}^k,$$

$$\text{for } l = 1, \ldots, N - 1,$$

$$(5.4)$$

where we define

$$F_{l-1/2,+}^k := \frac{1}{|Y_{l-1/2}^k|}\iint_{Y_{l-1/2}^k}f, \quad F_{l+1/2,-}^k := \frac{1}{|Z_{l+1/2}^k|}\iint_{Z_{l+1/2}^k}f.$$

The values $U_{1/2}^{k+1}$ and $V_{N-1/2}^{k+1}$ have to be supplied by the evaluation of the given boundary conditions.

In order to gain the lattice Boltzmann equations (5.1) and (5.2) by the application of FV methods, we have to make use of specific integration formulas for the integrals of the collision term $u - v$.

From the lattice Boltzmann variables U_l^k and V_l^k we switch to the macroscopic variables R_l^k and J_l^k by taking

$$R_l^k := U_l^k + V_l^k, \quad J_l^k := U_l^k - V_l^k.$$

We shall find the approximation properties

$$R_l^k \approx r(t_k, x_l), \quad -\frac{2\omega}{h} J_l^k \approx \partial_x r(t_k, x_l)$$

for the FD schemes and

$$R_l^k \approx \frac{1}{h} \int_{K_l} r(t_k, \cdot), \quad -\frac{2\omega}{h} J_l^k \approx \frac{1}{h} \int_{K_l} \partial_x r(t_k, \cdot)$$

for the FV schemes, where r is the solution of the heat equation with viscosity ν.

While the numerical viscosity of the FD lattice Boltzmann schemes (5.1) and (5.2) turns out to be

$$\nu_{FD} := \frac{1-\omega}{2\omega} \gamma,$$

we find the numerical viscosity

$$\nu_{FV} := \frac{1}{2\omega} \gamma$$

for the FV lattice Boltzmann schemes (5.3) and (5.4). This means that for a given viscosity ν and a given relaxation parameter ω the ratio $\gamma = h^2/\tau$ has to be adapted in order to achieve the correct numerical viscosity. Hence, for a given grid size h there is a prescribed value for the time step τ.

Discrete stability is proved directly only for the FD lattice Boltzmann schemes (5.1) and (5.2). We shall see that these schemes lack consistency with the equations of the advection system (3.1). This is no failure, because the approximation of the solutions of the advection system is not within our scope.

The stability of the FV lattice Boltzmann schemes (5.3) and (5.4) can be proved after a detailed Fourier analysis.

The choice $h = \gamma\epsilon$ and $\tau = \gamma\epsilon^2$ leads to a uniform hyperbolic CFL-condition (*Courant-Friedrichs-Levi-condition*, see Ref. [**22**, 7.2]). By using the characteristic gradient $a := 1/\epsilon$ we find $a\tau/h = 1$. On the other hand, we have parabolic dependence in the form $\gamma\tau/h^2 = 1$.

5.3. Finite Element Methods

For the technical analysis a variational formulation and the application of *Petrov-Galerkin finite element methods* are favorable; see Ref. [**40**]. This approach enables the introduction of a posteriori error estimation techniques.

The natural choice is the definition of space-time finite elements along the characteristic cells with the goal to achieve the explicit lattice Boltzmann schemes as presented in the previous section. In order to succeed, we have to choose ansatz functions that are piecewise constant in time or we have to apply specific quadrature formulas. These quadrature formulas and the displacement of the characteristic cells of the u-equation and the v-equation lead to additional difficulties.

The second choice is the combination of spatial finite elements and time integration methods like the explicit Euler scheme. In general, the application of finite element methods leads to nondiagonal mass matrices that are expected to be positive definite. As we shall see at the end of Section 6.3, the lattice Boltzmann schemes give rise to a system with the central second order schemes for the derivatives of the macroscopic quantities and a nonsymmetric mass matrix with non-negative eigenvalues. An interpretation of this structure is still missing.

The crucial point for the discretizations under consideration is the lattice Boltzmann type structure. Hence, we can neglect the knowledge of details on their respective derivations and study the lattice Boltzmann schemes in the presented form. But it is well-known that the disregard of Hilbert space structures limits the variety of analytical tools for the investigations of the convergence processes.

5.4. The Lattice Boltzmann Scaling

The lattice Boltzmann equations

$$U_{l+1}^{k+1} = U_l^k - \omega(U_l^k - V_l^k) \quad \text{for } l = 0, \ldots, N-1$$
$$V_{l-1}^{k+1} = V_l^k + \omega(U_l^k - V_l^k) \quad \text{for } l = 1, \ldots, N$$

can be viewed as discretizations of the equations

$$\partial_{\hat{t}}\hat{u} + \partial_{\hat{x}}\hat{u} + \omega(\hat{u} - \hat{v}) = 0,$$
$$\partial_{\hat{t}}\hat{v} - \partial_{\hat{x}}\hat{v} - \omega(\hat{u} - \hat{v}) = 0$$

for the unknowns $\hat{u} = \hat{u}(\hat{t}, \hat{x})$ and $\hat{v} = \hat{v}(\hat{t}, \hat{x})$ with the time step $\Delta\hat{t} = 1$ and the grid size $\Delta\hat{x} = 1$. Setting $u(t, x) = u(\tau\hat{t}, h\hat{x}) := \hat{u}(\hat{t}, \hat{x})$ and $v(t, x) = v(\tau\hat{t}, h\hat{x}) := \hat{v}(\hat{t}, \hat{x})$ with $t := \tau\hat{t}$ and $x := h\hat{x}$, we find

$$\tau\partial_t u + h\partial_x u + \omega(u - v) = 0,$$
$$\tau\partial_t v + h\partial_x v - \omega(u - v) = 0.$$

The diffusion scaling and the space-time coupling is attained by using $\tau := \gamma\epsilon^2$ and $h := \gamma\epsilon$. These definitions contain two kinds of information. The characteristic gradient $h/\tau = 1/\epsilon$ determines the structure of the characteristic grid. Secondly, $h^2/\tau = \gamma$ arranges the diffusive scaling. By rewriting the latter equations in the form

$$\epsilon\partial_t u + \partial_x u + \frac{\omega}{\gamma\epsilon}(u - v) = 0,$$
$$\epsilon\partial_t v + \partial_x v - \frac{\omega}{\gamma\epsilon}(u - v) = 0,$$

we can identify the Mach number $\epsilon = h/\gamma$ and the Knudsen number $\gamma\epsilon = h$. Hence, the parameter γ determines the space-time coupling, but it can also be seen as the ratio of the Knudsen number to the Mach number. As we shall see, the numerical viscosity of the lattice Boltzmann schemes depends on the relaxation parameter ω and on γ.

In the applications the viscosity ν and the discretization parameter N is given. Upon choosing the relaxation parameter ω, the space-time coupling γ and hence the time step is determined.

5.5. Boundary Conditions

The missing inflow values are computed by evaluating the boundary conditions. As in the continuous case, we prescribe periodic, density, flux or inflow boundary conditions.

For the VC lattice Boltzmann schemes (5.1) and (5.3) boundary values have to be prescribed for U_0^{k+1} and V_N^{k+1}. In the case of periodic boundary conditions (2.6) for the heat equation, periodic boundary conditions for the VC schemes are given by

$$U_0^{k+1} := U_N^{k+1},$$
$$V_N^{k+1} := V_0^{k+1}. \tag{5.5}$$

For Dirichlet boundary conditions (2.3) for the heat equation, we prescribe density boundary conditions on VC grids by

$$U_0^{k+1} := r_L^{k+1} - V_0^{k+1},$$
$$V_N^{k+1} := r_R^{k+1} - U_N^{k+1}, \tag{5.6}$$

with

$$r_L^k := r_L^{\mathrm{Di}}(t_k), \quad r_R^k := r_R^{\mathrm{Di}}(t_k). \tag{5.7}$$

Flux boundary conditions on VC grids are given in the case of Neumann boundary conditions (2.4) for the heat equation by choosing

$$
\begin{aligned}
U_0^{k+1} &:= j_L^{k+1} + V_0^{k+1}, \\
V_N^{k+1} &:= U_N^{k+1} - j_R^{k+1},
\end{aligned}
\tag{5.8}
$$

with

$$j_L^k := -\frac{h}{2\omega} r_L^{\mathrm{Neu}}(t_k), \quad j_R^k := -\frac{h}{2\omega} r_R^{\mathrm{Neu}}(t_k). \tag{5.9}$$

Flux boundary conditions with vanishing flux $j_L^{k+1} = j_R^{k+1} = 0$ are known as *bounce-back boundary conditions*, where the outflow value is used as the inflow value.

If Robin boundary conditions (2.5) are prescribed for the heat equation, we use inflow boundary conditions

$$
\begin{aligned}
U_0^{k+1} &:= u_L^{k+1}, \\
V_N^{k+1} &:= v_R^{k+1},
\end{aligned}
\tag{5.10}
$$

with

$$u_L^k := r_L^{\mathrm{Rob}}(t_k), \quad v_R^k := r_R^{\mathrm{Rob}}(t_k). \tag{5.11}$$

The parameter ϑ in (2.5) has to be chosen as $\vartheta := h/(2\omega)$ in order to gain the correct consistency of the boundary conditions.

On CC grids the boundary values have to be used to compute $U_{1/2}^{k+1}$ and $V_{N-1/2}^{k+1}$. We define the outflow values

$$
\begin{aligned}
V_{\mathrm{Out}}^{k+1} &:= V_{1/2}^k + \omega(U_{1/2}^k - V_{1/2}^k) + \frac{\tau}{2} F_{1/2,-}^k, \\
U_{\mathrm{Out}}^{k+1} &:= U_{N-1/2}^k - \omega(U_{N-1/2}^k - V_{N-1/2}^k) + \frac{\tau}{2} F_{N-1/2,+}^k.
\end{aligned}
$$

Then we define density boundary conditions for the CC lattice Boltzmann schemes (5.2) and (5.4), if Dirichlet boundary conditions (2.3) for the heat equation are prescribed. For a $\delta \in [0,1]$, we define $t_{k+\delta} := t_k + \delta\tau$ and then we put

$$
\begin{aligned}
U_{1/2}^{k+1} &:= r_L^{k+\delta} - V_{\mathrm{Out}}^{k+1}, \\
V_{N-1/2}^{k+1} &:= r_R^{k+\delta} - U_{\mathrm{Out}}^{k+1},
\end{aligned}
\tag{5.12}
$$

with r_L^k and r_R^k given in (5.7).

Flux boundary conditions on CCFD grids can be given in the case of Neumann boundary conditions (2.4) for the heat equation by choosing

$$
\begin{aligned}
U_{1/2}^{k+1} &:= (1-\omega)j_L^{k+\delta} + V_{\mathrm{Out}}^{k+1}, \\
V_{N-1/2}^{k+1} &:= U_{\mathrm{Out}}^{k+1} - (1-\omega)j_R^{k+\delta},
\end{aligned}
\tag{5.13}
$$

with j_L^k and j_R^k given in (5.9). On CCFV grids we use instead

$$
\begin{aligned}
U_{1/2}^{k+1} &:= j_L^{k+\delta} + V_{1/2}^k, \\
V_{N-1/2}^{k+1} &:= U_{N-1/2}^k - j_R^{k+\delta}.
\end{aligned}
\tag{5.14}
$$

The evaluation of the boundary values is chosen at time $t_{k+\delta} := t_k + \delta\tau$ for a $\delta \in [0,1]$.

Let the discrete mass m_V^k on VC and m_C^k on CC grids be defined by

$$m_V^k := \frac{h}{2} R_0^k + \sum_{l=1}^{N-1} h R_l^k + \frac{h}{2} R_N^k, \quad m_C^k := \sum_{l=1}^{N} h R_{l-1/2}^k.$$

For periodic boundary conditions (5.5) or flux boundary conditions (5.8) with vanishing flux (that is, bounce-back boundary conditions) and for vanishing right hand side, there is conservation of mass for the VC lattice Boltzmann schemes (5.1) and (5.3) of the form

$$m_V^k = m_V^0 \quad \text{for } k = 0, \dots, M - 1.$$

For vanishing flux boundary conditions (5.13) and vanishing right hand side we have conservation of mass for the CCFD lattice Boltzmann scheme (5.2), that is, we have

$$m_C^k = m_C^0 \quad \text{for } k = 0, \dots, M - 1.$$

For vanishing flux boundary conditions (5.14) and vanishing right hand side, we have conservation of mass for the CCFV lattice Boltzmann scheme (5.4).

For non-negative initial data U_l^0, V_l^0 and non-negative source terms $F_{l,+}^k$, $F_{l,-}^k$, the discrete solutions U_l^k, V_l^k of the VCFD lattice Boltzmann equations (5.1) with periodic boundary conditions (5.5) are non-negative. For non-negative data, the solutions of (5.1) in combination with density boundary conditions (5.6) are non-negative, provided $\omega \geq 1/2$ and $r_L^k, r_R^k \geq 0$. For the flux boundary conditions (5.8), we have to impose $j_L^k \geq 0$ and $j_R^k \leq 0$ in addition to the non-negativity of the initial data and the source terms in order to assure non-negativity of the solutions of (5.1). For the inflow boundary conditions (5.10), we have to provide $u_L^k, v_R^k \geq 0$.

For the CCFD lattice Boltzmann equations (5.3) in combination with flux boundary conditions (5.13) with $j_L^{k+\delta} \geq 0$, $j_R^{k+\delta} \leq 0$ and non-negative initial data and source terms, the solutions $U_{l-1/2}^k$ and $V_{l-1/2}^k$ are non-negative.

5.6. Matrix Formulations for the UV-Systems

For an analysis of the lattice Boltzmann equations we switch to matrix formulations. For the FD schemes (5.1) and (5.2) we can directly prove stability with respect to discrete norms. For the stability of the FV schemes (5.3) and (5.4) the analysis of Fourier representations is required.

The lattice Boltzmann equations (5.1) on VCFD grids and (5.3) on VCFV grids can be written in matrix formulation as equation

$$\boldsymbol{W}^{k+1} = \boldsymbol{M}\boldsymbol{W}^k + \boldsymbol{B}^k + \tau \boldsymbol{F}^k \quad \text{for } k = 0, \dots, M - 1 \tag{5.15}$$

for the unknowns $\boldsymbol{W}^k$, the boundary values $\boldsymbol{B}^k$ and the source terms $\boldsymbol{F}^k$. Here, we employ the time evolution matrix $\boldsymbol{M}$.

On VC grids we consider the unknowns $\boldsymbol{W}^k$ given by

$$\boldsymbol{W}^k := \left[V_0^k; U_1^k; V_1^k; \dots; U_{N-1}^k; V_{N-1}^k; U_N^k \right] \in \mathbb{R}^{2N}$$

and the right hand side

$$\boldsymbol{F}^k := \frac{1}{2} \left[F_{1,-}^k; F_{0,+}^k; F_{2,-}^k; F_{1,+}^k; F_{3,-}^k; \dots; F_{N-2,+}^k; F_{N,-}^k; h F_{N-1,+}^k \right] \in \mathbb{R}^{2N}.$$

The time evolution matrix $\boldsymbol{M} \in \mathbb{R}^{2N,2N}$ for the lattice Boltzmann equations (5.1) on VCFD grids in combination with periodic boundary conditions (5.5) is given by

$$
\boldsymbol{M} := \begin{bmatrix}
0 & \omega & 1-\omega & & & & & & & & & \\
\omega & 0 & 0 & & & & & & & & & 1-\omega \\
& 0 & 0 & \omega & 1-\omega & & & & & & & \\
& 1-\omega & \omega & 0 & 0 & & & & & & & \\
& & & 0 & 0 & \omega & 1-\omega & & & & & \\
& & & & & \ddots & & & & & & \\
& & & & & 1-\omega & \omega & 0 & 0 & & & \\
& & & & & & & 0 & 0 & \omega & 1-\omega & \\
& & & & & & & 1-\omega & \omega & 0 & 0 & \\
1-\omega & & & & & & & & & 0 & 0 & \omega \\
& & & & & & & & & 1-\omega & \omega & 0
\end{bmatrix} .
$$

$$\tag{5.16}$$

For the lattice Boltzmann equations (5.3) on VCFV grids with periodic boundary conditions (5.5) the time evolution matrix $\boldsymbol{M} \in \mathbb{R}^{2N,2N}$ results in

$$
\boldsymbol{M} := \frac{1}{2} \begin{bmatrix}
-\omega & \omega & 2-\omega & & & & & & & & & \omega \\
\omega & -\omega & \omega & & & & & & & & & 2-\omega \\
& \omega & -\omega & \omega & 2-\omega & & & & & & & \\
& 2-\omega & \omega & -\omega & \omega & & & & & & & \\
& & & \omega & -\omega & \omega & 2-\omega & & & & & \\
& & & & & \ddots & & & & & & \\
& & & & & 2-\omega & \omega & -\omega & \omega & & & \\
& & & & & & & \omega & -\omega & \omega & 2-\omega & \\
& & & & & & & 2-\omega & \omega & -\omega & \omega & \\
2-\omega & & & & & & & & & \omega & -\omega & \omega \\
\omega & & & & & & & & & 2-\omega & \omega & -\omega
\end{bmatrix} .
$$

$$\tag{5.17}$$

The boundary values are chosen as $\boldsymbol{B}^k := \boldsymbol{0}$ in both cases.

In the density, flux or inflow boundary case on VCFD grids the time evolution matrix $\boldsymbol{M} \in \mathbb{R}^{2N,2N}$ for the lattice Boltzmann equations (5.1) reads

$$
\boldsymbol{M} := \begin{bmatrix}
0 & \omega & 1-\omega & & & & & & & & \\
m(\omega) & 0 & 0 & 0 & & & & & & & \\
& 0 & 0 & \omega & 1-\omega & & & & & & \\
& 1-\omega & \omega & 0 & 0 & & & & & & \\
& & & 0 & 0 & \omega & 1-\omega & & & & \\
& & & & & \ddots & & & & & \\
& & & & & 1-\omega & \omega & 0 & 0 & & \\
& & & & & & & 0 & 0 & \omega & 1-\omega \\
& & & & & & & 1-\omega & \omega & 0 & 0 \\
& & & & & & & & & 0 & 0 & m(\omega) \\
& & & & & & & & & 1-\omega & \omega & 0
\end{bmatrix} .
$$

$$\tag{5.18}$$

For density boundary conditions (5.6) we choose

$$
m(\omega) := 2\omega - 1 \tag{5.19}
$$

and the boundary values are given by

$$
\boldsymbol{B}^k := (1-\omega)\big[0; r_L^k; 0; \ldots; 0; r_R^k; 0\big] \in \mathbb{R}^{2N}.
$$

For flux boundary conditions (5.8) we choose

$$m(\omega) := 1 \tag{5.20}$$

and the boundary values are given by

$$\boldsymbol{B}^k := (1 - \omega)\big[0; j_L^k; 0; \ldots; 0; -j_R^k; 0\big] \in \mathbb{R}^{2N}.$$

For inflow boundary (5.10) conditions we choose

$$m(\omega) := \omega$$

and the boundary values are given by

$$\boldsymbol{B}^k := (1 - \omega)\big[0; u_L^k; 0; \ldots; 0; v_R^k; 0\big] \in \mathbb{R}^{2N}.$$

In the density, flux or inflow boundary case on VCFV grids the time evolution matrix $\boldsymbol{M} \in \mathbb{R}^{2N,2N}$ for the lattice Boltzmann equations (5.3) reads

$$\boldsymbol{M} := \frac{1}{2}\begin{bmatrix}
m_1(\omega) & \omega & 2-\omega & & & & & & & \\
m_2(\omega) & -\omega & \omega & & & & & & & \\
& \omega & -\omega & \omega & 2-\omega & & & & & \\
& 2-\omega & \omega & -\omega & \omega & & & & & \\
& & & \omega & -\omega & \omega & 2-\omega & & & \\
& & & & & \ddots & & & & \\
& & & 2-\omega & \omega & -\omega & \omega & & & \\
& & & & & \omega & -\omega & \omega & 2-\omega & \\
& & & & & 2-\omega & \omega & -\omega & \omega & \\
& & & & & & & \omega & -\omega & m_2(\omega) \\
& & & & & & & 2-\omega & \omega & m_1(\omega)
\end{bmatrix}. \tag{5.21}$$

For density boundary conditions (5.6) we choose

$$m_1(\omega) := -2\omega, \quad m_2(\omega) := -(2 - 2\omega) \tag{5.22}$$

and the boundary values are given by

$$\boldsymbol{B}^k := \Big[\frac{\omega}{2}r_L^k; \Big(1 - \frac{\omega}{2}\Big)r_L^k; 0; \ldots; 0; \Big(1 - \frac{\omega}{2}\Big)r_R^k; \frac{\omega}{2}r_R^k\Big] \in \mathbb{R}^{2N}.$$

For flux boundary conditions (5.8) we choose

$$m_1(\omega) := 0, \quad m_2(\omega) := 2 \tag{5.23}$$

and the boundary values are given by

$$\boldsymbol{B}^k := \Big[\frac{\omega}{2}j_L^k; \Big(1 - \frac{\omega}{2}\Big)j_L^k; 0; \ldots; 0; -\Big(1 - \frac{\omega}{2}\Big)j_R^k; -\frac{\omega}{2}j_R^k\Big] \in \mathbb{R}^{2N}.$$

For inflow boundary (5.10) conditions we choose

$$m_1(\omega) := -\omega, \quad m_2(\omega) := \omega$$

and the boundary values are given by

$$\boldsymbol{B}^k := \Big[\frac{\omega}{2}u_L^k; \Big(1 - \frac{\omega}{2}\Big)u_L^k; 0; \ldots; 0; \Big(1 - \frac{\omega}{2}\Big)v_R^k; \frac{\omega}{2}v_R^k\Big] \in \mathbb{R}^{2N}.$$

On CC grids with density or flux boundary conditions the lattice Boltzmann equations (5.2) and (5.4) can be written as matrix equations

$$\boldsymbol{W}^{k+1} = \boldsymbol{M}\boldsymbol{W}^k + \boldsymbol{B}^{k+\delta} + \tau\boldsymbol{F}^k \quad \text{for } k = 0, \ldots, M - 1 \tag{5.24}$$

for the unknown quantities

$$\boldsymbol{W}^k := \big[U_{1/2}^k; V_{1/2}^k; U_{3/2}^k; V_{3/2}^k; \ldots; U_{N-1/2}^k; V_{N-1/2}^k\big] \in \mathbb{R}^{2N}.$$

The boundary values are now evaluated at time $t^{k+\delta} := t^k + \delta\tau$ for a $\delta \in [0,1]$. The time evolution matrix $\boldsymbol{M} \in \mathbb{R}^{2N,2N}$ for the lattice Boltzmann equations (5.2) is chosen as

$$
\boldsymbol{M} := \begin{bmatrix}
n_1(\omega) & n_2(\omega) & & & & & & \\
0 & 0 & \omega & 1-\omega & & & & \\
1-\omega & \omega & 0 & 0 & & & & \\
& & 0 & 0 & \omega & 1-\omega & & \\
& & & & \ddots & & & \\
& & 1-\omega & \omega & 0 & 0 & & \\
& & & & 0 & 0 & \omega & 1-\omega \\
& & & & 1-\omega & \omega & 0 & 0 \\
& & & & & & n_2(\omega) & n_1(\omega)
\end{bmatrix}. \tag{5.25}
$$

For density boundary conditions (5.12) on CCFD grids we choose

$$
n_1(\omega) := -\omega, \quad n_2(\omega) := -(1-\omega), \tag{5.26}
$$

the right hand side

$$
\boldsymbol{F}^k := \frac{1}{2}\Big[-F^k_{1/2,-}; F^k_{3/2,-}; F^k_{1/2,+}; F^k_{5/2,-}; F^k_{3/2,+}; F^k_{7/2,-}; \ldots
$$
$$
\ldots; F_{N-5/2,+}; F^k_{N-1/2,-}; F^k_{N-3/2,+}; -F^k_{N-1/2,+}\Big] \in \mathbb{R}^{2N}
$$

and boundary values

$$
\boldsymbol{B}^{k+\delta} := \big[r_L^{k+\delta}; 0; \ldots; 0; r_R^{k+\delta}\big] \in \mathbb{R}^{2N}.
$$

For flux boundary conditions (5.13) on CCFD grids we choose

$$
n_1(\omega) := \omega, \quad n_2(\omega) := 1-\omega, \tag{5.27}
$$

the right hand side

$$
\boldsymbol{F}^k := \frac{1}{2}\Big[F^k_{1/2,-}; F^k_{3/2,-}; F^k_{1/2,+}; F^k_{5/2,-}; F^k_{3/2,+}; F^k_{7/2,-}; \ldots
$$
$$
\ldots; F_{N-5/2,+}; F^k_{N-1/2,-}; F^k_{N-3/2,+}; F^k_{N-1/2,+}\Big] \in \mathbb{R}^{2N}
$$

and boundary values

$$
\boldsymbol{B}^{k+\delta} := (1-\omega)\big[j_L^{k+\delta}; 0; \ldots; 0; -j_R^{k+\delta}\big] \in \mathbb{R}^{2N}.
$$

The time evolution matrix $\boldsymbol{M} \in \mathbb{R}^{2N,2N}$ for the lattice Boltzmann equations (5.4) is chosen as

$$
\boldsymbol{M} := \frac{1}{2}\begin{bmatrix}
n_1(\omega) & n_2(\omega) & & & & & & \\
\omega & -\omega & \omega & 2-\omega & & & & \\
2-\omega & \omega & -\omega & \omega & & & & \\
& & \omega & -\omega & \omega & 2-\omega & & \\
& & & & \ddots & & & \\
& & 2-\omega & \omega & -\omega & \omega & & \\
& & & & \omega & -\omega & \omega & 2-\omega \\
& & & & 2-\omega & \omega & -\omega & \omega \\
& & & & & & n_2(\omega) & n_1(\omega)
\end{bmatrix}. \tag{5.28}
$$

For density boundary conditions (5.12) on CCFV grids we choose

$$
n_1(\omega) := -2\omega, \quad n_2(\omega) := -(2-2\omega), \tag{5.29}
$$

the right hand side

$$\boldsymbol{F}^k := \frac{1}{2}\Big[-F^k_{1/2,-}; F^k_{3/2,-}; F^k_{1/2,+}; F^k_{5/2,-}; F^k_{3/2,+}; F^k_{7/2,-}; \cdots$$

$$\cdots; F^k_{N-5/2,+}; F^k_{N-1/2,-}; F^k_{N-3/2,+}; -F^k_{N-1/2,+}\Big] \in \mathbb{R}^{2N}$$

and boundary values

$$\boldsymbol{B}^{k+\delta} := \big[r^{k+\delta}_L; 0; \ldots; 0; r^{k+\delta}_R\big] \in \mathbb{R}^{2N}.$$

For flux boundary conditions (5.14) on CCFV grids we choose

$$n_1(\omega) := 0, \quad n_2(\omega) := 2, \tag{5.30}$$

the right hand side

$$\boldsymbol{F}^k := \frac{1}{2}\Big[0; F^k_{3/2,-}; F^k_{1/2,+}; F^k_{5/2,-}; F^k_{3/2,+}; F^k_{7/2,-}; \cdots$$

$$\cdots; F^k_{N-5/2,+}; F^k_{N-1/2,-}; F^k_{N-3/2,+}; 0\Big] \in \mathbb{R}^{2N}$$

and boundary values

$$\boldsymbol{B}^{k+\delta} := \big[j^{k+\delta}_L; 0; \ldots; 0; -j^{k+\delta}_R\big] \in \mathbb{R}^{2N}.$$

5.7. Matrix Formulations for the *RJ*-Systems

For further examinations we switch to the variables R^k_l and J^k_l describing the density r and the flux j. This transformation is done upon multiplying the lattice Boltzmann systems (5.15) and (5.24) by nonsingular transformation matrices $\boldsymbol{T} \in \mathbb{R}^{2N,2N}$.

5.7.1. FD Lattice Boltzmann Schemes. In this section we consider the systems belonging to the FD lattice Boltzmann schemes (5.1) and (5.2).

For periodic boundary conditions (5.5) on VCFD grids we use the new variables $\boldsymbol{P}^k := [\boldsymbol{R}^k; \boldsymbol{J}^k] := \boldsymbol{T}\boldsymbol{W}^k$ with

$$\boldsymbol{R}^k := \big[R^k_0; \ldots; R^k_{N-1}\big] \in \mathbb{R}^N, \quad \boldsymbol{J}^k := \big[J^k_0; \ldots; J^k_{N-1}\big] \in \mathbb{R}^N. \tag{5.31}$$

A transformation of the basis results in the matrix equation

$$\boldsymbol{P}^{k+1} = \boldsymbol{K}\boldsymbol{P}^k + \tau \boldsymbol{G}^k \tag{5.32}$$

with $\boldsymbol{K} := \boldsymbol{T}\boldsymbol{M}\boldsymbol{T}^{-1}$ and $\boldsymbol{G}^k := [\boldsymbol{f}^k; \boldsymbol{g}^k] := \boldsymbol{T}\boldsymbol{F}^k$, where the transformation matrix $\boldsymbol{T} \in \mathbb{R}^{2N,2N}$ is defined by

$$\boldsymbol{T} := \begin{bmatrix} 1 & & & & & & & & 1 \\ & 1 & 1 & & & & & & \\ & & & \ddots & & & & & \\ & & & & 1 & 1 & & & \\ & & & & & & 1 & 1 & \\ -1 & & & & & & & & 1 \\ & 1 & -1 & & & & & & \\ & & & \ddots & & & & & \\ & & & & 1 & -1 & & & \\ & & & & & & 1 & -1 & \end{bmatrix}. \tag{5.33}$$

Then the time evolution matrix $\boldsymbol{K}$ is of block form. We obtain

$$\boldsymbol{K} = \frac{1}{2}\begin{bmatrix} \boldsymbol{A} & \theta\boldsymbol{D} \\ -\boldsymbol{D} & -\theta\boldsymbol{A} \end{bmatrix}, \tag{5.34}$$

where we define $\theta := 2\omega - 1$. The submatrices $\boldsymbol{A} \in \mathbb{R}^{N,N}$ and $\boldsymbol{D} \in \mathbb{R}^{N,N}$ are given by

$$
\boldsymbol{A} := \begin{bmatrix} 0 & 1 & & & 1 \\ 1 & 0 & 1 & & \\ & & \ddots & & \\ & & 1 & 0 & 1 \\ 1 & & & 1 & 0 \end{bmatrix}, \quad
\boldsymbol{D} := \begin{bmatrix} 0 & 1 & & & -1 \\ -1 & 0 & 1 & & \\ & & \ddots & & \\ & & -1 & 0 & 1 \\ 1 & & & -1 & 0 \end{bmatrix}. \tag{5.35}
$$

Furthermore, we get the transformed right hand sides

$$
\begin{aligned}
\boldsymbol{f}^k &= \frac{1}{2}\left[F^k_{N-1,+} + F^k_{1,-}; F^k_{0,+} + F^k_{2,-}; \ldots; F^k_{N-2,+} + F^k_{N,-}\right] \in \mathbb{R}^N, \\
\boldsymbol{g}^k &= \frac{1}{2}\left[F^k_{N-1,+} - F^k_{1,-}; F^k_{0,+} - F^k_{2,-}; \ldots; F^k_{N-2,+} - F^k_{N,-}\right] \in \mathbb{R}^N.
\end{aligned} \tag{5.36}
$$

On VCFD grids with density, flux or inflow boundary conditions we have to incorporate the boundary values in the form

$$
\boldsymbol{P}^k := \left[\boldsymbol{R}^k; \boldsymbol{J}^k\right] := \boldsymbol{T}\boldsymbol{W}^k + \boldsymbol{Q}^k.
$$

Hence, we get the matrix equation

$$
\boldsymbol{P}^{k+1} = \boldsymbol{K}\boldsymbol{P}^k + \boldsymbol{C}^k + \boldsymbol{Q}^{k+1} - \boldsymbol{K}\boldsymbol{Q}^k + \tau \boldsymbol{G}^k \tag{5.37}
$$

with $\boldsymbol{K} := \boldsymbol{T}\boldsymbol{M}\boldsymbol{T}^{-1}$ and $\boldsymbol{G}^k := [\boldsymbol{f}^k; \boldsymbol{g}^k] := \boldsymbol{T}\boldsymbol{F}^k$. Furthermore, we use $\boldsymbol{Q}^k := [\boldsymbol{p}^k; \boldsymbol{q}^k]$ and $\boldsymbol{C}^k := [\boldsymbol{c}^k; \boldsymbol{d}^k] := \boldsymbol{T}\boldsymbol{B}^k$.

For density boundary conditions (5.6) on VCFD grids the transformation matrix $\boldsymbol{T} \in \mathbb{R}^{2N,2N}$ is defined by

$$
\boldsymbol{T} := \begin{bmatrix} & 1 & 1 & & & & & & \\ & & & \ddots & & & & & \\ & & & & 1 & 1 & & & \\ & & & & & & 1 & 1 & \\ -2 & & & & & & & & \\ & 1 & -1 & & & & & & \\ & & & \ddots & & & & & \\ & & & & 1 & -1 & & & \\ & & & & & & 1 & -1 & \\ & & & & & & & & 2 \end{bmatrix}. \tag{5.38}
$$

We get the new variables

$$
\boldsymbol{R}^k = \left[R^k_1; \ldots; R^k_{N-1}\right] \in \mathbb{R}^{N-1}, \quad \boldsymbol{J}^k = \left[J^k_0; \ldots; J^k_N\right] \in \mathbb{R}^{N+1}, \tag{5.39}
$$

using

$$
\boldsymbol{p}^k := [0; \ldots; 0] \in \mathbb{R}^{N-1}, \quad \boldsymbol{q}^k := \left[r^k_L; 0; \ldots; 0; -r^k_R\right] \in \mathbb{R}^{N+1}.
$$

The time evolution matrix $\boldsymbol{K}$ then reads

$$
\boldsymbol{K} = \frac{1}{2}\begin{bmatrix} \boldsymbol{A} & \theta\boldsymbol{D} \\ -\boldsymbol{E} & -\theta\boldsymbol{B} \end{bmatrix} \tag{5.40}
$$

with $\theta := 2\omega - 1$ and $\boldsymbol{A} \in \mathbb{R}^{N-1,N-1}$, $\boldsymbol{B} \in \mathbb{R}^{N+1,N+1}$, $\boldsymbol{D} \in \mathbb{R}^{N-1,N+1}$ and $\boldsymbol{E} \in \mathbb{R}^{N+1,N-1}$ given by

$$
\boldsymbol{A} := \begin{bmatrix} 0 & 1 & & & \\ 1 & 0 & 1 & & \\ & & \ddots & & \\ & & 1 & 0 & 1 \\ & & & 1 & 0 \end{bmatrix}, \quad
\boldsymbol{D} := \begin{bmatrix} -1 & 0 & 1 & & & \\ & -1 & 0 & 1 & & \\ & & & \ddots & & \\ & & & -1 & 0 & 1 \\ & & & & -1 & 0 & 1 \end{bmatrix},
$$

$$
\tag{5.41}
$$

$$
\boldsymbol{E} := \begin{bmatrix} 2 & & \\ 0 & 1 & \\ -1 & 0 & 1 \\ & \ddots & \\ & -1 & 0 & 1 \\ & & -1 & 0 \\ & & & -2 \end{bmatrix}, \quad
\boldsymbol{B} := \begin{bmatrix} 0 & 2 & & & \\ 1 & 0 & 1 & & \\ & 1 & 0 & 1 & \\ & & \ddots & & \\ & & 1 & 0 & 1 \\ & & & 1 & 0 & 1 \\ & & & & 2 & 0 \end{bmatrix}.
$$

The right hand sides take the form

$$
\boldsymbol{f}^k = \frac{1}{2}\left[F_{0,+}^k + F_{2,-}^k ; F_{1,+}^k + F_{3,-}^k ; \ldots ; F_{N-2,+}^k + F_{N,-}^k \right] \in \mathbb{R}^{N-1},
$$

$$
\boldsymbol{g}^k = \frac{1}{2}\left[-2F_{1,-}^k ; F_{0,+}^k - F_{2,-}^k ; F_{1,+}^k - F_{3,-}^k ; \ldots ; F_{N-2,+}^k - F_{N,-}^k ; 2F_{N-1,+}^k \right] \in \mathbb{R}^{N+1}
$$

$$
\tag{5.42}
$$

and

$$
\boldsymbol{c}^k = (1-\omega)\left[r_L^k ; 0 ; \ldots ; 0 ; r_R^k \right] \in \mathbb{R}^{N-1},
$$

$$
\boldsymbol{d}^k = (1-\omega)\left[0 ; r_L^k ; 0 ; \ldots ; 0 ; -r_R^k ; 0 \right] \in \mathbb{R}^{N+1}.
$$

For flux boundary conditions (5.8) on VCFD grids the transformation matrix $\boldsymbol{T} \in \mathbb{R}^{2N,2N}$ is defined by

$$
\boldsymbol{T} := \begin{bmatrix} 2 & & & & & & & & \\ & 1 & & 1 & & & & & \\ & & \ddots & & & & & & \\ & & & 1 & & 1 & & & \\ & & & & 1 & & 1 & & \\ & & & & & & & 2 & \\ & 1 & -1 & & & & & & \\ & & \ddots & & & & & & \\ & & & 1 & -1 & & & & \\ & & & & 1 & -1 & & & \end{bmatrix}.
\tag{5.43}
$$

We get the new variables

$$
\boldsymbol{R}^k = \left[R_0^k ; \ldots ; R_N^k \right] \in \mathbb{R}^{N+1}, \quad \boldsymbol{J}^k = \left[J_1^k ; \ldots ; J_{N-1}^k \right] \in \mathbb{R}^{N-1},
\tag{5.44}
$$

using

$$
\boldsymbol{p}^k := \left[j_L^k ; 0 ; \ldots ; 0 ; -j_R^k \right] \in \mathbb{R}^{N+1}, \quad \boldsymbol{q}^k := \left[0 ; \ldots ; 0 \right] \in \mathbb{R}^{N-1}.
$$

The time evolution matrix $\boldsymbol{K}$ now reads

$$
\boldsymbol{K} = \frac{1}{2} \begin{bmatrix} \boldsymbol{B} & \theta\boldsymbol{E} \\ -\boldsymbol{D} & -\theta\boldsymbol{A} \end{bmatrix}
\tag{5.45}
$$

with $\theta := 2\omega - 1$ and $\boldsymbol{A} \in \mathbb{R}^{N-1,N-1}$, $\boldsymbol{B} \in \mathbb{R}^{N+1,N+1}$, $\boldsymbol{D} \in \mathbb{R}^{N-1,N+1}$ and $\boldsymbol{E} \in \mathbb{R}^{N+1,N-1}$ given in (5.41). We find the right hand sides

$$\boldsymbol{f}^k = \frac{1}{2}\left[2F^k_{1,-}; F^k_{0,+} + F^k_{2,-}; F^k_{1,+} + F^k_{3,-}; \ldots; F^k_{N-2,+} + F^k_{N,-}; 2F^k_{N-1,+}\right] \in \mathbb{R}^{N+1},$$

$$\boldsymbol{g}^k = \frac{1}{2}\left[F^k_{0,+} - F^k_{2,-}; F^k_{1,+} - F^k_{3,-}; \ldots; F^k_{N-2,+} - F^k_{N,-}\right] \in \mathbb{R}^{N-1}$$

$$(5.46)$$

and

$$\boldsymbol{c}^k = (1 - \omega)\left[0; j^k_L; 0; \ldots; 0; -j^k_R; 0\right] \in \mathbb{R}^{N+1},$$

$$\boldsymbol{d}^k = (1 - \omega)\left[j^k_L; 0; \ldots; 0; j^k_R\right] \in \mathbb{R}^{N-1}.$$

For inflow boundary conditions (5.10) on VCFD grids the transformation matrix $\boldsymbol{T} \in \mathbb{R}^{2N,2N}$ is defined by

$$\boldsymbol{T} := \begin{bmatrix} 1 & & & & & & & & & \\ & 1 & 1 & & & & & & & \\ & & & \ddots & & & & & & \\ & & & & 1 & 1 & & & & \\ & & & & & & 1 & 1 & & \\ & & & & & & & & 1 & \\ & 1 & -1 & & & & & & & \\ & & & \ddots & & & & & & \\ & & & & 1 & -1 & & & & \\ & & & & & & 1 & -1 & & \end{bmatrix}.$$

$$(5.47)$$

We get the new variables

$$\boldsymbol{R}^k = \left[R^k_0; \ldots; R^k_N\right] \in \mathbb{R}^{N+1}, \quad \boldsymbol{J}^k = \left[J^k_1; \ldots; J^k_{N-1}\right] \in \mathbb{R}^{N-1}, \qquad (5.48)$$

using

$$\boldsymbol{p}^k := \left[u^k_L; 0; \ldots; 0; v^k_R\right] \in \mathbb{R}^{N+1}, \quad \boldsymbol{q}^k := \left[0; \ldots; 0\right] \in \mathbb{R}^{N-1}.$$

The time evolution matrix $\boldsymbol{K}$ then reads

$$\boldsymbol{K} = \frac{1}{2}\begin{bmatrix} \boldsymbol{A} & \theta\boldsymbol{D} \\ -\boldsymbol{E} & -\theta\boldsymbol{B} \end{bmatrix} \qquad (5.49)$$

with $\theta := 2\omega - 1$ and $\boldsymbol{A} \in \mathbb{R}^{N+1,N+1}$, $\boldsymbol{B} \in \mathbb{R}^{N-1,N-1}$, $\boldsymbol{D} \in \mathbb{R}^{N+1,N-1}$ and $\boldsymbol{E} \in \mathbb{R}^{N-1,N+1}$ given by

$$\boldsymbol{A} := \begin{bmatrix} 0 & 1 & & & & & \\ 2\omega & 0 & 1 & & & & \\ & 1 & 0 & 1 & & & \\ & & & \ddots & & & \\ & & & 1 & 0 & 1 & \\ & & & & 1 & 0 & 2\omega \\ & & & & & 1 & 0 \end{bmatrix}, \quad \boldsymbol{D} := \begin{bmatrix} 1 & & & & \\ 0 & 1 & & & \\ -1 & 0 & 1 & & \\ & & \ddots & & \\ & -1 & 0 & 1 \\ & & -1 & 0 \\ & & & -1 \end{bmatrix},$$

$$(5.50)$$

$$\boldsymbol{E} := \begin{bmatrix} -2\omega & 0 & 1 & & & \\ & -1 & 0 & 1 & & \\ & & \ddots & & & \\ & & & -1 & 0 & 1 \\ & & & & -1 & 0 & 2\omega \end{bmatrix}, \quad \boldsymbol{B} := \begin{bmatrix} 0 & 1 & & & \\ 1 & 0 & 1 & & \\ & & \ddots & & \\ & & 1 & 0 & 1 \\ & & & 1 & 0 \end{bmatrix}.$$

We find the right hand sides

$$\boldsymbol{f}^k = \frac{1}{2}\big[F_{1,-}^k; F_{0,+}^k + F_{2,-}^k; F_{1,+}^k + F_{3,-}^k; \ldots; F_{N-2,+}^k + F_{N,-}^k; F_{N-1,+}^k\big] \in \mathbb{R}^{N+1},$$

$$\boldsymbol{g}^k = \frac{1}{2}\big[F_{0,+}^k - F_{2,-}^k; F_{1,+}^k - F_{3,-}^k; \ldots; F_{N-2,+}^k - F_{N,-}^k\big] \in \mathbb{R}^{N-1}$$

$$(5.51)$$

and

$$\boldsymbol{c}^k = (1-\omega)\big[0; u_L^k; 0; \ldots; 0; v_R^k; 0\big] \in \mathbb{R}^{N+1},$$

$$\boldsymbol{d}^k = (1-\omega)\big[u_L^k; 0; \ldots, 0; -v_R^k\big] \in \mathbb{R}^{N-1}.$$

On CCFD grids we use

$$\boldsymbol{P}^k := \big[\boldsymbol{R}^k; \boldsymbol{J}^k\big] := \boldsymbol{T}\boldsymbol{W}^k.$$

Hence, we get the matrix equation

$$\boldsymbol{P}^{k+1} = \boldsymbol{K}\boldsymbol{P}^k + \boldsymbol{C}^{k+\delta} + \tau\boldsymbol{G}^k \tag{5.52}$$

with $\boldsymbol{K} := \boldsymbol{T}\boldsymbol{M}\boldsymbol{T}^{-1}$, $\boldsymbol{C}^{k+\delta} := [\boldsymbol{c}^{k+\delta}; \boldsymbol{d}^{k+\delta}] := \boldsymbol{T}\boldsymbol{B}^{k+\delta}$ and $\boldsymbol{G}^k := [\boldsymbol{f}^k; \boldsymbol{g}^k] := \boldsymbol{T}\boldsymbol{F}^k$. For the density boundary conditions (5.12) and for the flux boundary conditions (5.13) we use the transformation matrix $\boldsymbol{T} \in \mathbb{R}^{2N,2N}$ defined by

$$\boldsymbol{T} := \begin{bmatrix} 1 & 1 & & & & & & \\ & & \ddots & & & & & \\ & & & 1 & 1 & & & \\ & & & & & 1 & 1 & \\ 1 & -1 & & & & & & \\ & & \ddots & & & & & \\ & & & 1 & -1 & & & \\ & & & & & 1 & -1 & \end{bmatrix}. \tag{5.53}$$

We get the unknowns

$$\boldsymbol{R}^k = \big[R_{1/2}^k; \ldots; R_{N-1/2}^k\big] \in \mathbb{R}^N, \quad \boldsymbol{J}^k = \big[J_{1/2}^k; \ldots; J_{N-1/2}^k\big] \in \mathbb{R}^N. \tag{5.54}$$

For density boundary conditions (5.12) on CCFD grids the time evolution matrix $\boldsymbol{K} \in \mathbb{R}^{2N,2N}$ reads

$$\boldsymbol{K} = \frac{1}{2}\begin{bmatrix} \boldsymbol{A} & \theta\boldsymbol{D} \\ -\boldsymbol{E} & -\theta\boldsymbol{B} \end{bmatrix} \tag{5.55}$$

with $\theta := 2\omega - 1$ and $\boldsymbol{A}, \boldsymbol{B}, \boldsymbol{D}, \boldsymbol{E} \in \mathbb{R}^{N,N}$ given by

$$\boldsymbol{A} := \begin{bmatrix} -1 & 1 & & & \\ 1 & 0 & 1 & & \\ & & \ddots & & \\ & & 1 & 0 & 1 \\ & & & 1 & -1 \end{bmatrix}, \quad \boldsymbol{D} := \begin{bmatrix} -1 & 1 & & & \\ -1 & 0 & 1 & & \\ & & \ddots & & \\ & & -1 & 0 & 1 \\ & & & -1 & 1 \end{bmatrix},$$

$$(5.56)$$

$$\boldsymbol{E} := \begin{bmatrix} 1 & 1 & & & \\ -1 & 0 & 1 & & \\ & & \ddots & & \\ & & -1 & 0 & 1 \\ & & & -1 & -1 \end{bmatrix}, \quad \boldsymbol{B} := \begin{bmatrix} 1 & 1 & & & \\ 1 & 0 & 1 & & \\ & & \ddots & & \\ & & 1 & 0 & 1 \\ & & & 1 & 1 \end{bmatrix}.$$

We find the right hand sides

$$\boldsymbol{f}^k = \frac{1}{2}\Big[-F^k_{1/2,-} + F^k_{3/2,-}; F^k_{1/2,+} + F^k_{5/2,-}; F^k_{3/2,+} + F^k_{7/2,-}; \cdots$$
$$\cdots; F^k_{N-5/2,+} + F^k_{N-1/2,-}; F^k_{N-3/2,+} - F^k_{N-1/2,+}\Big] \in \mathbb{R}^N,$$
$$\boldsymbol{g}^k = \frac{1}{2}\Big[-F^k_{1/2,-} - F^k_{3/2,-}; F^k_{1/2,+} - F^k_{5/2,-}; F^k_{3/2,+} - F^k_{7/2,-}; \cdots$$
$$\cdots; F^k_{N-5/2,+} - F^k_{N-1/2,-}; F^k_{N-3/2,+} + F^k_{N-1/2,+}\Big] \in \mathbb{R}^N \tag{5.57}$$

and

$$\boldsymbol{c}^{k+\delta} = \big[r^{k+\delta}_L; 0; \ldots, 0; r^{k+\delta}_R\big] \in \mathbb{R}^N, \quad \boldsymbol{d}^{k+\delta} = \big[r^{k+\delta}_L; 0; \ldots; 0; -r^{k+\delta}_R\big] \in \mathbb{R}^N.$$

For flux boundary conditions (5.13) on CCFD grids the time evolution matrix $\boldsymbol{K} \in \mathbb{R}^{2N,2N}$ reads

$$\boldsymbol{K} = \frac{1}{2}\begin{bmatrix} \boldsymbol{B} & \theta\boldsymbol{E} \\ -\boldsymbol{D} & -\theta\boldsymbol{A} \end{bmatrix} \tag{5.58}$$

with $\theta := 2\omega - 1$ and $\boldsymbol{A}, \boldsymbol{B}, \boldsymbol{D}, \boldsymbol{E} \in \mathbb{R}^{N,N}$ given in (5.56). We find the right hand sides

$$\boldsymbol{f}^k = \frac{1}{2}\Big[F^k_{1/2,-} + F^k_{3/2,-}; F^k_{1/2,+} + F^k_{5/2,-}; F^k_{3/2,+} + F^k_{7/2,-}; \cdots$$
$$\cdots; F^k_{N-5/2,+} + F^k_{N-1/2,-}; F^k_{N-3/2,+} + F^k_{N-1/2,+}\Big] \in \mathbb{R}^N,$$
$$\boldsymbol{g}^k = \frac{1}{2}\Big[F^k_{1/2,-} - F^k_{3/2,-}; F^k_{1/2,+} - F^k_{5/2,-}; F^k_{3/2,+} - F^k_{7/2,-}; \cdots$$
$$\cdots; F^k_{N-5/2,+} - F^k_{N-1/2,-}; F^k_{N-3/2,+} - F^k_{N-1/2,+}\Big] \in \mathbb{R}^N \tag{5.59}$$

and

$$\boldsymbol{c}^{k+\delta} = (1-\omega)\big[j^{k+\delta}_L; 0; \ldots; 0; -j^{k+\delta}_R\big] \in \mathbb{R}^N,$$
$$\boldsymbol{d}^{k+\delta} = (1-\omega)\big[j^{k+\delta}_L; 0; \ldots; 0; j^{k+\delta}_R\big] \in \mathbb{R}^N.$$

5.7.2. FV Lattice Boltzmann Schemes. In this section we examine the discrete systems belonging to the FV lattice Boltzmann schemes (5.3) and (5.4). While the variables $\boldsymbol{R}^k$ and $\boldsymbol{J}^k$ and their transformations remain unchanged in comparison to the previous section, the time evolution matrix $\boldsymbol{K}$ and parts of the right hand sides differ.

For periodic boundary conditions (5.5) on VCFV grids we use $\boldsymbol{T}$ as in (5.33). We get the matrix equation (5.32) for the variables (5.31) with $\boldsymbol{K}$ given by

$$\boldsymbol{K} = \frac{1}{2}\begin{bmatrix} \boldsymbol{A} & -(1-\omega)\boldsymbol{D} \\ -\boldsymbol{D} & (1-\omega)\boldsymbol{A} - 2\omega\boldsymbol{Id}_N \end{bmatrix} \tag{5.60}$$

with $\boldsymbol{A} \in \mathbb{R}^{N,N}$ and $\boldsymbol{D} \in \mathbb{R}^{N,N}$ given in (5.35). Furthermore, $\boldsymbol{Id}_N$ denotes the identity matrix in $\mathbb{R}^{N,N}$. The transformed right hand sides are the same as in (5.36).

On VCFV grids with density, flux or inflow boundary conditions we get the matrix equation (5.37). For density boundary conditions (5.6) on VCFV grids we use the transformation matrix $\boldsymbol{T}$ as in (5.38) and the variables (5.39). The time evolution matrix $\boldsymbol{K}$ then reads

$$\boldsymbol{K} = \frac{1}{2}\begin{bmatrix} \boldsymbol{A} & -(1-\omega)\boldsymbol{D} \\ -\boldsymbol{E} & (1-\omega)\boldsymbol{B} - 2\omega\boldsymbol{Id}_{N+1} \end{bmatrix} \tag{5.61}$$

with $A \in \mathbb{R}^{N-1,N-1}$, $B \in \mathbb{R}^{N+1,N+1}$, $D \in \mathbb{R}^{N-1,N+1}$ and $E \in \mathbb{R}^{N+1,N-1}$ given in (5.41). The right hand sides are given by (5.42) and

$$c^k = \left(1 - \frac{\omega}{2}\right) \left[r_L^k; 0; \ldots; 0; r_R^k\right] \in \mathbb{R}^{N-1},$$

$$d^k = \left[-\omega r_L^k; \left(1 - \frac{\omega}{2}\right) r_L^k; 0; \ldots; 0; -\left(1 - \frac{\omega}{2}\right) r_R^k; \omega r_R^k\right] \in \mathbb{R}^{N+1}.$$

For flux boundary conditions (5.8) on VCFV grids we use $T \in \mathbb{R}^{2N,2N}$ as in (5.43) and the variables (5.44). The time evolution matrix K then reads

$$K = \frac{1}{2} \begin{bmatrix} B & -(1-\omega)E \\ -D & (1-\omega)A - 2\omega Id_{N-1} \end{bmatrix} \tag{5.62}$$

with $A \in \mathbb{R}^{N-1,N-1}$, $B \in \mathbb{R}^{N+1,N+1}$, $D \in \mathbb{R}^{N-1,N+1}$ and $E \in \mathbb{R}^{N+1,N-1}$ given in (5.41). We find the right hand sides given by (5.46) and

$$c^k = \left[\omega j_L^k; \left(1 - \frac{\omega}{2}\right) j_L^k; 0; \ldots; 0; -\left(1 - \frac{\omega}{2}\right) j_R^k; -\omega j_R^k\right] \in \mathbb{R}^{N+1},$$

$$d^k = \left(1 - \frac{\omega}{2}\right) \left[j_L^k; 0; \ldots; 0; j_R^k\right] \in \mathbb{R}^{N-1}.$$

For inflow boundary conditions (5.10) on VCFV grids we use T as in (5.47) and the variables (5.48). The time evolution matrix K then reads

$$K = \frac{1}{2} \begin{bmatrix} A & -(1-\omega)D \\ -E & (1-\omega)B - 2\omega Id_{N-1} \end{bmatrix} \tag{5.63}$$

with $A \in \mathbb{R}^{N+1,N+1}$ and $E \in \mathbb{R}^{N-1,N+1}$ given by

$$A := \begin{bmatrix} -\omega & 1 & & & & & \\ \omega & 0 & 1 & & & & \\ & 1 & 0 & 1 & & & \\ & & & \ddots & & & \\ & & & 1 & 0 & 1 & \\ & & & & 1 & 0 & \omega \\ & & & & & 1 & -\omega \end{bmatrix}, \quad E := \begin{bmatrix} -\omega & 0 & 1 & & & \\ & -1 & 0 & 1 & & \\ & & & \ddots & & \\ & & & -1 & 0 & 1 \\ & & & & -1 & 0 & \omega \end{bmatrix}$$

and $B \in \mathbb{R}^{N-1,N-1}$, $D \in \mathbb{R}^{N+1,N-1}$ given in (5.50). We find the right hand sides given in (5.51) and

$$c^k = \left[\frac{\omega}{2} u_L^k; \left(1 - \frac{\omega}{2}\right) u_L^k; 0; \ldots; 0; \left(1 - \frac{\omega}{2}\right) v_R^k; \frac{\omega}{2} v_R^k\right] \in \mathbb{R}^{N+1},$$

$$d^k = \left(1 - \frac{\omega}{2}\right) \left[u_L^k; 0; \ldots, 0; -v_R^k\right] \in \mathbb{R}^{N-1}.$$

On CCFV grids we get the matrix equation (5.52). For the density boundary conditions (5.12) and for the flux boundary conditions (5.14) we use the transformation matrix T defined by (5.53) and the variables (5.54).

For density boundary conditions (5.12) on CCFV grids the time evolution matrix $K \in \mathbb{R}^{2N,2N}$ reads

$$K = \frac{1}{2} \begin{bmatrix} A & -(1-\omega)D \\ -E & (1-\omega)B - 2\omega Id_N \end{bmatrix} \tag{5.64}$$

with $A, B, D, E \in \mathbb{R}^{N,N}$ given by (5.56). We find the right hand sides by (5.57) and

$$c^{k+\delta} = \left[r_L^{k+\delta}; 0; \ldots; 0; r_R^{k+\delta}\right] \in \mathbb{R}^N, \quad d^{k+\delta} = \left[r_L^{k+\delta}; 0; \ldots; 0; -r_R^{k+\delta}\right] \in \mathbb{R}^N.$$

For flux boundary conditions (5.14) on CCFV grids the time evolution matrix $\boldsymbol{K} \in \mathbb{R}^{2N,2N}$ reads

$$\boldsymbol{K} = \frac{1}{2} \begin{bmatrix} \boldsymbol{B} & -(1-\omega)\boldsymbol{E} \\ -\boldsymbol{D} & (1-\omega)\boldsymbol{A} - 2\omega\boldsymbol{Id}_N \end{bmatrix} \tag{5.65}$$

with $\boldsymbol{A}, \boldsymbol{B}, \boldsymbol{D}, \boldsymbol{E} \in \mathbb{R}^{N,N}$ given in (5.56). We find the right hand sides

$$\boldsymbol{f}^k = \frac{1}{2}\Big[F_{3/2,-}^k; F_{1/2,+}^k + F_{5/2,-}^k; F_{3/2,+}^k + F_{7/2,-}^k; \ldots$$
$$\ldots; F_{N-5/2,+}^k + F_{N-1/2,-}^k; F_{N-3/2,+}^k\Big] \in \mathbb{R}^N,$$

$$\boldsymbol{g}^k = \frac{1}{2}\Big[-F_{3/2,-}^k; F_{1/2,+}^k - F_{5/2,-}^k; F_{3/2,+}^k - F_{7/2,-}^k; \ldots$$
$$\ldots; F_{N-5/2,+}^k - F_{N-1/2,-}^k; F_{N-3/2,+}^k\Big] \in \mathbb{R}^N$$

and

$$\boldsymbol{c}^{k+\delta} = \big[j_L^{k+\delta}; 0; \ldots; 0; -j_R^{k+\delta}\big] \in \mathbb{R}^N, \quad \boldsymbol{d}^{k+\delta} = \big[j_L^{k+\delta}; 0; \ldots; 0; j_R^{k+\delta}\big] \in \mathbb{R}^N.$$

5.8. Reduction of the Boundary Values

As in the continuous case, boundary values have to be subtracted for the consideration of stability properties and the usage of Fourier series.

For density boundary conditions (5.6) on VC grids we solve

$$-(\boldsymbol{K} - \boldsymbol{Id}_{2N})\boldsymbol{P}_B^k = \boldsymbol{C}^k - (\boldsymbol{K} - \boldsymbol{Id}_{2N})\boldsymbol{Q}^k.$$

We take $\boldsymbol{P}_B^k := [\boldsymbol{R}_B^k; \boldsymbol{J}_B^k]$ and we find in the VCFD case and in the VCFV case

$$\boldsymbol{R}_B^k := r_R^k \left[\frac{i}{N}\right]_{i=1,\ldots,N-1} + r_L^k \left[\frac{N-i}{N}\right]_{i=1,\ldots,N-1} \in \mathbb{R}^{N-1},$$
$$\boldsymbol{J}_B^k := -\frac{1}{2\omega}(r_R^k - r_L^k)\left[\frac{1}{N}\right]_{i=0,\ldots,N} \in \mathbb{R}^{N+1}. \tag{5.66}$$

With

$$\boldsymbol{G}_{LR}^k := \boldsymbol{G}^k - \frac{1}{\tau}(\boldsymbol{P}_B^{k+1} - \boldsymbol{P}_B^k) + \frac{1}{\tau}(\boldsymbol{Q}^{k+1} - \boldsymbol{Q}^k) \tag{5.67}$$

and $\boldsymbol{P}_{LR}^k := \boldsymbol{P}^k - \boldsymbol{P}_B^k$ the system (5.37) transforms to

$$\boldsymbol{P}_{LR}^{k+1} = \boldsymbol{K}\boldsymbol{P}_{LR}^k + \tau\boldsymbol{G}_{LR}^k. \tag{5.68}$$

In the case of flux boundary conditions (5.8) on VC grids we observe that $\boldsymbol{C}^k$ is not in the range of $-(\boldsymbol{K} - \boldsymbol{Id}_{2N})$. But for

$$\boldsymbol{v}_0 := \big[[1]_{i=0,\ldots,N}; [0]_{i=1,\ldots,N-1}\big] \in \mathbb{R}^{2N}$$

with $-(\boldsymbol{K} - \boldsymbol{Id}_{2N})\boldsymbol{v}_0 = \boldsymbol{0}$ and

$$\boldsymbol{C}_0^k := -(1-\omega)\frac{j_R^k - j_L^k}{N}\boldsymbol{v}_0 \quad \text{in the VCFD case,}$$

$$\boldsymbol{C}_0^k := -\frac{j_R^k - j_L^k}{N}\boldsymbol{v}_0 \qquad \text{in the VCFV case}$$

we can solve

$$-(\boldsymbol{K} - \boldsymbol{Id}_{2N})\boldsymbol{P}_B^k = \boldsymbol{C}^k - \boldsymbol{C}_0^k - (\boldsymbol{K} - \boldsymbol{Id}_{2N})\boldsymbol{Q}^k.$$

With $\boldsymbol{P}_B^k := [\boldsymbol{R}_B^k; \boldsymbol{J}_B^k]$ we find in the VCFD and in the VCFV case

$$
\begin{aligned}
\boldsymbol{R}_B^k &:= -\omega j_R^k \left[\frac{i^2}{N}\right]_{i=0,\ldots,N} + \omega j_L^k \left[\frac{(N-i)^2}{N}\right]_{i=0,\ldots,N} \in \mathbb{R}^{N+1}, \\
\boldsymbol{J}_B^k &:= j_R^k \left[\frac{i}{N}\right]_{i=1,\ldots,N-1} + j_L^k \left[\frac{N-i}{N}\right]_{i=1,\ldots,N-1} \in \mathbb{R}^{N-1}.
\end{aligned}
\tag{5.69}
$$

Now we get

$$
\boldsymbol{P}_{LR}^{k+1} = \boldsymbol{K}\boldsymbol{P}_{LR}^k + \tau \boldsymbol{G}_{LR}^k + \boldsymbol{C}_0^k,
\tag{5.70}
$$

where $\boldsymbol{G}_{LR}^k$ is given in (5.67).

For inflow boundary conditions (5.10) on VC grids we solve

$$
-(\boldsymbol{K} - \boldsymbol{Id}_{2N})\boldsymbol{P}_B^k = \boldsymbol{C}^k - (\boldsymbol{K} - \boldsymbol{Id}_{2N})\boldsymbol{Q}^k.
$$

We take $\boldsymbol{P}_B^k := [\boldsymbol{R}_B^k; \boldsymbol{J}_B^k]$ and we find in the VCFD and the VCFV case

$$
\begin{aligned}
\boldsymbol{R}_B^k &:= v_R^k \left[\frac{1+2\omega i}{1+N\omega}\right]_{i=0,\ldots,N} + u_L^k \left[\frac{1+2\omega(N-i)}{1+N\omega}\right]_{i=0,\ldots,N} \in \mathbb{R}^{N+1}, \\
\boldsymbol{J}_B^k &:= -(v_R^k - u_L^k) \left[\frac{1}{1+N\omega}\right]_{i=1,\ldots,N-1} \in \mathbb{R}^{N-1}.
\end{aligned}
\tag{5.71}
$$

We find the equation (5.68), where $\boldsymbol{G}_{LR}^k$ is given in (5.67).

For density boundary conditions (5.12) on CC grids we solve

$$
-(\boldsymbol{K} - \boldsymbol{Id}_{2N})\boldsymbol{P}_B^k = \boldsymbol{C}^k.
$$

We take $\boldsymbol{P}_B^k := [\boldsymbol{R}_B^k; \boldsymbol{J}_B^k]$ and find in the VCFD and in the VCFV case

$$
\begin{aligned}
\boldsymbol{R}_B^k &:= r_R^k \left[\frac{2i-1}{2N}\right]_{i=1,\ldots,N} + r_L^k \left[\frac{2(N-i)+1}{2N}\right]_{i=1,\ldots,N} \in \mathbb{R}^{N}, \\
\boldsymbol{J}_B^k &:= -\frac{1}{2\omega}(r_R^k - r_L^k) \left[\frac{1}{N}\right]_{i=1,\ldots,N} \in \mathbb{R}^{N}.
\end{aligned}
\tag{5.72}
$$

With

$$
\boldsymbol{G}_{LR}^k := \boldsymbol{G}^k - \frac{1}{\tau}(\boldsymbol{P}_B^{k+1} - \boldsymbol{P}_B^k) + \frac{1}{\tau}(\boldsymbol{C}^{k+\delta} - \boldsymbol{C}^k)
\tag{5.73}
$$

and $\boldsymbol{P}_{LR}^k := \boldsymbol{P}^k - \boldsymbol{P}_B^k$ equation (5.52) transforms to (5.68).

In the case of flux boundary conditions (5.13) on CCFD grids and (5.14) on CCFV grids we observe again that $\boldsymbol{C}^k$ is not in the range of $-(\boldsymbol{K} - \boldsymbol{Id}_{2N})$. But for

$$
\boldsymbol{v}_0 := \left[[1]_{i=1,\ldots,N}; [0]_{i=1,\ldots,N}\right] \in \mathbb{R}^{2N}
$$

with $-(\boldsymbol{K} - \boldsymbol{Id}_{2N})\boldsymbol{v}_0 = \boldsymbol{0}$ and

$$
\begin{aligned}
\boldsymbol{C}_0^k &:= -(1-\omega)\frac{j_R^k - j_L^k}{N}\boldsymbol{v}_0 \quad \text{in the VCFD case,} \\
\boldsymbol{C}_0^k &:= -\frac{j_R^k - j_L^k}{N}\boldsymbol{v}_0 \qquad\quad \text{in the VCFV case}
\end{aligned}
$$

we can solve

$$
-(\boldsymbol{K} - \boldsymbol{Id}_{2N})\boldsymbol{P}_B^k = \boldsymbol{C}^k - \boldsymbol{C}_0^k.
$$

With $\boldsymbol{P}_B^k := [\boldsymbol{R}_B^k ; \boldsymbol{J}_B^k]$ we find in both cases

$$\boldsymbol{R}_B^k := -\omega j_R^k \left[\frac{(2i-1)^2}{4N} \right]_{i=1,\ldots,N} + \omega j_L^k \left[\frac{(2(N-i)+1)^2}{4N} \right]_{i=1,\ldots,N} \in \mathbb{R}^N,$$

$$\boldsymbol{J}_B^k := j_R^k \left[\frac{2i-1}{2N} \right]_{i=1,\ldots,N} + j_L^k \left[\frac{2(N-i)+1}{2N} \right]_{i=1,\ldots,N} \in \mathbb{R}^N. \tag{5.74}$$

Now we obtain equation (5.70), where $\boldsymbol{G}_{LR}^k$ is given in (5.73).

5.9. Discrete Stability of the FD Lattice Boltzmann Solutions

In this section we consider solutions for the periodic problem on VCFD grids, solutions for the density and flux problem on VCFD grids and CCFD grids, and solutions for the inflow problem on VCFD grids. Stability can be proved directly only for these schemes. We exploit the fact that in the periodic case $\boldsymbol{K}^\dagger \boldsymbol{K}$ is a diagonal matrix of a special structure. For the FV schemes this property lacks.

For vectors $\boldsymbol{x} = [x_i]_i$ and $\boldsymbol{y} = [y_i]_i \in \mathbb{R}^K$, $K \in \mathbb{N}$, we define the scalar product

$$\boldsymbol{x} \cdot \boldsymbol{y} := \sum_{i=1}^{K} x_i y_i .$$

All vectors in this work are understood as column vectors. For $h := |\Omega|/K$ we introduce the discrete L^1-, L^∞- and L^2-norms

$$\|\boldsymbol{x}\|_1 := \sum_{i=1}^{K} h|x_i|, \quad \|\boldsymbol{x}\|_\infty := \max_{i=1,\ldots,K} |x_i|,$$

$$\|\boldsymbol{x}\|_2 := \left(\sum_{i=1}^{K} h|x_i|^2 \right)^{1/2} = (h\boldsymbol{x} \cdot \boldsymbol{x})^{1/2} .$$

For a matrix $\boldsymbol{N} = [n_{ij}]_{i,j} \in \mathbb{R}^{K,L}$, $K, L \in \mathbb{N}$, we define its transposed matrix $\boldsymbol{N}^\dagger := [n_{ji}]_{i,j} \in \mathbb{R}^{L,K}$. For $L = K$ we define the spectral radius of $\boldsymbol{N} \in \mathbb{R}^{K,K}$

$$\rho(\boldsymbol{N}) := \max\{|\lambda| : \lambda \text{ is an eigenvalue of } \boldsymbol{N}\}$$

and we use the following matrixnorms

$$\|\boldsymbol{N}\|_2 := \sqrt{\rho(\boldsymbol{N}^\dagger \boldsymbol{N})},$$

$$\|\boldsymbol{N}\|_1 := \max_{j=1,\ldots,K} \sum_{i=1}^{K} |n_{ij}|,$$

$$\|\boldsymbol{N}\|_\infty := \max_{i=1,\ldots,K} \sum_{j=1}^{K} |n_{ij}|.$$

We have the following compatibilities

$$\|\boldsymbol{N}\boldsymbol{x}\|_l \leq \|\boldsymbol{N}\|_l \|\boldsymbol{x}\|_l \quad \text{for } \boldsymbol{x} \in \mathbb{R}^K \text{ and } l \in \{1, 2, \infty\}.$$

For the FD lattice Boltzmann schemes in Section 5.6 we find by simple computations

$$\|\boldsymbol{M}\|_2 \leq \|\boldsymbol{M}\|_1 = \|\boldsymbol{M}\|_\infty = 1.$$

Here, we used the estimate $\rho(\boldsymbol{N}) \leq \|\boldsymbol{N}\|_2 \leq \sqrt{\|\boldsymbol{N}\|_1 \|\boldsymbol{N}\|_\infty}$ for all $\boldsymbol{N} \in \mathbb{R}^{K,K}$. Hence, we can derive stability estimates in the discrete L^1-, L^2- or $L^\infty-$norm in the variables $\boldsymbol{U}^k$ and $\boldsymbol{V}^k$. But if we switch to the variables $\boldsymbol{R}^k$ and $\boldsymbol{J}^k$ we find

$$\|\boldsymbol{K}\|_1 = 2, \quad \|\boldsymbol{K}\|_\infty = 2\omega.$$

Hence, we cannot directly prove stability estimates in the discrete L^1- or L^∞-norm for $\omega \geq 1/2$.

In the periodic case (5.5) and for density boundary conditions (5.12) or flux boundary conditions (5.13) on CCFD grids we find

$$\boldsymbol{K}^\dagger \boldsymbol{K} = \mathrm{diag}_{2N}(1,\ldots,1,\theta^2,\ldots,\theta^2)$$

with $\theta := 2\omega - 1$. So we have $\|\boldsymbol{K}\|_2 = 1$. Here, $\mathrm{diag}_K(d_1,\ldots,d_K) \in \mathbb{R}^{K,K}$ is a diagonal matrix with the diagonal entries $d_1,\ldots,d_K$.

The solution of the lattice Boltzmann systems (5.68) and (5.70) can be represented by

$$\boldsymbol{P}^k = \boldsymbol{K}^k \boldsymbol{P}^0 + \sum_{l=0}^{k-1} \tau \boldsymbol{K}^{k-1-l} \boldsymbol{G}^l + \sum_{l=0}^{k-1} \boldsymbol{C}_0^l,$$

where we omit the indices LR and the last expression only appears in the flux case. Hence, we get estimates of the form

$$\|\boldsymbol{R}^k\|_2 + \|\boldsymbol{J}^k\|_2 \leq \sqrt{2}\left(\|\boldsymbol{R}^k\|_2^2 + \|\boldsymbol{J}^k\|_2^2\right)^{1/2} = \left(2h\boldsymbol{P}^k \cdot \boldsymbol{P}^k\right)^{1/2}$$

$$\leq \sqrt{2}\left(\|\boldsymbol{R}^0\|_2^2 + \|\boldsymbol{J}^0\|_2^2\right)^{1/2} + \sqrt{2}\sum_{l=0}^{k-1}\tau\left(\|\boldsymbol{f}^l\|_2^2 + \|\boldsymbol{g}^l\|_2^2\right)^{1/2}$$

$$\leq \sqrt{2}\left(\|\boldsymbol{R}^0\|_2 + \|\boldsymbol{J}^0\|_2 + \sum_{l=0}^{k-1}\tau\|\boldsymbol{f}^l\|_2 + \sum_{l=0}^{k-1}\tau\|\boldsymbol{g}^l\|_2\right).$$

For flux boundary conditions (5.13) on CCFD grids we have to add the bounded expression

$$(1-\omega)\frac{\gamma}{\sqrt{|\Omega|}}\sum_{l=0}^{k-1}\tau\frac{|j_R^l - j_L^l|}{h}$$

to the right hand side of the estimate. An analogous expression does not appear in the continuous case (confer Lemma 3.1).

We shall see that in the general case we have to adapt the choice of the discrete norms to the underlying grid situation. For density, flux or inflow boundary conditions on VCFD grids we observe that $\boldsymbol{K}^\dagger \boldsymbol{K}$ is not diagonal and that in the first two cases there are two eigenvalues larger than 1. Hence, we have $\|\boldsymbol{K}\|_2 > 1$. But in all three cases we can find diagonal matrices $\boldsymbol{H} \in \mathbb{R}^{2N,2N}$ with positive elements such that $\boldsymbol{K}^\dagger \boldsymbol{H} \boldsymbol{K}$ is a diagonal matrix with positive elements. The choice of $\boldsymbol{H}$ implies the discrete norms that have to be applied to prove discrete stability for the given schemes.

In the periodic case or on CCFD grids we define the discrete norms for $\boldsymbol{R}, \boldsymbol{J} \in \mathbb{R}^N$ by

$$\|\boldsymbol{R}\|_2 := (h\boldsymbol{R} \cdot \boldsymbol{R})^{1/2}, \quad \|\boldsymbol{J}\|_2 := (h\boldsymbol{J} \cdot \boldsymbol{J})^{1/2} \tag{5.75}$$

with $h := |\Omega|/N$.

For density or flux boundary conditions on VCFD grids we use

$$\boldsymbol{d} := \mathrm{diag}_{N+1}(1/2, 1, \ldots, 1, 1/2) \in \mathbb{R}^{N+1,N+1}. \tag{5.76}$$

For density boundary conditions (5.6) on VCFD grids we define for $\boldsymbol{R} \in \mathbb{R}^{N-1}$ and $\boldsymbol{J} \in \mathbb{R}^{N+1}$

$$\|\boldsymbol{R}\|_2 := (h\boldsymbol{R} \cdot \boldsymbol{R})^{1/2}, \quad \|\boldsymbol{J}\|_2 := (h\boldsymbol{J} \cdot \boldsymbol{d}\boldsymbol{J})^{1/2}. \tag{5.77}$$

For flux boundary conditions (5.8) on VCFD grids we define for $\boldsymbol{R} \in \mathbb{R}^{N+1}$ and $\boldsymbol{J} \in \mathbb{R}^{N-1}$

$$\|\boldsymbol{R}\|_2 := (h\boldsymbol{R} \cdot d\boldsymbol{R})^{1/2}, \quad \|\boldsymbol{J}\|_2 := (h\boldsymbol{J} \cdot \boldsymbol{J})^{1/2}. \tag{5.78}$$

For inflow boundary conditions (5.10) on VCFD grids we use

$$\boldsymbol{e} := \mathrm{diag}_{N+1}(2, 1, \ldots, 1, 2) \in \mathbb{R}^{N+1,N+1}, \tag{5.79}$$

and we define for $\boldsymbol{R} \in \mathbb{R}^{N+1}$ and $\boldsymbol{J} \in \mathbb{R}^{N-1}$

$$\|\boldsymbol{R}\|_2 := (h\boldsymbol{R} \cdot \boldsymbol{e}\boldsymbol{R})^{1/2}, \quad \|\boldsymbol{J}\|_2 := (h\boldsymbol{J} \cdot \boldsymbol{J})^{1/2}. \tag{5.80}$$

In all situations we choose $h := |\Omega|/N$.

Now we find

Theorem 5.1. *Let* $[\boldsymbol{R}^l; \boldsymbol{J}^l]$, $l = 0, \ldots, M - 1$, *be the solution of*

$$\begin{bmatrix} \boldsymbol{R}^{k+1} \\ \boldsymbol{J}^{k+1} \end{bmatrix} = \frac{1}{2} \begin{bmatrix} \boldsymbol{A}_1 & \boldsymbol{D}_1 \\ \boldsymbol{D}_2 & \boldsymbol{A}_2 \end{bmatrix} \begin{bmatrix} \boldsymbol{R}^k \\ \boldsymbol{J}^k \end{bmatrix} + \tau \begin{bmatrix} \boldsymbol{f}_0^k \\ \boldsymbol{g}_0^k \end{bmatrix} \quad \textit{for } k = 0, \ldots, M - 1. \tag{5.81}$$

Let the discrete norms be defined by

$$\|\boldsymbol{R}\|_2 := (h\boldsymbol{R} \cdot \boldsymbol{d}_1 \boldsymbol{R})^{1/2}, \quad \|\boldsymbol{J}\|_2 := (h\boldsymbol{J} \cdot \boldsymbol{d}_2 \boldsymbol{J})^{1/2}$$

with symmetric and positive definite matrices $\boldsymbol{d}_1$ and $\boldsymbol{d}_2$ of appropriate sizes. We assume equivalence of the norms, that is,

$$\boldsymbol{R} \cdot \boldsymbol{R} \le c\boldsymbol{R} \cdot \boldsymbol{d}_1 \boldsymbol{R} \le C\boldsymbol{R} \cdot \boldsymbol{R},$$

$$\boldsymbol{J} \cdot \boldsymbol{J} \le d\boldsymbol{J} \cdot \boldsymbol{d}_2 \boldsymbol{J} \le D\boldsymbol{J} \cdot \boldsymbol{J},$$

and

$$\frac{h}{4} \begin{bmatrix} \boldsymbol{R} \\ \boldsymbol{J} \end{bmatrix} \cdot \begin{bmatrix} \boldsymbol{A}_1^\dagger & \boldsymbol{D}_2^\dagger \\ \boldsymbol{D}_1^\dagger & \boldsymbol{A}_2^\dagger \end{bmatrix} \begin{bmatrix} \boldsymbol{d}_1 & \\ & \boldsymbol{d}_2 \end{bmatrix} \begin{bmatrix} \boldsymbol{A}_1 & \boldsymbol{D}_1 \\ \boldsymbol{D}_2 & \boldsymbol{A}_2 \end{bmatrix} \begin{bmatrix} \boldsymbol{R} \\ \boldsymbol{J} \end{bmatrix} \le \|\boldsymbol{R}\|_2^2 + \theta^2 \|\boldsymbol{J}\|_2^2$$

for $h := |\Omega|/N$, $c, C, d, D > 0$ and $\theta := 2\omega - 1$. Then we have

$$\|\boldsymbol{R}^k\|_2^2 + \|\boldsymbol{J}^k\|_2^2 + 2\omega(1 - \omega)\frac{1}{\tau} \sum_{l=0}^{k-1} \tau \|\boldsymbol{J}^l\|_2^2$$

$$\le 2\|\boldsymbol{R}^0\|_2^2 + 2\|\boldsymbol{J}^0\|_2^2 + 2\tau \sum_{l=0}^{k-1} \tau \|\boldsymbol{f}_0^l\|_2^2 + 2\tau \sum_{l=0}^{k-1} \tau \|\boldsymbol{g}_0^l\|_2^2 \tag{5.82}$$

$$+ 5c^2 t_k \sum_{l=0}^{k-1} \tau \|\boldsymbol{f}_*^l\|_2^2 + \frac{d^2}{\omega(1 - \omega)} \tau \sum_{l=0}^{k-1} \tau \|\boldsymbol{g}_*^l\|_2^2,$$

where we use

$$\boldsymbol{f}_*^k := \frac{1}{2}\boldsymbol{A}_1^\dagger \boldsymbol{d}_1 \boldsymbol{f}_0^k + \frac{1}{2}\boldsymbol{D}_2^\dagger \boldsymbol{d}_2 \boldsymbol{g}_0^k,$$

$$\boldsymbol{g}_*^k := \frac{1}{2}\boldsymbol{D}_1^\dagger \boldsymbol{d}_1 \boldsymbol{f}_0^k + \frac{1}{2}\boldsymbol{A}_2^\dagger \boldsymbol{d}_2 \boldsymbol{g}_0^k.$$

PROOF. Multiply (5.81) by

$$h \begin{bmatrix} \boldsymbol{R}^{k+1} \\ \boldsymbol{J}^{k+1} \end{bmatrix} \cdot \begin{bmatrix} \boldsymbol{d}_1 & \\ & \boldsymbol{d}_2 \end{bmatrix}$$

We gain

$$\|\boldsymbol{R}^{k+1}\|_2^2 + \|\boldsymbol{J}^{k+1}\|_2^2 \le \|\boldsymbol{R}^k\|_2^2 + \theta^2 \|\boldsymbol{J}^k\|_2^2 + \tau^2 \|\boldsymbol{f}_0^k\|_2^2 + \tau^2 \|\boldsymbol{g}_0^k\|_2^2$$

$$+ 2h\tau \boldsymbol{R}^k \cdot \boldsymbol{f}_*^k + 2h\tau \boldsymbol{J}^k \cdot \boldsymbol{g}_*^k.$$

By observing $\theta^2 = 1 - 4\omega(1 - \omega)$ and

$$2h\tau \boldsymbol{J}^k \cdot \boldsymbol{g}_*^k \le 2d\tau \|\boldsymbol{J}^k\|_2 \|\boldsymbol{g}_*^k\|_2 \le 2\omega(1 - \omega)\|\boldsymbol{J}^k\|_2^2 + \frac{d^2}{2\omega(1 - \omega)}\tau^2\|\boldsymbol{g}_*^k\|_2^2$$

we get

$$\|\boldsymbol{R}^{k+1}\|_2^2 + \|\boldsymbol{J}^{k+1}\|_2^2 + 2\omega(1 - \omega)\|\boldsymbol{J}^k\|_2^2$$
$$\le \|\boldsymbol{R}^k\|_2^2 + \|\boldsymbol{J}^k\|_2^2 + \tau^2\|\boldsymbol{f}_0^k\|_2^2 + \tau^2\|\boldsymbol{g}_0^k\|_2^2$$
$$+ \frac{d^2}{2\omega(1 - \omega)}\tau^2\|\boldsymbol{g}_*^k\|_2^2 + 2c\tau\|\boldsymbol{R}^k\|_2\|\boldsymbol{f}_*^k\|_2.$$

Summation over the time intervals renders

$$\|\boldsymbol{R}^k\|_2^2 + \|\boldsymbol{J}^k\|_2^2 + 2\omega(1 - \omega)\frac{1}{\tau}\sum_{l=0}^{k-1}\tau\|\boldsymbol{J}^l\|_2^2$$
$$\le \|\boldsymbol{R}^0\|_2^2 + \|\boldsymbol{J}^0\|_2^2 + \tau\sum_{l=0}^{k-1}\tau\|\boldsymbol{f}_0^l\|_2^2 + \tau\sum_{l=0}^{k-1}\tau\|\boldsymbol{g}_0^l\|_2^2 \tag{5.83}$$
$$+ \frac{d^2}{2\omega(1 - \omega)}\tau\sum_{l=0}^{k-1}\tau\|\boldsymbol{g}_*^l\|_2^2 + 2c\left(\sum_{l=0}^{k-1}\tau\|\boldsymbol{R}^l\|_2^2\right)^{1/2}\left(\sum_{l=0}^{k-1}\tau\|\boldsymbol{f}_*^l\|_2^2\right)^{1/2}.$$

We apply a Gronwall-like argument. We gain

$$\|\boldsymbol{R}^i\|_2^2 \le 2c\left(\sum_{l=0}^{k-1}\tau\|\boldsymbol{R}^l\|_2^2\right)^{1/2}\left(\sum_{l=0}^{k-1}\tau\|\boldsymbol{f}_*^l\|_2^2\right)^{1/2} + S^2 \quad \text{for } i = 0, \ldots, k,$$

where S^2 consists of all the unmentioned terms of the right hand side in (5.83). Adding up these terms, we get

$$\sum_{l=0}^{k-1}\tau\|\boldsymbol{R}^l\|_2^2 \le 2ct_k\left(\sum_{l=0}^{k-1}\tau\|\boldsymbol{R}^l\|_2^2\right)^{1/2}\left(\sum_{l=0}^{k-1}\tau\|\boldsymbol{f}_*^l\|_2^2\right)^{1/2} + t_k S^2.$$

This yields

$$\left(\sum_{l=0}^{k-1}\tau\|\boldsymbol{R}^l\|_2^2\right)^{1/2} \le 2ct_k\left(\sum_{l=0}^{k-1}\tau\|\boldsymbol{f}_*^l\|_2^2\right)^{1/2} + t_k^{1/2}S.$$

By inserting this expression into (5.83), we end up with the estimate (5.82). $\qquad\square$

The discrete stability estimate (5.82) in Theorem 5.1 is the discrete counterpart of the a priori estimate in Lemma 3.1.

In the next step we apply the result of Theorem 5.1 to the different grid and boundary situations. For the boundary value problems we assume reduced boundary conditions with appropriately adapted right hand sides (see Section 5.8).

Corollary 5.2. *In the case of periodic boundary conditions (5.5) on VCFD grids with $\boldsymbol{A}_1 = \boldsymbol{A}$, $\boldsymbol{A}_2 = -\theta\boldsymbol{A}$, $\boldsymbol{D}_1 = \theta\boldsymbol{D}$ and $\boldsymbol{D}_2 = -\boldsymbol{D}$ with $\boldsymbol{A}$ and $\boldsymbol{D}$ given in (5.35) we use $\boldsymbol{f}_0^k := \boldsymbol{f}^k$ and $\boldsymbol{g}_0^k := \boldsymbol{g}^k$ given in (5.36). We find*

$$\frac{1}{4}\begin{bmatrix} \boldsymbol{A} & \theta\boldsymbol{D} \\ -\boldsymbol{D} & -\theta\boldsymbol{A} \end{bmatrix}^\dagger \begin{bmatrix} \boldsymbol{A} & \theta\boldsymbol{D} \\ -\boldsymbol{D} & -\theta\boldsymbol{A} \end{bmatrix} = \begin{bmatrix} \boldsymbol{Id}_N & \\ & \theta^2\boldsymbol{Id}_N \end{bmatrix}.$$

So we can apply (5.82) with $\boldsymbol{d}_1 = \boldsymbol{d}_2 = \boldsymbol{Id}_N$ and the discrete norms (5.75) with $c = d = 1$. We get

$$\boldsymbol{f}_*^k = \frac{1}{2}[F_{0,-}^k + F_{0,+}^k; \ldots; F_{N,-}^k + F_{N,+}^k],$$

$$\boldsymbol{g}_*^k = \frac{\theta}{2}[F_{0,-}^k - F_{0,+}^k; \ldots; F_{N,-}^k - F_{N,+}^k].$$

Hence, the solutions of the VCFD lattice Boltzmann equations (5.1) with periodic boundary conditions (5.5) are stable.

Corollary 5.3. *In the case of density boundary conditions (5.6) on VCFD grids with $\boldsymbol{A}_1 = \boldsymbol{A}$, $\boldsymbol{A}_2 = -\theta \boldsymbol{B}$, $\boldsymbol{D}_1 = \theta \boldsymbol{D}$ and $\boldsymbol{D}_2 = -\boldsymbol{E}$ with $\boldsymbol{A}$, $\boldsymbol{B}$, $\boldsymbol{D}$ and $\boldsymbol{E}$ given in (5.41) we use*

$$\boldsymbol{f}_0^k := \boldsymbol{f}^k - \frac{1}{\tau}(\boldsymbol{R}_B^{k+1} - \boldsymbol{R}_B^k),$$

$$\boldsymbol{g}_0^k := \boldsymbol{g}^k - \frac{1}{\tau}(\boldsymbol{J}_B^{k+1} - \boldsymbol{J}_B^k) + \frac{1}{\tau}(r_L^{k+1} - r_L^k)\boldsymbol{e}_1 - \frac{1}{\tau}(r_R^{k+1} - r_R^k)\boldsymbol{e}_{N+1},$$

with $\boldsymbol{f}^k$, $\boldsymbol{g}^k$ given in (5.42) and $\boldsymbol{R}_B^k$, $\boldsymbol{J}_B^k$ as in (5.66). We find

$$\frac{1}{4}\begin{bmatrix} \boldsymbol{A} & \theta \boldsymbol{D} \\ -\boldsymbol{E} & -\theta \boldsymbol{B} \end{bmatrix}^{\dagger} \begin{bmatrix} \boldsymbol{Id}_{N-1} & \\ & \boldsymbol{d} \end{bmatrix} \begin{bmatrix} \boldsymbol{A} & \theta \boldsymbol{D} \\ -\boldsymbol{E} & -\theta \boldsymbol{B} \end{bmatrix} = \begin{bmatrix} \boldsymbol{Id}_{N-1} & \\ & \theta^2 \boldsymbol{d} \end{bmatrix}$$

with $\boldsymbol{d} := \mathrm{diag}_{N+1}(1/2, 1, \ldots, 1, 1/2)$. So we can apply (5.82) with $\boldsymbol{d}_1 = \boldsymbol{Id}_{N-1}$ and $\boldsymbol{d}_2 = \boldsymbol{d}$ to $\boldsymbol{R}_{LR}^k$ and $\boldsymbol{J}_{LR}^k$ with the discrete norms (5.77) and $c = 1$ and $d = 2$. We get

$$\boldsymbol{f}_*^k = \frac{1}{2}[F_{1,-}^k + F_{1,+}^k; \ldots; F_{N-1,-}^k + F_{N-1,+}^k] - \frac{1}{\tau}(\boldsymbol{R}_B^{k+1} - \boldsymbol{R}_B^k),$$

$$\boldsymbol{g}_*^k = \frac{\theta}{2}[-F_{0,+}^k; F_{1,-}^k - F_{1,+}^k; \ldots; F_{N-1,-}^k - F_{N-1,+}^k; F_{N,-}^k]$$

$$- \frac{\theta^2}{\tau}\boldsymbol{d}(\boldsymbol{J}_B^{k+1} - \boldsymbol{J}_B^k) + \frac{\theta}{2\tau}(r_L^{k+1} - r_L^k)\boldsymbol{e}_1 - \frac{\theta}{2\tau}(r_R^{k+1} - r_R^k)\boldsymbol{e}_{N+1}.$$

Hence, the solutions of the VCFD lattice Boltzmann equations (5.1) with density boundary conditions (5.6) are stable.

Corollary 5.4. *In the case of flux boundary conditions (5.8) on VCFD grids with $\boldsymbol{A}_1 = \boldsymbol{B}$, $\boldsymbol{A}_2 = -\theta \boldsymbol{A}$, $\boldsymbol{D}_1 = \theta \boldsymbol{E}$ and $\boldsymbol{D}_2 = -\boldsymbol{D}$ with $\boldsymbol{A}$, $\boldsymbol{B}$, $\boldsymbol{D}$ and $\boldsymbol{E}$ given in (5.41) we use*

$$\boldsymbol{f}_0^k := \boldsymbol{f}^k - \frac{1}{\tau}(\boldsymbol{R}_B^{k+1} - \boldsymbol{R}_B^k) - \frac{1}{\tau}\frac{j_R^k - j_L^k}{N}(1 - \omega)[1; \ldots; 1]$$

$$+ \frac{1}{\tau}(j_L^{k+1} - j_L^k)\boldsymbol{e}_1 - \frac{1}{\tau}(j_R^{k+1} - j_R^k)\boldsymbol{e}_{N+1},$$

$$\boldsymbol{g}_0^k := \boldsymbol{g}^k - \frac{1}{\tau}(\boldsymbol{J}_B^{k+1} - \boldsymbol{J}_B^k),$$

with $\boldsymbol{f}^k$, $\boldsymbol{g}^k$ given in (5.46) and $\boldsymbol{R}_B^k$, $\boldsymbol{J}_B^k$ as in (5.69). We find

$$\frac{1}{4}\begin{bmatrix} \boldsymbol{B} & \theta \boldsymbol{E} \\ -\boldsymbol{D} & -\theta \boldsymbol{A} \end{bmatrix}^{\dagger} \begin{bmatrix} \boldsymbol{d} & \\ & \boldsymbol{Id}_{N-1} \end{bmatrix} \begin{bmatrix} \boldsymbol{B} & \theta \boldsymbol{E} \\ -\boldsymbol{D} & -\theta \boldsymbol{A} \end{bmatrix} = \begin{bmatrix} \boldsymbol{d} & \\ & \theta^2 \boldsymbol{Id}_{N-1} \end{bmatrix}$$

with $\boldsymbol{d} := \mathrm{diag}_{N+1}(1/2, 1, \ldots, 1, 1/2)$. So we can apply (5.82) with $\boldsymbol{d}_1 = \boldsymbol{d}$ and $\boldsymbol{d}_2 = \boldsymbol{Id}_{N-1}$ to $\boldsymbol{R}_{LR}^k$ and $\boldsymbol{J}_{LR}^k$ with the discrete norms (5.78) and $c = 2$ and $d = 1$.

We get

$$\boldsymbol{f}_*^k = \frac{1}{2}[F_{0,+}^k; F_{1,-}^k + F_{1,+}^k; \ldots; F_{N-1,-}^k + F_{N-1,+}^k; F_{N,-}^k] - \frac{1}{\tau}\boldsymbol{d}(\boldsymbol{R}_B^{k+1} - \boldsymbol{R}_B^k)$$

$$- \frac{1-\omega}{\tau}\frac{j_R^{k+1} - j_L^{k+1}}{N}[1/2; 1; \ldots; 1; 1/2]$$

$$+ \frac{\theta}{2\tau}(j_L^{k+1} - j_L^k)\boldsymbol{e}_1 - \frac{\theta}{2\tau}(j_R^{k+1} - j_R^k)\boldsymbol{e}_{N+1},$$

$$\boldsymbol{g}_*^k = \frac{\theta}{2}[F_{1,-}^k - F_{1,+}^k; \ldots; F_{N-1,-}^k - F_{N-1,+}^k] - \frac{\theta^2}{\tau}(\boldsymbol{J}_B^{k+1} - \boldsymbol{J}_B^k).$$

Hence, the solutions of the VCFD lattice Boltzmann equations (5.1) with flux boundary conditions (5.8) are stable.

Corollary 5.5. *In the case of inflow boundary conditions (5.10) on VCFD grids with $\boldsymbol{A}_1 = \boldsymbol{A}$, $\boldsymbol{A}_2 = -\theta\boldsymbol{B}$, $\boldsymbol{D}_1 = \theta\boldsymbol{D}$ and $\boldsymbol{D}_2 = -\boldsymbol{E}$ with $\boldsymbol{A}$, $\boldsymbol{B}$, $\boldsymbol{D}$ and $\boldsymbol{E}$ given in (5.50) we use*

$$\boldsymbol{f}_0^k := \boldsymbol{f}^k - \frac{1}{\tau}(\boldsymbol{R}_B^{k+1} - \boldsymbol{R}_B^k) + \frac{1}{\tau}(u_L^{k+1} - u_L^k)\boldsymbol{e}_1 + \frac{1}{\tau}(v_R^{k+1} - v_R^k)\boldsymbol{e}_{N+1},$$

$$\boldsymbol{g}_0^k := \boldsymbol{g}^k - \frac{1}{\tau}(\boldsymbol{J}_B^{k+1} - \boldsymbol{J}_B^k),$$

with $\boldsymbol{f}^k$, $\boldsymbol{g}^k$ given in (5.51) and $\boldsymbol{R}_B^k$, $\boldsymbol{J}_B^k$ as in (5.71). We find

$$\frac{1}{4}\begin{bmatrix} \boldsymbol{A} & \theta\boldsymbol{D} \\ -\boldsymbol{E} & -\theta\boldsymbol{B} \end{bmatrix}^\dagger \begin{bmatrix} \boldsymbol{e} & \\ & \boldsymbol{Id}_{N-1} \end{bmatrix}\begin{bmatrix} \boldsymbol{A} & \theta\boldsymbol{D} \\ -\boldsymbol{E} & -\theta\boldsymbol{B} \end{bmatrix} = \begin{bmatrix} \boldsymbol{f} & \\ & \theta^2\boldsymbol{Id}_{N-1} \end{bmatrix}$$

with $\boldsymbol{e} := \mathrm{diag}_{N+1}(2,1,\ldots,1,2)$ and $\boldsymbol{f} := \mathrm{diag}_{N+1}(2\omega^2, 1, \ldots, 1, 2\omega^2)$. So we can apply (5.82) with $\boldsymbol{d}_1 = \boldsymbol{e}$ and $\boldsymbol{d}_2 = \boldsymbol{Id}_{N-1}$ to $\boldsymbol{R}_{LR}^k$ and $\boldsymbol{J}_{LR}^k$ with the discrete norms (5.80) and $c = 1$ and $d = 1$. In addition we can add the outflow values

$$2(1-\omega^2)\frac{h}{\tau}\sum_{l=0}^{k-1}\tau\left(|U_N^l|^2 + |V_0^l|^2\right)$$

to the right hand side of the stability estimate (5.82). This is in accordance with the continuous case (confer Lemma 3.1). We get

$$\boldsymbol{f}_*^k = \frac{1}{2}[2\omega F_{0,+}^k; F_{1,-}^k + F_{1,+}^k; \ldots, F_{N-1,-}^k + F_{N-1,+}^k; 2\omega F_{N,-}^k]$$

$$- \frac{1}{\tau}\mathrm{diag}_{N+1}(0, 1, \ldots, 1, 0)(\boldsymbol{R}_B^{k+1} - \boldsymbol{R}_B^k)$$

$$- \frac{2\omega}{\tau}\left(\frac{\omega}{1+N\omega}(v_R^{k+1} - v_R^k) + \frac{1+(N-1)\omega}{1+N\omega}(u_L^{k+1} - u_L^k)\right)\boldsymbol{e}_1$$

$$- \frac{2\omega}{\tau}\left(\frac{1+(N-1)\omega}{1+N\omega}(v_R^{k+1} - v_R^k) + \frac{\omega}{1+N\omega}(u_L^{k+1} - u_L^k)\right)\boldsymbol{e}_{N+1},$$

$$\boldsymbol{g}_*^k = \frac{\theta}{2}[F_{1,-}^k - F_{1,+}^k; \ldots; F_{N-1,-}^k - F_{N-1,+}^k] - \frac{\theta^2}{\tau}(\boldsymbol{J}_B^{k+1} - \boldsymbol{J}_B^k).$$

Hence, the solutions of the VCFD lattice Boltzmann equations (5.1) with inflow boundary conditions (5.10) are stable.

Corollary 5.6. *In the case of density boundary conditions (5.12) on CCFD grids with $\boldsymbol{A}_1 = \boldsymbol{A}$, $\boldsymbol{A}_2 = -\theta\boldsymbol{B}$, $\boldsymbol{D}_1 = \theta\boldsymbol{D}$ and $\boldsymbol{D}_2 = -\boldsymbol{E}$ with $\boldsymbol{A}$, $\boldsymbol{B}$, $\boldsymbol{D}$ and $\boldsymbol{E}$ given in (5.56) we use*

$$\boldsymbol{f}_0^k := \boldsymbol{f}^k - \frac{1}{\tau}(\boldsymbol{R}_B^{k+1} - \boldsymbol{R}_B^k) + \frac{1}{\tau}(r_L^{k+\delta} - r_L^k)\boldsymbol{e}_1 + \frac{1}{\tau}(r_R^{k+\delta} - r_R^k)\boldsymbol{e}_N,$$

$$\boldsymbol{g}_0^k := \boldsymbol{g}^k - \frac{1}{\tau}(\boldsymbol{J}_B^{k+1} - \boldsymbol{J}_B^k) + \frac{1}{\tau}(r_L^{k+\delta} - r_L^k)\boldsymbol{e}_1 - \frac{1}{\tau}(r_R^{k+\delta} - r_R^k)\boldsymbol{e}_N,$$

with $\boldsymbol{f}^k$, $\boldsymbol{g}^k$ given in (5.57) and $\boldsymbol{R}_B^k$, $\boldsymbol{J}_B^k$ as in (5.72). We find

$$\frac{1}{4}\begin{bmatrix} \boldsymbol{A} & \theta\boldsymbol{D} \\ -\boldsymbol{E} & -\theta\boldsymbol{B} \end{bmatrix}^{\dagger}\begin{bmatrix} \boldsymbol{A} & \theta\boldsymbol{D} \\ -\boldsymbol{E} & -\theta\boldsymbol{B} \end{bmatrix} = \begin{bmatrix} \boldsymbol{Id}_N & \\ & \theta^2\boldsymbol{Id}_N \end{bmatrix}.$$

So we can apply (5.82) with $\boldsymbol{d}_1 = \boldsymbol{d}_2 = \boldsymbol{Id}_N$ to $\boldsymbol{R}_{LR}^k$ and $\boldsymbol{J}_{LR}^k$ with the discrete norms (5.75) and $c = 1$ and $d = 1$. We get

$$\boldsymbol{f}_*^k = \frac{1}{2}[F_{1/2,-}^k + F_{1/2,+}^k; \ldots; F_{N-1/2,-}^k + F_{N-1/2,+}^k] - \frac{1}{\tau}(\boldsymbol{R}_B^{k+1} - \boldsymbol{R}_B^k)$$

$$+ \frac{1}{\tau}(r_L^{k+1} - r_L^{k+\delta})e_1 + \frac{1}{\tau}(r_R^{k+1} - r_R^{k+\delta})e_N,$$

$$\boldsymbol{g}_*^k = \frac{\theta}{2}[F_{1/2,-}^k - F_{1/2,+}^k; \ldots; F_{N-1/2,-}^k - F_{N-1/2,+}^k] - \frac{\theta^2}{\tau}(\boldsymbol{J}_B^{k+1} - \boldsymbol{J}_B^k)$$

$$+ \frac{\theta}{\tau}(r_L^{k+1} - r_L^{k+\delta})e_1 - \frac{\theta}{\tau}(r_R^{k+1} - r_R^{k+\delta})e_N.$$

Hence, the solutions of the CCFD lattice Boltzmann equations (5.2) with density boundary conditions (5.12) are stable.

Corollary 5.7. In the case of flux boundary conditions (5.13) on CCFD grids with $\boldsymbol{A}_1 = \boldsymbol{B}$, $\boldsymbol{A}_2 = -\theta\boldsymbol{A}$, $\boldsymbol{D}_1 = \theta\boldsymbol{E}$ and $\boldsymbol{D}_2 = -\boldsymbol{D}$ with $\boldsymbol{A}$, $\boldsymbol{B}$, $\boldsymbol{D}$ and $\boldsymbol{E}$ given in (5.56) we use

$$\boldsymbol{f}_0^k := \boldsymbol{f}^k - \frac{1}{\tau}(\boldsymbol{R}_B^{k+1} - \boldsymbol{R}_B^k) - \frac{1-\omega}{\tau}\frac{j_R^k - j_L^k}{N}[1; \ldots; 1]$$

$$+ \frac{1-\omega}{\tau}(j_L^{k+\delta} - j_L^k)e_1 - \frac{1-\omega}{\tau}(j_R^{k+\delta} - j_R^k)e_N,$$

$$\boldsymbol{g}_0^k := \boldsymbol{g}^k - \frac{1}{\tau}(\boldsymbol{J}_B^{k+1} - \boldsymbol{J}_B^k)$$

$$+ \frac{1-\omega}{\tau}(j_L^{k+\delta} - j_L^k)e_1 + \frac{1-\omega}{\tau}(j_R^{k+\delta} - j_R^k)e_N,$$

with $\boldsymbol{f}^k$, $\boldsymbol{g}^k$ given in (5.59) and $\boldsymbol{R}_B^k$, $\boldsymbol{J}_B^k$ as in (5.74). We find

$$\frac{1}{4}\begin{bmatrix} \boldsymbol{B} & \theta\boldsymbol{E} \\ -\boldsymbol{D} & -\theta\boldsymbol{A} \end{bmatrix}^{\dagger}\begin{bmatrix} \boldsymbol{B} & \theta\boldsymbol{E} \\ -\boldsymbol{D} & -\theta\boldsymbol{A} \end{bmatrix} = \begin{bmatrix} \boldsymbol{Id}_N & \\ & \theta^2\boldsymbol{Id}_N \end{bmatrix}.$$

So we can apply (5.82) with $\boldsymbol{d}_1 = \boldsymbol{d}_2 = \boldsymbol{Id}_N$ to $\boldsymbol{R}_{LR}^k$ and $\boldsymbol{J}_{LR}^k$ with the discrete norms (5.75) and $c = 1$ and $d = 1$. We get

$$\boldsymbol{f}_*^k = \frac{1}{2}[F_{1/2,-}^k + F_{1/2,+}^k; \ldots; F_{N-1/2,-}^k + F_{N-1/2,+}^k] - \frac{1}{\tau}(\boldsymbol{R}_B^{k+1} - \boldsymbol{R}_B^k)$$

$$- \frac{1-\omega}{\tau}\frac{j_R^{k+1} - j_L^{k+1}}{N}[1; \ldots; 1]$$

$$- \frac{1-\omega}{\tau}(j_L^{k+1} - j_L^{k+\delta})e_1 + \frac{1-\omega}{\tau}(j_R^{k+1} - j_R^{k+\delta})e_N,$$

$$\boldsymbol{g}_*^k = \frac{\theta}{2}[F_{1/2,-}^k - F_{1/2,+}^k; \ldots; F_{N-1/2,-}^k - F_{N-1/2,+}^k] - \frac{\theta^2}{\tau}(\boldsymbol{J}_B^{k+1} - \boldsymbol{J}_B^k)$$

$$- \frac{1-\omega}{\tau}\theta(j_L^{k+1} - j_L^{k+\delta})e_1 - \frac{1-\omega}{\tau}\theta(j_R^{k+1} - j_R^{k+\delta})e_N.$$

Hence, the solutions of the CCFD lattice Boltzmann equations (5.2) with flux boundary conditions (5.13) are stable.

For the FV lattice Boltzmann schemes in Section 5.6 we find

$$\|\boldsymbol{M}\|_1 = \|\boldsymbol{M}\|_\infty = 1 + \omega > 1, \quad \|\boldsymbol{K}\|_1 = \|\boldsymbol{K}\|_\infty = 2.$$

Yet we have not found a symmetric and positive definite matrix $\boldsymbol{H}$ such that $\boldsymbol{K}^{\dagger}\boldsymbol{H}\boldsymbol{K}$ is a diagonal matrix, as in the FD case. Hence, the stability considerations have to be postponed until the examination of the discrete Fourier solutions of the underlying problems.

5.10. Discrete Fourier Solutions for the FD Lattice Boltzmann Schemes

For the computation of discrete Fourier solutions we consider orthonormal bases in $\mathbb{R}^{2N}$. We define the scaled grid points $y_i := \pi i / N$ for $i = 0, \ldots, N$ and the intermediate scaled grid points $y_{i-1/2} := \pi(2i - 1)/(2N)$ for $i = 1, \ldots, N$. For $N = 2n - 1$ or $N = 2n$ we introduce

$$
\widetilde{S}_c^0 := \frac{1}{\sqrt{2N}} \left[\begin{bmatrix} 1 \\ 1 \end{bmatrix} \right]_{i=1,\ldots,N}, \qquad
\widetilde{T}_c^0 := \frac{1}{\sqrt{2N}} \left[\begin{bmatrix} 1 \\ -1 \end{bmatrix} \right]_{i=1,\ldots,N},
$$

$$
\widetilde{S}_c^l := \frac{1}{\sqrt{N}} \left[\begin{bmatrix} \cos(2ly_{i-1/2}) \\ \cos(2ly_{i-1/2}) \end{bmatrix} \right]_{i=1,\ldots,N}, \qquad
\widetilde{T}_s^l := \frac{1}{\sqrt{N}} \left[\begin{bmatrix} \sin(2ly_{i-1/2}) \\ -\sin(2ly_{i-1/2}) \end{bmatrix} \right]_{i=1,\ldots,N},
$$

$$
\widetilde{S}_s^l := \frac{1}{\sqrt{N}} \left[\begin{bmatrix} \sin(2ly_{i-1/2}) \\ \sin(2ly_{i-1/2}) \end{bmatrix} \right]_{i=1,\ldots,N}, \qquad
\widetilde{T}_c^l := \frac{1}{\sqrt{N}} \left[\begin{bmatrix} \cos(2ly_{i-1/2}) \\ -\cos(2ly_{i-1/2}) \end{bmatrix} \right]_{i=1,\ldots,N}, \qquad (5.84)
$$

$$
\text{for } l = 1, \ldots, n - 1.
$$

For even $N = 2n$ we add the two basis vectors

$$
\widetilde{S}_s^n := \frac{1}{\sqrt{2N}} \left[\begin{bmatrix} (-1)^{i+1} \\ (-1)^{i+1} \end{bmatrix} \right]_{i=1,\ldots,N}, \qquad
\widetilde{T}_s^n := \frac{1}{\sqrt{2N}} \left[\begin{bmatrix} (-1)^{i+1} \\ -(-1)^{i+1} \end{bmatrix} \right]_{i=1,\ldots,N}.
$$

The column vectors above are composed by the indicated 2-by-1 blocks in sequence. Using these vectors we define an orthogonal matrix $\boldsymbol{S} \in \mathbb{R}^{2N,2N}$ with $\boldsymbol{S}^\dagger \boldsymbol{S} = \boldsymbol{Id}_{2N}$ by

$$
\begin{aligned}
\boldsymbol{S} := [\widetilde{S}_c^0, \widetilde{S}_c^1, \widetilde{T}_s^1, &\ldots, \widetilde{S}_c^{n-1}, \widetilde{T}_s^{n-1}, \widetilde{T}_s^n, \\
&\widetilde{T}_c^0, \widetilde{S}_s^1, \widetilde{T}_c^1, \ldots, \widetilde{S}_s^{n-1}, \widetilde{T}_c^{n-1}, \widetilde{S}_s^n], \quad \text{if } N = 2n, \\[4pt]
\boldsymbol{S} := [\widetilde{S}_c^0, \widetilde{S}_c^1, \widetilde{T}_s^1, &\ldots, \widetilde{S}_c^{n-1}, \widetilde{T}_s^{n-1}, \\
&\widetilde{T}_c^0, \widetilde{S}_s^1, \widetilde{T}_c^1, \ldots, \widetilde{S}_s^{n-1}, \widetilde{T}_c^{n-1}], \qquad \text{if } N = 2n - 1.
\end{aligned} \qquad (5.85)
$$

Furthermore, we define for $\omega \in (0, 1)$

$$
\alpha_l := (1 - \omega) \sin(2y_l), \quad \beta_l := (1 - \omega) \cos(2y_l), \quad V_l := \sqrt{\omega^2 - \alpha_l^2} \qquad (5.86)
$$

for $l = 1, \ldots, n - 1$. For $\omega \in [1/2, 1)$, we find that V_l is a real number for all $l = 1, \ldots, n - 1$. For $\omega < 1/2$, the V_l are real for small values of l and for values close to n. For values in between, the V_l are imaginary.

We find that $\boldsymbol{S}^\dagger \boldsymbol{M} \boldsymbol{S}$ is a block-diagonal matrix with diagonal elements ± 1 and $\pm \theta$ ($\theta = 2\omega - 1$) and with the 2-by-2 blocks

$$
\begin{bmatrix} \beta_l + \omega & \alpha_l \\ -\alpha_l & \beta_l - \omega \end{bmatrix}, \quad
\begin{bmatrix} \beta_l + \omega & -\alpha_l \\ \alpha_l & \beta_l - \omega \end{bmatrix} \quad \text{for } l = 1, \ldots, n - 1
$$

on the diagonal. Here, $\boldsymbol{M}$ is the time evolution matrix given in (5.16), belonging to the VCFD lattice Boltzmann equations (5.1) subject to periodic boundary conditions (5.5). The eigenvalues of $\boldsymbol{M}$ are found to read

$$
\begin{aligned}
&\lambda_0^+ = 1, \quad \lambda_0^- = -\theta, \\
&\left. \begin{aligned} \lambda_l^+ &= \beta_l + V_l, \\ \lambda_l^- &= \beta_l - V_l, \end{aligned} \right\} \quad \text{for } l = 1, \ldots, n - 1, \\
&\lambda_n^+ = \theta, \quad \lambda_n^- = -1 \quad \text{if } N = 2n.
\end{aligned} \qquad (5.87)
$$

See Figure 5.3 for the cases $\omega > 1/2$ and $\omega = 1/2$. Furthermore, we get the

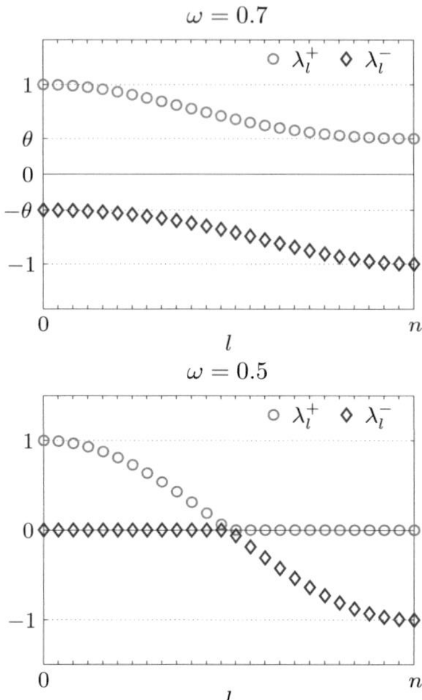

FIGURE 5.3. Eigenvalues of $\boldsymbol{M}$ and $\boldsymbol{K}$ in the periodic VCFD case for $\omega = 0.5$, $\omega = 0.7$ and $N = 50$.

eigenvectors of $\boldsymbol{M}$ in the form

$$
\widetilde{\boldsymbol{V}}_{+}^{0} := \widetilde{\boldsymbol{S}}_{c}^{0}, \qquad \widetilde{\boldsymbol{W}}_{-}^{0} := \widetilde{\boldsymbol{T}}_{c}^{0},
$$

$$
\left.
\begin{aligned}
\widetilde{\boldsymbol{V}}_{+}^{l} &:= \alpha_l \widetilde{\boldsymbol{S}}_{c}^{l} - (\omega - V_l)\widetilde{\boldsymbol{T}}_{s}^{l}, \\
\widetilde{\boldsymbol{V}}_{-}^{l} &:= \alpha_l \widetilde{\boldsymbol{S}}_{c}^{l} - (\omega + V_l)\widetilde{\boldsymbol{T}}_{s}^{l}, \\
\widetilde{\boldsymbol{W}}_{+}^{l} &:= \alpha_l \widetilde{\boldsymbol{S}}_{s}^{l} + (\omega - V_l)\widetilde{\boldsymbol{T}}_{c}^{l}, \\
\widetilde{\boldsymbol{W}}_{-}^{l} &:= \alpha_l \widetilde{\boldsymbol{S}}_{s}^{l} + (\omega + V_l)\widetilde{\boldsymbol{T}}_{c}^{l},
\end{aligned}
\right\} \quad \text{for } l = 1, \ldots, n-1,
$$

$$
\widetilde{\boldsymbol{V}}_{-}^{n} := \widetilde{\boldsymbol{T}}_{s}^{n}, \qquad \widetilde{\boldsymbol{W}}_{+}^{n} := \widetilde{\boldsymbol{S}}_{s}^{n}, \qquad \text{if } N = 2n,
$$

with

$$
\boldsymbol{M}\widetilde{\boldsymbol{V}}_{+}^{0} = \lambda_{0}^{+}\widetilde{\boldsymbol{V}}_{+}^{0}, \quad \boldsymbol{M}\widetilde{\boldsymbol{W}}_{-}^{0} = \lambda_{n}^{-}\widetilde{\boldsymbol{W}}_{-}^{0},
$$

$$
\left.
\begin{aligned}
\boldsymbol{M}\widetilde{\boldsymbol{V}}_{+}^{l} &= \lambda_{l}^{+}\widetilde{\boldsymbol{V}}_{+}^{l}, \quad \boldsymbol{M}\widetilde{\boldsymbol{W}}_{+}^{l} = \lambda_{l}^{+}\widetilde{\boldsymbol{W}}_{+}^{l}, \\
\boldsymbol{M}\widetilde{\boldsymbol{V}}_{-}^{l} &= \lambda_{l}^{-}\widetilde{\boldsymbol{V}}_{-}^{l}, \quad \boldsymbol{M}\widetilde{\boldsymbol{W}}_{-}^{l} = \lambda_{l}^{-}\widetilde{\boldsymbol{W}}_{-}^{l},
\end{aligned}
\right\} \quad \text{for } l = 1, \ldots, n-1,
$$

$$
\boldsymbol{M}\widetilde{\boldsymbol{V}}_{-}^{n} = \lambda_{n}^{-}\widetilde{\boldsymbol{V}}_{-}^{n}, \quad \boldsymbol{M}\widetilde{\boldsymbol{W}}_{+}^{n} = \lambda_{n}^{+}\widetilde{\boldsymbol{W}}_{+}^{n} \qquad \text{if } N = 2n.
$$

As V_l turns imaginary for $\omega < 1/2$ for some values of l, the corresponding eigenvalues and eigenvectors turn complex. See Figure 5.4 for the case $\omega < 1/2$. For $l \leq l^*$ and $l \geq n - l^*$ with $l^* := \lfloor N \arcsin(\omega/(1-\omega))/(2\pi) \rfloor$, the eigenvalues are real. For the remaining values of l, the eigenvalues are located on a circle in the complex plane with midpoint 0 and radius $\sqrt{1 - 2\omega}$.

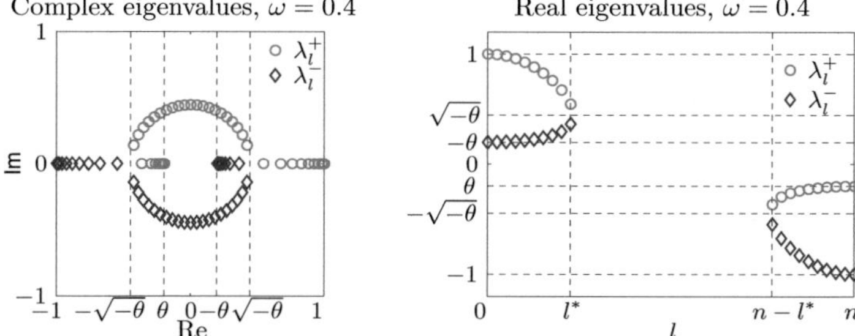

FIGURE 5.4. Eigenvalues of $\boldsymbol{M}$ and $\boldsymbol{K}$ in the periodic VCFD case for $\omega = 0.4$ and $N = 80$.

The eigenvectors are not orthogonal, especially we see for $\omega \geq 1/2$

$$\widetilde{\boldsymbol{V}}^l_+ \cdot \widetilde{\boldsymbol{V}}^l_- = 2\alpha_l^2, \quad \widetilde{\boldsymbol{W}}^l_+ \cdot \widetilde{\boldsymbol{W}}^l_- = 2\alpha_l^2 \quad \text{for } l = 1, \ldots, n-1.$$

For further examinations we introduce an orthonormal basis in $\mathbb{R}^N$. For even $N = 2n$ we use

$$\widetilde{\boldsymbol{t}}^0 := \frac{1}{\sqrt{N}} \, [1]_{i=0,\ldots,N-1} \, , \qquad \widetilde{\boldsymbol{t}}^n := \frac{1}{\sqrt{N}} \, [(-1)^i]_{i=0,\ldots,N-1} \, ,$$

$$\widetilde{\boldsymbol{s}}^l := \sqrt{\frac{2}{N}} \, [\sin(2ly_i)]_{i=0,\ldots,N-1} \, , \qquad \widetilde{\boldsymbol{t}}^l := \sqrt{\frac{2}{N}} \, [\cos(2ly_i)]_{i=0,\ldots,N-1} \, ,$$
$$\text{for } l = 1, \ldots, n-1.$$

(5.88)

For odd $N = 2n - 1$ we omit the vector $\widetilde{\boldsymbol{t}}^n$.

Employing the orthogonal transformation via $\widetilde{\boldsymbol{T}} := \boldsymbol{T}/\sqrt{2}$ with $\boldsymbol{T}$ as in (5.33), we get the eigenvectors of $\boldsymbol{K}$ (given in (5.34)) in the form

$$\widetilde{\boldsymbol{P}}^0_+ := \widetilde{\boldsymbol{T}}\widetilde{\boldsymbol{V}}^0_+ = \begin{bmatrix} \widetilde{\boldsymbol{t}}^0 \\ \boldsymbol{0} \end{bmatrix}, \qquad \widetilde{\boldsymbol{Q}}^0_- := \widetilde{\boldsymbol{T}}\widetilde{\boldsymbol{W}}^0_- = \begin{bmatrix} \boldsymbol{0} \\ -\widetilde{\boldsymbol{t}}^0 \end{bmatrix},$$

$$\widetilde{\boldsymbol{P}}^l_+ := \widetilde{\boldsymbol{T}}\widetilde{\boldsymbol{V}}^l_+ = \begin{bmatrix} (\alpha_l \cos(y_l) - (\omega - V_l)\sin(y_l))\widetilde{\boldsymbol{t}}^l \\ (\alpha_l \sin(y_l) + (\omega - V_l)\cos(y_l))\widetilde{\boldsymbol{s}}^l \end{bmatrix},$$

$$\widetilde{\boldsymbol{P}}^l_- := \widetilde{\boldsymbol{T}}\widetilde{\boldsymbol{V}}^l_- = \begin{bmatrix} (\alpha_l \cos(y_l) - (\omega + V_l)\sin(y_l))\widetilde{\boldsymbol{t}}^l \\ (\alpha_l \sin(y_l) + (\omega + V_l)\cos(y_l))\widetilde{\boldsymbol{s}}^l \end{bmatrix},$$

$$\widetilde{\boldsymbol{Q}}^l_+ := \widetilde{\boldsymbol{T}}\widetilde{\boldsymbol{W}}^l_+ = \begin{bmatrix} (\alpha_l \cos(y_l) - (\omega - V_l)\sin(y_l))\widetilde{\boldsymbol{s}}^l \\ -(\alpha_l \sin(y_l) + (\omega - V_l)\cos(y_l))\widetilde{\boldsymbol{t}}^l \end{bmatrix},$$

(5.89)

$$\widetilde{\boldsymbol{Q}}^l_- := \widetilde{\boldsymbol{T}}\widetilde{\boldsymbol{W}}^l_- = \begin{bmatrix} (\alpha_l \cos(y_l) - (\omega + V_l)\sin(y_l))\widetilde{\boldsymbol{s}}^l \\ -(\alpha_l \sin(y_l) + (\omega + V_l)\cos(y_l))\widetilde{\boldsymbol{t}}^l \end{bmatrix},$$

$$\widetilde{\boldsymbol{P}}^n_- := \widetilde{\boldsymbol{T}}\widetilde{\boldsymbol{V}}^n_- = \begin{bmatrix} \widetilde{\boldsymbol{t}}^n \\ \boldsymbol{0} \end{bmatrix}, \qquad \widetilde{\boldsymbol{Q}}^n_+ := \widetilde{\boldsymbol{T}}\widetilde{\boldsymbol{W}}^n_+ = \begin{bmatrix} \boldsymbol{0} \\ -\widetilde{\boldsymbol{t}}^n \end{bmatrix}, \quad \text{if } N = 2n$$

with the same eigenvalues as for $\boldsymbol{M}$. If there is a l with $(1 - \omega)\sin(2y_l) = \omega$ for $\omega \leq 1/2$, that is, $V_l = 0$, then the rank of the presented eigenvalues is only $2N - 2$ (for $\omega = 1/2$ and $N = 4m$, $m \in \mathbb{N}$, or $\omega < 1/2$ and $N = 2n - 1$) or $2N - 4$ (for $\omega < 1/2$ and $N = 2n$). In the latter case we get $V_l = V_{n-l} = 0$, and hence, we have to consider l and $n-l$. In all these cases we use in addition

$$\widetilde{\boldsymbol{P}}_J^l := \frac{1}{2}\begin{bmatrix}(\sin(y_l)+\cos(y_l))\,\widetilde{\boldsymbol{t}}^l\\(\sin(y_l)-\cos(y_l))\,\widetilde{\boldsymbol{s}}^l\end{bmatrix}, \qquad \widetilde{\boldsymbol{Q}}_J^l := \frac{1}{2}\begin{bmatrix}(\sin(y_l)+\cos(y_l))\,\widetilde{\boldsymbol{s}}^l\\-(\sin(y_l)-\cos(y_l))\,\widetilde{\boldsymbol{t}}^l\end{bmatrix}$$

with $(\boldsymbol{K}-\beta_l\boldsymbol{Id})\widetilde{\boldsymbol{P}}_J^l = \widetilde{\boldsymbol{P}}_+^l = \widetilde{\boldsymbol{P}}_-^l$ and $(\boldsymbol{K}-\beta_l\boldsymbol{Id})\widetilde{\boldsymbol{Q}}_J^l = \widetilde{\boldsymbol{Q}}_+^l = \widetilde{\boldsymbol{Q}}_-^l$. This renders two Jordan blocks of length two belonging to the eigenvalue $\lambda_l^+ = \lambda_l^- = \beta_l$. In the case $\omega = 1/2$ and $N = 4m$ we find

$$\boldsymbol{K}\begin{bmatrix}\widetilde{\boldsymbol{t}}^m\\0\end{bmatrix} = \begin{bmatrix}0\\\widetilde{\boldsymbol{s}}^m\end{bmatrix} = \sqrt{2}\widetilde{\boldsymbol{P}}_+^m = \sqrt{2}\widetilde{\boldsymbol{P}}_-^m, \qquad \boldsymbol{K}^2\begin{bmatrix}\widetilde{\boldsymbol{t}}^m\\0\end{bmatrix} = 0,$$

$$\boldsymbol{K}\begin{bmatrix}\widetilde{\boldsymbol{s}}^m\\0\end{bmatrix} = \begin{bmatrix}0\\-\widetilde{\boldsymbol{t}}^m\end{bmatrix} = \sqrt{2}\widetilde{\boldsymbol{Q}}_+^m = \sqrt{2}\widetilde{\boldsymbol{Q}}_-^m, \qquad \boldsymbol{K}^2\begin{bmatrix}\widetilde{\boldsymbol{s}}^m\\0\end{bmatrix} = 0.$$

For $l = 1,\ldots,n-1$, we define

$$\gamma_l^\pm := \alpha_l\cos(y_l) - (\omega\mp V_l)\sin(y_l) = (2(1-\omega)\cos^2(y_l) - (\omega\mp V_l))\sin(y_l),$$

$$\delta_l^\pm := \alpha_l\sin(y_l) + (\omega\mp V_l)\cos(y_l) = (2(1-\omega)\sin^2(y_l) + (\omega\mp V_l))\cos(y_l),$$

and for $l = 0,\ldots,n$, we use

$$\widetilde{\boldsymbol{S}}_D^l := \begin{bmatrix}\widetilde{\boldsymbol{s}}^l\\0\end{bmatrix}, \quad \widetilde{\boldsymbol{S}}_N^l := \begin{bmatrix}0\\\widetilde{\boldsymbol{s}}^l\end{bmatrix}, \quad \widetilde{\boldsymbol{T}}_N^l := \begin{bmatrix}\widetilde{\boldsymbol{t}}^l\\0\end{bmatrix}, \quad \widetilde{\boldsymbol{T}}_D^l := \begin{bmatrix}0\\\widetilde{\boldsymbol{t}}^l\end{bmatrix}. \tag{5.90}$$

Basis transformations are then given by

$$\widetilde{\boldsymbol{P}}_+^l = \gamma_l^+\widetilde{\boldsymbol{T}}_N^l + \delta_l^+\widetilde{\boldsymbol{S}}_N^l, \qquad \widetilde{\boldsymbol{P}}_-^l = \gamma_l^-\widetilde{\boldsymbol{T}}_N^l + \delta_l^-\widetilde{\boldsymbol{S}}_N^l,$$

$$\widetilde{\boldsymbol{Q}}_+^l = \gamma_l^+\widetilde{\boldsymbol{S}}_D^l - \delta_l^+\widetilde{\boldsymbol{T}}_D^l, \qquad \widetilde{\boldsymbol{Q}}_-^l = \gamma_l^-\widetilde{\boldsymbol{S}}_D^l - \delta_l^-\widetilde{\boldsymbol{T}}_D^l, \tag{5.91}$$

and

$$\widetilde{\boldsymbol{S}}_N^l = \frac{-\gamma_l^-}{\gamma_l^+\delta_l^- - \gamma_l^-\delta_l^+}\widetilde{\boldsymbol{P}}_+^l + \frac{\gamma_l^+}{\gamma_l^+\delta_l^- - \gamma_l^-\delta_l^+}\widetilde{\boldsymbol{P}}_-^l,$$

$$\widetilde{\boldsymbol{T}}_N^l = \frac{\delta_l^-}{\gamma_l^+\delta_l^- - \gamma_l^-\delta_l^+}\widetilde{\boldsymbol{P}}_+^l - \frac{\delta_l^+}{\gamma_l^+\delta_l^- - \gamma_l^-\delta_l^+}\widetilde{\boldsymbol{P}}_-^l,$$

$$\widetilde{\boldsymbol{S}}_D^l = \frac{\delta_l^-}{\gamma_l^+\delta_l^- - \gamma_l^-\delta_l^+}\widetilde{\boldsymbol{Q}}_+^l - \frac{\delta_l^+}{\gamma_l^+\delta_l^- - \gamma_l^-\delta_l^+}\widetilde{\boldsymbol{Q}}_-^l, \tag{5.92}$$

$$\widetilde{\boldsymbol{T}}_D^l = \frac{\gamma_l^-}{\gamma_l^+\delta_l^- - \gamma_l^-\delta_l^+}\widetilde{\boldsymbol{Q}}_+^l - \frac{\gamma_l^+}{\gamma_l^+\delta_l^- - \gamma_l^-\delta_l^+}\widetilde{\boldsymbol{Q}}_-^l$$

for $l = 1,\ldots,n-1$. Furthermore, we find

$$\frac{\gamma_l^+\delta_l^-}{\gamma_l^+\delta_l^- - \gamma_l^-\delta_l^+} = \frac{1}{2}(1+A_l), \qquad \frac{\gamma_l^+\gamma_l^-}{\gamma_l^+\delta_l^- - \gamma_l^-\delta_l^+} = \frac{1-2\omega}{2\omega}B_l,$$

$$\frac{-\gamma_l^-\delta_l^+}{\gamma_l^+\delta_l^- - \gamma_l^-\delta_l^+} = \frac{1}{2}(1-A_l), \qquad \frac{\delta_l^+\delta_l^-}{\gamma_l^+\delta_l^- - \gamma_l^-\delta_l^+} = \frac{1}{2\omega}B_l, \tag{5.93}$$

where we use the definitions

$$A_l := \frac{\omega}{V_l}\cos(2y_l), \quad B_l := \frac{\omega}{V_l}\sin(2y_l) \quad \text{for } l = 1,\ldots,n-1. \tag{5.94}$$

The behavior of A_l and B_l depending on ω and l is plotted in Figure 5.5. For $\omega < 1/2$ the values of A_l and B_l turn imaginary for $l^* < l < N/2 - l^*$ with $l^* := \lfloor N\arcsin(\omega/(1-\omega))/(2\pi)\rfloor$. There are poles at l^* and $N/2 - l^*$. We find

$$\lim_{\omega\to 1/2+} A_l = \begin{cases}1 & \text{if } l < N/4,\\-1 & \text{if } l > N/4,\end{cases} \qquad \lim_{\omega\to 1/2+} B_l = \begin{cases}\tan(2y_l) & \text{if } l < N/4,\\-\tan(2y_l) & \text{if } l > N/4.\end{cases}$$

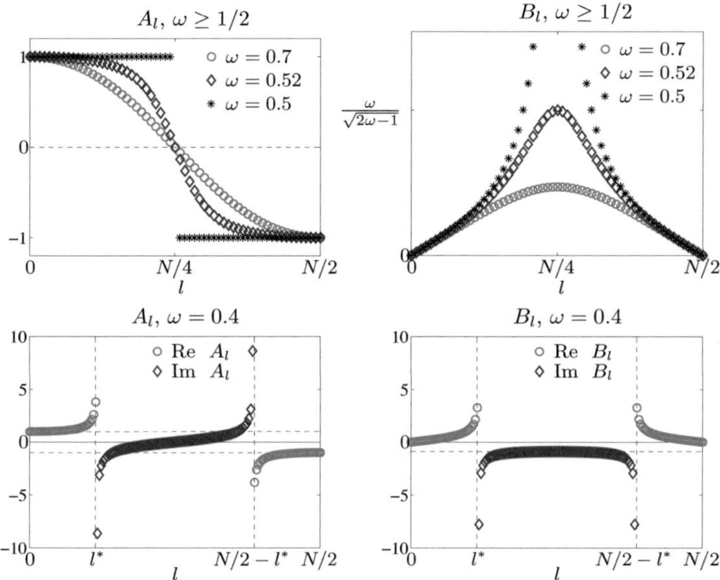

FIGURE 5.5. Behavior of A_l and B_l for $l = 0, \ldots, N/2$.

The transformation formulas only fail in the case $V_l = 0$, which is impossible for $\omega > 1/2$. For $\omega = 1/2$ and $N = 4m$ we get $V_m = 0$. For $\omega < 1/2$ we have $V_l = 0$, if l fulfills $\sin(2y_l) = \omega/(1 - \omega)$. Then the periodic matrices $\boldsymbol{M}$ and $\boldsymbol{K}$ are not diagonalizable and there are Jordan-blocks of length two belonging to the eigenvalue $\lambda_l^+ = \lambda_l^-$.

Now we find the Fourier representations for the FD lattice Boltzmann solutions.

Theorem 5.8. *Assume that* $\omega \in (1/2, 1)$ *and* $N = 2n$. *Let the initial values be given by* $\boldsymbol{P}^0 = [\boldsymbol{R}^0; \boldsymbol{J}^0]$ *and let the right hand side be given by* $\boldsymbol{G}^l = [\boldsymbol{f}^l; \boldsymbol{g}^l]$ *for* $l = 0, \ldots, k-1$. *Then the discrete Fourier solution* $\boldsymbol{P}^k = [\boldsymbol{R}^k; \boldsymbol{J}^k]$ *of the system*

$$\boldsymbol{P}^{l+1} = \boldsymbol{K}\boldsymbol{P}^l + \tau\boldsymbol{G}^l \quad \text{for } l = 0, \ldots, k-1$$

for the VCFD lattice Boltzmann equations (5.1) *with periodic boundary conditions* (5.5) *can be represented in terms of the orthogonal basis* (5.88) *for* $k \geq 0$ *as*

$$\boldsymbol{R}^k = \left(\boldsymbol{R}^0 \cdot \widetilde{\boldsymbol{t}}^0\right)\widetilde{\boldsymbol{t}}^0 + (-1)^k \left(\boldsymbol{R}^0 \cdot \widetilde{\boldsymbol{t}}^n\right)\widetilde{\boldsymbol{t}}^n$$

$$+ \sum_{j=1}^{n-1} \left(\frac{1}{2}(1 + A_j)(\lambda_j^+)^k + \frac{1}{2}(1 - A_j)(\lambda_j^-)^k\right)\left(\left(\boldsymbol{R}^0 \cdot \widetilde{\boldsymbol{t}}^j\right)\widetilde{\boldsymbol{t}}^j + \left(\boldsymbol{R}^0 \cdot \widetilde{\boldsymbol{s}}^j\right)\widetilde{\boldsymbol{s}}^j\right)$$

$$+ \frac{\theta}{2\omega}\sum_{j=1}^{n-1} B_j \left((\lambda_j^+)^k - (\lambda_j^-)^k\right)\left(\left(\boldsymbol{J}^0 \cdot \widetilde{\boldsymbol{s}}^j\right)\widetilde{\boldsymbol{t}}^j - \left(\boldsymbol{J}^0 \cdot \widetilde{\boldsymbol{t}}^j\right)\widetilde{\boldsymbol{s}}^j\right)$$

$$+ \tau\sum_{l=0}^{k-1} \left(\boldsymbol{f}^l \cdot \widetilde{\boldsymbol{t}}^0\right)\widetilde{\boldsymbol{t}}^0 + \tau\sum_{l=0}^{k-1}(-1)^{k-l-1}\left(\boldsymbol{f}^l \cdot \widetilde{\boldsymbol{t}}^n\right)\widetilde{\boldsymbol{t}}^n$$

$$+ \tau \sum_{l=0}^{k-1} \sum_{j=1}^{n-1} \frac{1}{2}(1 + A_j)(\lambda_j^+)^{k-l-1} \left(\left(\boldsymbol{f}^l \cdot \widetilde{\boldsymbol{t}}^j \right) \widetilde{\boldsymbol{t}}^j + \left(\boldsymbol{f}^l \cdot \widetilde{\boldsymbol{s}}^j \right) \widetilde{\boldsymbol{s}}^j \right)$$

$$+ \tau \sum_{l=0}^{k-1} \sum_{j=1}^{n-1} \frac{1}{2}(1 - A_j)(\lambda_j^-)^{k-l-1} \left(\left(\boldsymbol{f}^l \cdot \widetilde{\boldsymbol{t}}^j \right) \widetilde{\boldsymbol{t}}^j + \left(\boldsymbol{f}^l \cdot \widetilde{\boldsymbol{s}}^j \right) \widetilde{\boldsymbol{s}}^j \right)$$

$$+ \frac{\theta}{2\omega} \tau \sum_{l=0}^{k-2} \sum_{j=1}^{n-1} B_j \left((\lambda_j^+)^{k-l-1} - (\lambda_j^-)^{k-l-1} \right) \left(\left(\boldsymbol{g}^l \cdot \widetilde{\boldsymbol{s}}^j \right) \widetilde{\boldsymbol{t}}^j - \left(\boldsymbol{g}^l \cdot \widetilde{\boldsymbol{t}}^j \right) \widetilde{\boldsymbol{s}}^j \right),$$

$$\boldsymbol{J}^k = (-\theta)^k \left(\boldsymbol{J}^0 \cdot \widetilde{\boldsymbol{t}}^0 \right) \widetilde{\boldsymbol{t}}^0 + \theta^k \left(\boldsymbol{J}^0 \cdot \widetilde{\boldsymbol{t}}^n \right) \widetilde{\boldsymbol{t}}^n$$

$$+ \sum_{j=1}^{n-1} \left(\frac{1}{2}(1 - A_j)(\lambda_j^+)^k + \frac{1}{2}(1 + A_j)(\lambda_j^-)^k \right) \left(\left(\boldsymbol{J}^0 \cdot \widetilde{\boldsymbol{t}}^j \right) \widetilde{\boldsymbol{t}}^j + \left(\boldsymbol{J}^0 \cdot \widetilde{\boldsymbol{s}}^j \right) \widetilde{\boldsymbol{s}}^j \right)$$

$$- \frac{1}{2\omega} \sum_{j=1}^{n-1} B_j \left((\lambda_j^+)^k - (\lambda_j^-)^k \right) \left(\left(\boldsymbol{R}^0 \cdot \widetilde{\boldsymbol{s}}^j \right) \widetilde{\boldsymbol{t}}^j - \left(\boldsymbol{R}^0 \cdot \widetilde{\boldsymbol{t}}^j \right) \widetilde{\boldsymbol{s}}^j \right)$$

$$+ \tau \sum_{l=0}^{k-1} (-\theta)^{k-l-1} \left(\boldsymbol{g}^l \cdot \widetilde{\boldsymbol{t}}^0 \right) \widetilde{\boldsymbol{t}}^0 + \tau \sum_{l=0}^{k-1} \theta^{k-l-1} \left(\boldsymbol{g}^l \cdot \widetilde{\boldsymbol{t}}^n \right) \widetilde{\boldsymbol{t}}^n$$

$$+ \tau \sum_{l=0}^{k-1} \sum_{j=1}^{n-1} \frac{1}{2}(1 - A_j)(\lambda_j^+)^{k-l-1} \left(\left(\boldsymbol{g}^l \cdot \widetilde{\boldsymbol{t}}^j \right) \widetilde{\boldsymbol{t}}^j + \left(\boldsymbol{g}^l \cdot \widetilde{\boldsymbol{s}}^j \right) \widetilde{\boldsymbol{s}}^j \right)$$

$$+ \tau \sum_{l=0}^{k-1} \sum_{j=1}^{n-1} \frac{1}{2}(1 + A_j)(\lambda_j^-)^{k-l-1} \left(\left(\boldsymbol{g}^l \cdot \widetilde{\boldsymbol{t}}^j \right) \widetilde{\boldsymbol{t}}^j + \left(\boldsymbol{g}^l \cdot \widetilde{\boldsymbol{s}}^j \right) \widetilde{\boldsymbol{s}}^j \right)$$

$$- \frac{\tau}{2\omega} \sum_{l=0}^{k-2} \sum_{j=1}^{n-1} B_j \left((\lambda_j^+)^{k-l-1} - (\lambda_j^-)^{k-l-1} \right) \left(\left(\boldsymbol{f}^l \cdot \widetilde{\boldsymbol{s}}^j \right) \widetilde{\boldsymbol{t}}^j - \left(\boldsymbol{f}^l \cdot \widetilde{\boldsymbol{t}}^j \right) \widetilde{\boldsymbol{s}}^j \right),$$

where A_j and B_j are defined in (5.94) and the eigenvalues $\lambda_j^\pm$ are given in (5.87) with (5.86). For odd $N = 2n - 1$ the parts with $\widetilde{\boldsymbol{t}}^n$ disappear.

PROOF. The initial value $\boldsymbol{P}^0$ and the right hand side $\boldsymbol{G}^k$ of the problem can be written as

$$\boldsymbol{P}^0 = \left(\boldsymbol{R}^0 \cdot \widetilde{\boldsymbol{t}}^0 \right) \widetilde{\boldsymbol{T}}_N^0 + \sum_{j=1}^{n-1} \left(\left(\boldsymbol{R}^0 \cdot \widetilde{\boldsymbol{t}}^j \right) \widetilde{\boldsymbol{T}}_N^j + \left(\boldsymbol{R}^0 \cdot \widetilde{\boldsymbol{s}}^j \right) \widetilde{\boldsymbol{S}}_D^j \right) + \left(\boldsymbol{R}^0 \cdot \widetilde{\boldsymbol{t}}^n \right) \widetilde{\boldsymbol{T}}_N^n$$

$$+ \left(\boldsymbol{J}^0 \cdot \widetilde{\boldsymbol{t}}^0 \right) \widetilde{\boldsymbol{T}}_D^0 + \sum_{j=1}^{n-1} \left(\left(\boldsymbol{J}^0 \cdot \widetilde{\boldsymbol{t}}^j \right) \widetilde{\boldsymbol{T}}_D^j + \left(\boldsymbol{J}^0 \cdot \widetilde{\boldsymbol{s}}^j \right) \widetilde{\boldsymbol{S}}_N^j \right) + \left(\boldsymbol{J}^0 \cdot \widetilde{\boldsymbol{t}}^n \right) \widetilde{\boldsymbol{T}}_D^n,$$

$$\boldsymbol{G}^k = \left(\boldsymbol{f}^k \cdot \widetilde{\boldsymbol{t}}^0 \right) \widetilde{\boldsymbol{T}}_N^0 + \sum_{j=1}^{n-1} \left(\left(\boldsymbol{f}^k \cdot \widetilde{\boldsymbol{t}}^j \right) \widetilde{\boldsymbol{T}}_N^j + \left(\boldsymbol{f}^k \cdot \widetilde{\boldsymbol{s}}^j \right) \widetilde{\boldsymbol{S}}_D^j \right) + \left(\boldsymbol{f}^k \cdot \widetilde{\boldsymbol{t}}^n \right) \widetilde{\boldsymbol{T}}_N^n$$

$$+ \left(\boldsymbol{g}^k \cdot \widetilde{\boldsymbol{t}}^0 \right) \widetilde{\boldsymbol{T}}_D^0 + \sum_{j=1}^{n-1} \left(\left(\boldsymbol{g}^k \cdot \widetilde{\boldsymbol{t}}^j \right) \widetilde{\boldsymbol{T}}_D^j + \left(\boldsymbol{g}^k \cdot \widetilde{\boldsymbol{s}}^j \right) \widetilde{\boldsymbol{S}}_N^j \right) + \left(\boldsymbol{g}^k \cdot \widetilde{\boldsymbol{t}}^n \right) \widetilde{\boldsymbol{T}}_D^n.$$

By applying the transformation formula (5.92), we switch to an eigenvector representation. So we can easily apply the time evolution matrix $\boldsymbol{K}$ given in (5.34) and compute $\boldsymbol{P}^1$ and all further solutions. Upon applying the transformation (5.91) we gain the provided representation of the solution with regard to (5.93). $\square$

In the case $\omega \in (0, 1/2)$ the solution representation in Theorem 5.8 remains valid under the constraint

$$\sin(2y_l) \neq \frac{\omega}{1 - \omega} \quad \text{for } l = 1, \ldots, n.$$

The solution representation remains real although V_l and hence, the eigenvalues and eigenvectors turn complex for some l. If there is a j with $\sin(2y_j) = \omega/(1 - \omega)$, then the jth summand takes the form

$$\left(\beta_j^k + \omega\cos(2y_j)k\beta_j^{k-1}\right)\left(\left(\boldsymbol{R}^0 \cdot \widetilde{\boldsymbol{t}}^j\right)\widetilde{\boldsymbol{t}}^j + \left(\boldsymbol{R}^0 \cdot \widetilde{\boldsymbol{s}}^j\right)\widetilde{\boldsymbol{s}}^j\right)$$
$$+ \theta\sin(2y_j)k\beta_j^{k-1}\left(\left(\boldsymbol{J}^0 \cdot \widetilde{\boldsymbol{s}}^j\right)\widetilde{\boldsymbol{t}}^j - \left(\boldsymbol{J}^0 \cdot \widetilde{\boldsymbol{t}}^j\right)\widetilde{\boldsymbol{s}}^j\right)$$

for the $\boldsymbol{R}$-equation and

$$\left(\beta_j^k - \omega\cos(2y_j)k\beta_j^{k-1}\right)\left(\left(\boldsymbol{J}^0 \cdot \widetilde{\boldsymbol{t}}^j\right)\widetilde{\boldsymbol{t}}^j + \left(\boldsymbol{J}^0 \cdot \widetilde{\boldsymbol{s}}^j\right)\widetilde{\boldsymbol{s}}^j\right)$$
$$- \sin(2y_j)k\beta_j^{k-1}\left(\left(\boldsymbol{R}^0 \cdot \widetilde{\boldsymbol{s}}^j\right)\widetilde{\boldsymbol{t}}^j - \left(\boldsymbol{R}^0 \cdot \widetilde{\boldsymbol{t}}^j\right)\widetilde{\boldsymbol{s}}^j\right)$$

for the $\boldsymbol{J}$-equation. Accordingly, we have to add corresponding terms for the right hand sides.

For the low-frequency coefficients in the solution representation we find

$$A_j = 1 - \frac{\theta}{2\omega^2}(2y_j)^2 + \theta\mathcal{O}(y_j^4) \qquad \text{for } j \ll N,$$
$$B_j = 2y_j + \frac{3(1 - \omega)^2 - \omega^2}{6\omega^2}(2y_j)^3 + \mathcal{O}(y_j^5) \quad \text{for } j \ll N.$$

For the corresponding eigenvalues we find

$$\lambda_j^+ = 1 - \frac{1 - \omega}{2\omega}(2y_j)^2 + \mathcal{O}(y_j^4) \quad \text{for } j \ll N,$$
$$(\lambda_j^+)^k \approx e^{-4\nu j^2\pi^2 t_k/|\Omega|^2} \qquad \text{for } j \ll N,$$

where we take $\nu := (1 - \omega)\gamma/(2\omega)$ and $t_k := k\tau = k|\Omega|^2/(\gamma N^2)$. Hence, the leading order terms

$$\frac{1}{2}(1 + A_j)(\lambda_j^+)^k \approx e^{-4\nu j^2\pi^2 t_k/|\Omega|^2} \quad \text{for } j \ll N$$

show the right approximation property. In contrast, we have $1 - A_j = \mathcal{O}(y_j^2)$ and $(\lambda_j^-)^k$ is oscillating and it decays in a fast way. Since B_j and $\boldsymbol{J}^0$ are of order $\mathcal{O}(h)$, the terms induced by $\boldsymbol{J}^0$ in the $\boldsymbol{R}$-representation are only of order $\mathcal{O}(h^2)$. If we choose the initial data $j_0 := -h\partial_x r_0/(2\omega)$, we find

$$\left(\left(\boldsymbol{J}^0 \cdot \widetilde{\boldsymbol{s}}^j\right)\widetilde{\boldsymbol{t}}^j - \left(\boldsymbol{J}^0 \cdot \widetilde{\boldsymbol{t}}^j\right)\widetilde{\boldsymbol{s}}^j\right) = \frac{1}{2\omega}2y_j\left(\left(\boldsymbol{R}^0 \cdot \widetilde{\boldsymbol{t}}^j\right)\widetilde{\boldsymbol{t}}^j + \left(\boldsymbol{R}^0 \cdot \widetilde{\boldsymbol{s}}^j\right)\widetilde{\boldsymbol{s}}^j\right) + \mathcal{O}(y_j^3).$$

Then the leading order coefficient in the $\boldsymbol{R}$-representation is of the form

$$\frac{1}{2}(1 + A_j) + \frac{\theta}{4\omega^2}2y_jB_j = 1 + \mathcal{O}(y_j^4).$$

In fact this does not lead to an improvement, since the middle- and high-frequency terms are only of order h^2.

The crucial term in the $\boldsymbol{J}$-representation is $B_j(\lambda_j^+)^k$. With the background of partial integration we find

$$-\frac{B_j}{2\omega}\left(\left(\boldsymbol{R}^0 \cdot \widetilde{\boldsymbol{s}}^j\right) - \left(\boldsymbol{R}^0 \cdot \widetilde{\boldsymbol{t}}^j\right)\right) \approx -\frac{h}{2\omega}\left(\left(D_h\boldsymbol{R}^0 \cdot \widetilde{\boldsymbol{t}}^j\right) + \left(D_h\boldsymbol{R}^0 \cdot \widetilde{\boldsymbol{s}}^j\right)\right),$$

where we define $D_h\boldsymbol{R}^0 := [\partial_x r_0(x_i)]_i$. Hence, $\boldsymbol{J}^k$ is an approximation of $-h\partial_x r/(2\omega)$. The choice of $\boldsymbol{J}^0$ plays a subordinate role.

In the $\boldsymbol{R}$-representation the oscillating parts $(\lambda_j^-)^k$ are hidden in an order $\mathcal{O}(h^2)$-term. In the $\boldsymbol{J}$-representation the oscillating terms $B_j(\lambda_j^-)^k$ are apparent. Hence, we get the wrong viscosity in the first steps. After some iterations the oscillations faded away and the correct viscosity is reached. This can be seen in numerical experiments (see Section 7.3).

For the special choice $\omega = (3 - \sqrt{3})/2$ and vanishing right hand side, we find an improvement of the convergence properties in the numerical experiments. In this case we have $3(1 - \omega)^2 = \omega^2$ and hence we get $B_j = 2y_j + \mathcal{O}(y_j^5)$.

The behavior of the time-dependent coefficients

$$R_j^k := \frac{1}{2}(1 + A_j)(\lambda_j^+)^k + \frac{1}{2}(1 - A_j)(\lambda_j^-)^k$$

is examined in Figure 5.6. The poles in A_j for $\omega < 1/2$ are deleted. For large time steps only the coefficients with j close to 0 and $N/2$ are of significant size.

Corollary 5.9. *In the case $\omega = 1/2$, $N = 2n$, the solution representation in Theorem 5.8 for $k \geq 1$ simplifies to*

$$\boldsymbol{R}^k = \left(\boldsymbol{R}^0 \cdot \widetilde{\boldsymbol{t}}^0\right)\widetilde{\boldsymbol{t}}^0 + (-1)^k \left(\boldsymbol{R}^0 \cdot \widetilde{\boldsymbol{t}}^n\right)\widetilde{\boldsymbol{t}}^n$$

$$+ \sum_{j=1}^{n-1} \cos^k(2y_j)\left(\left(\boldsymbol{R}^0 \cdot \widetilde{\boldsymbol{t}}^j\right)\widetilde{\boldsymbol{t}}^j + \left(\boldsymbol{R}^0 \cdot \widetilde{\boldsymbol{s}}^j\right)\widetilde{\boldsymbol{s}}^j\right)$$

$$+ \tau \sum_{l=0}^{k-1} \left(\boldsymbol{f}^l \cdot \widetilde{\boldsymbol{t}}^0\right)\widetilde{\boldsymbol{t}}^0 + \tau \sum_{l=0}^{k-1}(-1)^{k-l-1}\left(\boldsymbol{f}^l \cdot \widetilde{\boldsymbol{t}}^n\right)\widetilde{\boldsymbol{t}}^n$$

$$+ \tau \sum_{l=0}^{k-1}\sum_{j=1}^{n-1} \cos^{k-l-1}(2y_j)\left(\left(\boldsymbol{f}^l \cdot \widetilde{\boldsymbol{t}}^j\right)\widetilde{\boldsymbol{t}}^j + \left(\boldsymbol{f}^l \cdot \widetilde{\boldsymbol{s}}^j\right)\widetilde{\boldsymbol{s}}^j\right),$$

$$\boldsymbol{J}^k = - \sum_{j=1}^{n-1} \sin(2y_j)\cos^{k-1}(2y_j)\left(\left(\boldsymbol{R}^0 \cdot \widetilde{\boldsymbol{s}}^j\right)\widetilde{\boldsymbol{t}}^j - \left(\boldsymbol{R}^0 \cdot \widetilde{\boldsymbol{t}}^j\right)\widetilde{\boldsymbol{s}}^j\right)$$

$$- \tau \sum_{l=0}^{k-2}\sum_{j=1}^{n-1} \sin(2y_j)\cos^{k-l-2}(2y_j)\left(\left(\boldsymbol{f}^l \cdot \widetilde{\boldsymbol{s}}^j\right)\widetilde{\boldsymbol{t}}^j - \left(\boldsymbol{f}^j \cdot \widetilde{\boldsymbol{t}}^j\right)\widetilde{\boldsymbol{s}}^j\right) + \tau \boldsymbol{g}^{k-1}.$$

For odd $N = 2n - 1$ the parts with $\widetilde{\boldsymbol{t}}^n$ disappear.

PROOF. We find $(1 + A_j)/2 = 1$ for $j < N/4$, $(1 + A_j)/2 = 0$ for $j > N/4$, $(1 - A_j)/2 = 0$ for $j < N/4$, $(1 - A_j)/2 = 1$ for $j > N/4$, $B_j = \tan(2y_j)$ for $j < N/4$ and $B_j = -\tan(2y_j)$ for $j > N/4$. In the case $N = 4m$, $m \in \mathbb{N}$, the solution representation remains unchanged for $j = m = N/4$. $\square$

In the case $\omega = 1/2$ the solution for the discrete density matches the solution obtained from an explicit Euler scheme. We observe that for $\omega = 1/2$ the initial values $\boldsymbol{J}^0$ have no influence on the solution. The right hand side $\boldsymbol{g}^k$ of the flux equation only has an effect in the following time step.

The structure of the basis vectors $\widetilde{\boldsymbol{t}}^l$ and $\widetilde{\boldsymbol{s}}^l$ implies to interpret the representations of the solutions in the context of VCFD grids. It turns out that the representation of $\boldsymbol{R}^k$ in terms of the odd basis vectors $\widetilde{\boldsymbol{s}}^l$ and the representation of $\boldsymbol{J}^k$ in terms of the even basis vectors $\widetilde{\boldsymbol{t}}^l$ is the representation of solutions of the problem subject to vanishing density boundary conditions. By exchanging the roles of $\widetilde{\boldsymbol{t}}^l$ and $\widetilde{\boldsymbol{s}}^l$ we gain solutions of the vanishing flux boundary value problem.

For an interpretation in the context of CC grids we slightly modify our approach. All the other previous computations can be applied without changes. Now

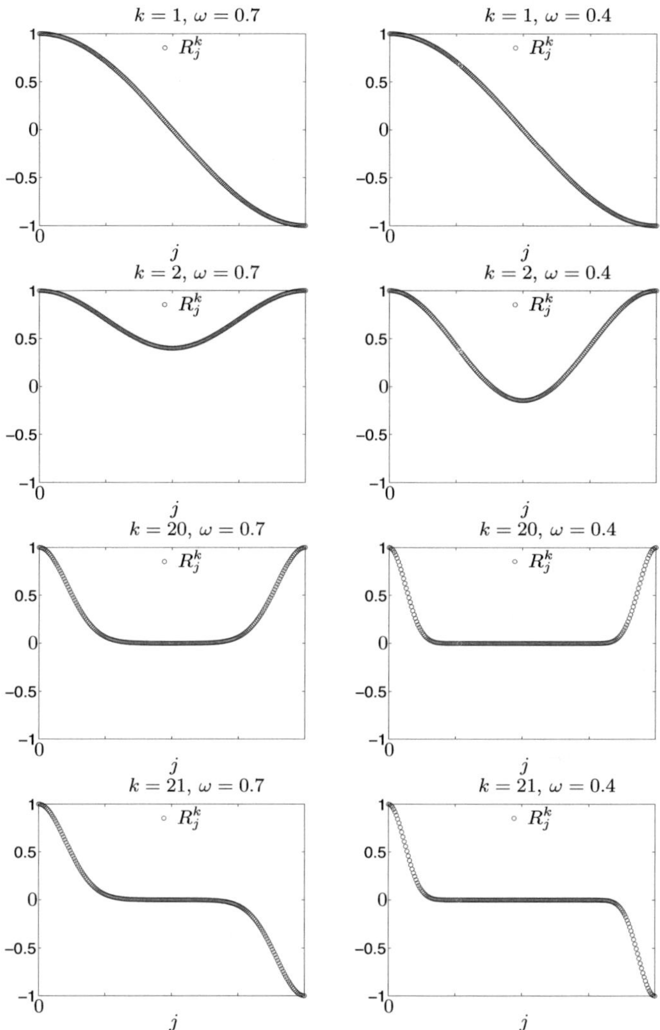

FIGURE 5.6. Behavior of R_j^k for $N = 200$ and selected time steps k.

we define for $N = 2n - 1$ or $N = 2n$

$$\widehat{\boldsymbol{S}}_c^0 := \frac{1}{\sqrt{2N}} \left[\begin{bmatrix} 1 \\ 1 \end{bmatrix} \right]_{i=1,\ldots,N}, \qquad \widehat{\boldsymbol{T}}_c^0 := \frac{1}{\sqrt{2N}} \left[\begin{bmatrix} 1 \\ -1 \end{bmatrix} \right]_{i=1,\ldots,N},$$

$$\widehat{\boldsymbol{S}}_c^l := \frac{1}{\sqrt{N}} \left[\begin{bmatrix} \cos(2ly_i) \\ \cos(2ly_i) \end{bmatrix} \right]_{i=1,\ldots,N}, \qquad \widehat{\boldsymbol{T}}_s^l := \frac{1}{\sqrt{N}} \left[\begin{bmatrix} \sin(2ly_i) \\ -\sin(2ly_i) \end{bmatrix} \right]_{i=1,\ldots,N},$$

$$\tag{5.95}$$

$$\widehat{\boldsymbol{S}}_s^l := \frac{1}{\sqrt{N}} \left[\begin{bmatrix} \sin(2ly_i) \\ \sin(2ly_i) \end{bmatrix} \right]_{i=1,\ldots,N}, \qquad \widehat{\boldsymbol{T}}_c^l := \frac{1}{\sqrt{N}} \left[\begin{bmatrix} \cos(2ly_i) \\ -\cos(2ly_i) \end{bmatrix} \right]_{i=1,\ldots,N},$$

$$\text{for } l = 1, \ldots, n - 1.$$

For even $N = 2n$ we further define

$$\widehat{S}_c^n := \frac{1}{\sqrt{2N}} \left[\begin{bmatrix} (-1)^i \\ (-1)^i \end{bmatrix} \right]_{i=1,\dots,N}, \qquad \widehat{T}_c^n := \frac{1}{\sqrt{2N}} \left[\begin{bmatrix} (-1)^i \\ -(-1)^i \end{bmatrix} \right]_{i=1,\dots,N}.$$

By applying the same procedure as before, we obtain the eigenvectors of M (given in (5.16)) in the form

$$\widehat{V}_+^0 := \widehat{S}_c^0, \qquad \widehat{W}_-^0 := \widehat{T}_c^0,$$

$$\left.\begin{aligned}
\widehat{V}_+^l &:= \alpha_l \widehat{S}_c^l - (\omega - V_l)\widehat{T}_s^l, \\
\widehat{V}_-^l &:= \alpha_l \widehat{S}_c^l - (\omega + V_l)\widehat{T}_s^l, \\
\widehat{W}_+^l &:= \alpha_l \widehat{S}_s^l + (\omega - V_l)\widehat{T}_c^l, \\
\widehat{W}_-^l &:= \alpha_l \widehat{S}_s^l + (\omega + V_l)\widehat{T}_c^l,
\end{aligned}\right\} \quad \text{for } l = 1, \dots, n-1,$$

$$\widehat{V}_+^n := \widehat{S}_c^n, \qquad \widehat{W}_-^n := \widehat{T}_c^n, \qquad \text{if } N = 2n,$$

with

$$\mathbf{M}\widehat{V}_+^0 = \lambda_0^+ \widehat{V}_+^0, \quad \mathbf{M}\widehat{W}_-^0 = \lambda_n^- \widehat{W}_-^0,$$

$$\left.\begin{aligned}
\mathbf{M}\widehat{V}_+^l &= \lambda_l^+ \widehat{V}_+^l, \quad \mathbf{M}\widehat{W}_+^l = \lambda_l^+ \widehat{W}_+^l, \\
\mathbf{M}\widehat{V}_-^l &= \lambda_l^- \widehat{V}_-^l, \quad \mathbf{M}\widehat{W}_-^l = \lambda_l^- \widehat{W}_-^l,
\end{aligned}\right\} \quad \text{for } l = 1, \dots, n-1,$$

$$\mathbf{M}\widehat{V}_+^n = \lambda_n^+ \widehat{V}_+^n, \quad \mathbf{M}\widehat{W}_-^n = \lambda_n^- \widehat{W}_-^n \qquad \text{if } N = 2n.$$

The eigenvalues $\lambda_l^{\pm}$ are the same as in (5.87). An orthonormal basis in $\mathbb{R}^N$ for $N = 2n$ is now defined by

$$\widehat{t}^0 := \frac{1}{\sqrt{N}} [1]_{i=1,\dots,N}, \qquad\qquad \widehat{t}^l := \sqrt{\frac{2}{N}} \left[\cos(2l y_{i-1/2})\right]_{i=1,\dots,N},$$

$$\widehat{s}^l := \sqrt{\frac{2}{N}} \left[\sin(2l y_{i-1/2})\right]_{i=1,\dots,N}, \qquad \widehat{s}^n := \frac{1}{\sqrt{N}} \left[(-1)^{i+1}\right]_{i=1,\dots,N},$$
$$\text{for } l = 1, \dots, n-1. \tag{5.96}$$

For odd $N = 2n - 1$ we omit the vector $\widehat{s}^n$. We get the eigenvectors of $\mathbf{K}$ (given in (5.34)) by transformation via $\widehat{T} := T/\sqrt{2}$ with T given in (5.33) in the form

$$\widehat{P}_+^0 := \widehat{T}\widehat{V}_+^0 = \begin{bmatrix} \widehat{t}^0 \\ 0 \end{bmatrix}, \qquad \widehat{Q}_-^0 := \widehat{T}\widehat{W}_-^0 = \begin{bmatrix} 0 \\ -\widehat{t}^0 \end{bmatrix},$$

$$\widehat{P}_+^l := \widehat{T}\widehat{V}_+^l = \begin{bmatrix} (\alpha_l \cos(y_l) - (\omega - V_l)\sin(y_l))\,\widehat{t}^l \\ (\alpha_l \sin(y_l) + (\omega - V_l)\cos(y_l))\,\widehat{s}^l \end{bmatrix},$$

$$\widehat{P}_-^l := \widehat{T}\widehat{V}_-^l = \begin{bmatrix} (\alpha_l \cos(y_l) - (\omega + V_l)\sin(y_l))\,\widehat{t}^l \\ (\alpha_l \sin(y_l) + (\omega + V_l)\cos(y_l))\,\widehat{s}^l \end{bmatrix},$$

$$\widehat{Q}_+^l := \widehat{T}\widehat{W}_+^l = \begin{bmatrix} (\alpha_l \cos(y_l) - (\omega - V_l)\sin(y_l))\,\widehat{s}^l \\ -(\alpha_l \sin(y_l) + (\omega - V_l)\cos(y_l))\,\widehat{t}^l \end{bmatrix},$$

$$\widehat{Q}_-^l := \widehat{T}\widehat{W}_-^l = \begin{bmatrix} (\alpha_l \cos(y_l) - (\omega + V_l)\sin(y_l))\,\widehat{s}^l \\ -(\alpha_l \sin(y_l) + (\omega + V_l)\cos(y_l))\,\widehat{t}^l \end{bmatrix},$$

$$\widehat{P}_+^n := \widehat{T}\widehat{V}_+^n = \begin{bmatrix} 0 \\ \widehat{s}^n \end{bmatrix}, \qquad \widehat{Q}_-^n := \widehat{T}\widehat{W}_-^n = \begin{bmatrix} -\widehat{s}^n \\ 0 \end{bmatrix}, \quad \text{if } N = 2n.$$

Now we find

Theorem 5.10. *Assume that $\omega \in (1/2, 1)$ and $N = 2n$. Let the initial values be given by $\boldsymbol{P}^0 = [\boldsymbol{R}^0; \boldsymbol{J}^0]$ and let the right hand side be given by $\boldsymbol{G}^l = [\boldsymbol{f}^l; \boldsymbol{g}^l]$ for $l = 0, \ldots, k - 1$. Then the discrete Fourier solution $\boldsymbol{P}^k = [\boldsymbol{R}^k; \boldsymbol{J}^k]$ of the system*

$$\boldsymbol{P}^{l+1} = \boldsymbol{K}\boldsymbol{P}^l + \tau\boldsymbol{G}^l \quad \text{for } l = 0, \ldots, k - 1$$

for the VCFD lattice Boltzmann equation (5.1) with periodic boundary conditions (5.5) can be represented in terms of the orthonormal basis (5.96). The representation is the same as in Theorem 5.8 with $\widetilde{\boldsymbol{s}}^l$ replaced by $\widehat{\boldsymbol{s}}^l$ for $l = 1, \ldots n - 1$, $\widetilde{\boldsymbol{t}}^l$ replaced by $\widehat{\boldsymbol{t}}^l$ for $l = 0, \ldots, n - 1$ and $\widetilde{\boldsymbol{t}}^n$ replaced by $\widehat{\boldsymbol{s}}^n$.

For the examination of discrete Fourier solutions for the density boundary value problem on VCFD grids we use the orthonormal basis

$$\mathring{\boldsymbol{T}}_c^0 := \frac{1}{\sqrt{2N}} \left[\begin{bmatrix} 1 \\ -1 \end{bmatrix}\right]_{i=1,\ldots,N}, \qquad \mathring{\boldsymbol{S}}_s^N := \frac{1}{\sqrt{2N}} \left[\begin{bmatrix} (-1)^{i+1} \\ (-1)^{i+1} \end{bmatrix}\right]_{i=1,\ldots,N},$$

$$\mathring{\boldsymbol{S}}_s^l := \frac{1}{\sqrt{N}} \left[\begin{bmatrix} \sin(ly_{i-1/2}) \\ \sin(ly_{i-1/2}) \end{bmatrix}\right]_{i=1,\ldots,N}, \qquad \mathring{\boldsymbol{T}}_c^l := \frac{1}{\sqrt{N}} \left[\begin{bmatrix} \cos(ly_{i-1/2}) \\ -\cos(ly_{i-1/2}) \end{bmatrix}\right]_{i=1,\ldots,N}, \quad (5.97)$$

$$\text{for } l = 1, \ldots, N - 1.$$

The eigenvectors of the time evolution matrix $\boldsymbol{M}$ (given in (5.18) and (5.19)) are then

$$\mathring{\boldsymbol{W}}_-^0 = \mathring{\boldsymbol{T}}_c^0, \quad \mathring{\boldsymbol{W}}_+^N = \mathring{\boldsymbol{S}}_s^N,$$

$$\left. \begin{array}{l} \mathring{\boldsymbol{W}}_+^l = \alpha_l \mathring{\boldsymbol{S}}_s^l + (\omega - V_l)\mathring{\boldsymbol{T}}_c^l, \\[2mm] \mathring{\boldsymbol{W}}_-^l = \alpha_l \mathring{\boldsymbol{S}}_s^l + (\omega + V_l)\mathring{\boldsymbol{T}}_c^l, \end{array} \right\} \quad \text{for} \quad l = 1, \ldots, N - 1.$$

For $\omega \in (0, 1)$ we define

$$\alpha_l := (1 - \omega)\sin(y_l), \quad \beta_l := (1 - \omega)\cos(y_l), \quad V_l := \sqrt{\omega^2 - \alpha_l^2}. \quad (5.98)$$

We find the corresponding eigenvalues

$$\begin{aligned} \lambda_0^- &= -\theta, \qquad \lambda_N^+ = \theta, \\[2mm] \lambda_l^+ &= \beta_l + V_l, \quad \lambda_l^- = \beta_l - V_l, \quad \text{for } l = 1, \ldots, N - 1. \end{aligned} \quad (5.99)$$

For $\omega \in (0, 1/2)$ the values V_l turn imaginary for some l.

On VC grids we define two sets of basis functions:

$$\mathring{\boldsymbol{s}}^l = \sqrt{\frac{2}{N}} \left[\sin(ly_i)\right]_{i=1,\ldots,N-1}, \quad l = 1, \ldots, N - 1 \quad (5.100)$$

and

$$\mathring{\boldsymbol{t}}^0 := \frac{1}{\sqrt{N}} [1]_{i=0,\ldots,N}, \quad \mathring{\boldsymbol{t}}^N := \frac{1}{\sqrt{N}} \left[(-1)^i\right]_{i=0,\ldots,N},$$

$$\mathring{\boldsymbol{t}}^l := \sqrt{\frac{2}{N}} \left[\cos(ly_i)\right]_{i=0,\ldots,N}, \quad l = 1, \ldots, N - 1. \quad (5.101)$$

While $\{\boldsymbol{s}^l\}_{l=1,\ldots,N-1} \subseteq \mathbb{R}^{N-1}$ is an orthonormal set, we have to define $\boldsymbol{d} := \text{diag}_{N+1}(1/2, 1, \ldots, 1, 1/2)$ in order to achieve

$$\mathring{\boldsymbol{t}}^i \cdot \boldsymbol{d}\mathring{\boldsymbol{t}}^j = \begin{cases} 1 & \text{if } i = j, \\ 0 & \text{if } i \neq j, \end{cases} \quad \text{for } i, j = 0, \ldots, N.$$

The eigenvectors of the time evolution matrix $\boldsymbol{K}$ (given in (5.40)) are gained via the transformation $\mathring{\boldsymbol{T}} := \boldsymbol{T}/\sqrt{2}$ with $\boldsymbol{T}$ given in (5.38)

$$\mathring{\boldsymbol{Q}}_-^0 = \mathring{\boldsymbol{T}}\mathring{\boldsymbol{W}}_-^0 = \begin{bmatrix} \boldsymbol{0} \\ -\mathring{\boldsymbol{t}}^0 \end{bmatrix}, \quad \mathring{\boldsymbol{Q}}_+^N = \mathring{\boldsymbol{T}}\mathring{\boldsymbol{W}}_+^N = \begin{bmatrix} \boldsymbol{0} \\ -\mathring{\boldsymbol{t}}^N \end{bmatrix},$$

$$\mathring{\boldsymbol{Q}}_+^l = \mathring{\boldsymbol{T}}\mathring{\boldsymbol{W}}_+^l = \begin{bmatrix} (\alpha_l \cos{(y_l/2)} - (\omega - V_l)\sin{(y_l/2)})\,\mathring{\boldsymbol{s}}^l \\ -(\alpha_l \sin{(y_l/2)} + (\omega - V_l)\cos{(y_l/2)})\,\mathring{\boldsymbol{t}}^l \end{bmatrix},$$

$$\mathring{\boldsymbol{Q}}_-^l = \mathring{\boldsymbol{T}}\mathring{\boldsymbol{W}}_-^l = \begin{bmatrix} (\alpha_l \cos{(y_l/2)} - (\omega + V_l)\sin{(y_l/2)})\,\mathring{\boldsymbol{s}}^l \\ -(\alpha_l \sin{(y_l/2)} + (\omega + V_l)\cos{(y_l/2)})\,\mathring{\boldsymbol{t}}^l \end{bmatrix}.$$

For the representation of the discrete Fourier solutions we define

$$A_l := \frac{\omega}{V_l}\cos(y_l), \quad B_l := \frac{\omega}{V_l}\sin(y_l) \quad \text{for } l = 1,\ldots,N-1. \tag{5.102}$$

Theorem 5.11. *Assume that $\omega \in (1/2, 1)$. Let the initial values be given by $\boldsymbol{P}^0 = [\boldsymbol{R}^0; \boldsymbol{J}^0]$ and let the right hand side be given by $\boldsymbol{G}^l = [\boldsymbol{f}^l; \boldsymbol{g}^l]$ for $l = 0,\ldots,k-1$. Then the discrete Fourier solution $\boldsymbol{P}_{LR}^k = [\boldsymbol{R}^k; \boldsymbol{J}^k]$ of the system*

$$\boldsymbol{P}_{LR}^{l+1} = \boldsymbol{K}\boldsymbol{P}_{LR}^l + \tau \boldsymbol{G}_{LR}^l \quad \text{for } l = 0,\ldots,k-1$$

for the VCFD lattice Boltzmann equations (5.1) with reduced density boundary conditions (5.6) can be represented in terms of the bases (5.100) and (5.101) by

$$\boldsymbol{R}^k = \sum_{j=1}^{N-1} \left(\frac{1}{2}(1 + A_j)(\lambda_j^+)^k + \frac{1}{2}(1 - A_j)(\lambda_j^-)^k \right) \left(\boldsymbol{R}^0 \cdot \mathring{\boldsymbol{s}}^j \right) \mathring{\boldsymbol{s}}^j$$

$$- \frac{\theta}{2\omega} \sum_{j=1}^{N-1} B_j \left((\lambda_j^+)^k - (\lambda_j^-)^k \right) \left(\boldsymbol{J}^0 \cdot \mathring{\boldsymbol{dt}}^j \right) \mathring{\boldsymbol{s}}^j$$

$$+ \tau \sum_{l=0}^{k-1} \sum_{j=1}^{N-1} \left(\frac{1}{2}(1 + A_j)(\lambda_j^+)^{k-l-1} + \frac{1}{2}(1 - A_j)(\lambda_j^-)^{k-l-1} \right) \left(\boldsymbol{f}^l \cdot \mathring{\boldsymbol{s}}^j \right) \mathring{\boldsymbol{s}}^j$$

$$- \frac{\theta}{2\omega}\tau \sum_{l=0}^{k-2} \sum_{j=1}^{N-1} B_j \left((\lambda_j^+)^{k-l-1} - (\lambda_j^-)^{k-l-1} \right) \left(\boldsymbol{g}^l \cdot \mathring{\boldsymbol{dt}}^j \right) \mathring{\boldsymbol{s}}^j,$$

$$\boldsymbol{J}^k = (-\theta)^k \left(\boldsymbol{J}^0 \cdot \mathring{\boldsymbol{dt}}^0 \right) \mathring{\boldsymbol{t}}^0 + \theta^k \left(\boldsymbol{J}^0 \cdot \mathring{\boldsymbol{dt}}^N \right) \mathring{\boldsymbol{t}}^N$$

$$+ \sum_{j=1}^{N-1} \left(\frac{1}{2}(1 - A_j)(\lambda_j^+)^k + \frac{1}{2}(1 + A_j)(\lambda_j^-)^k \right) \left(\boldsymbol{J}^0 \cdot \mathring{\boldsymbol{dt}}^j \right) \mathring{\boldsymbol{t}}^j$$

$$- \frac{1}{2\omega} \sum_{j=1}^{N-1} B_j \left((\lambda_j^+)^k - (\lambda_j^-)^k \right) \left(\boldsymbol{R}^0 \cdot \mathring{\boldsymbol{s}}^j \right) \mathring{\boldsymbol{t}}^j$$

$$+ \tau \sum_{l=0}^{k-1} (-\theta)^{k-l-1} \left(\boldsymbol{g}^l \cdot \mathring{\boldsymbol{dt}}^0 \right) \mathring{\boldsymbol{t}}^0 + \tau \sum_{l=0}^{k-1} \theta^{k-l-1} \left(\boldsymbol{g}^l \cdot \mathring{\boldsymbol{dt}}^N \right) \mathring{\boldsymbol{t}}^N$$

$$+ \tau \sum_{l=0}^{k-1} \sum_{j=1}^{N-1} \left(\frac{1}{2}(1 - A_j)(\lambda_j^+)^{k-l-1} + \frac{1}{2}(1 + A_j)(\lambda_j^-)^{k-l-1} \right) \left(\boldsymbol{g}^l \cdot \mathring{\boldsymbol{dt}}^j \right) \mathring{\boldsymbol{t}}^j$$

$$- \frac{\tau}{2\omega} \sum_{l=0}^{k-2} \sum_{j=1}^{N-1} B_j \left((\lambda_j^+)^{k-l-1} - (\lambda_j^-)^{k-l-1} \right) \left(\boldsymbol{f}^l \cdot \mathring{\boldsymbol{s}}^j \right) \mathring{\boldsymbol{t}}^j,$$

where A_j and B_j are defined in (5.102) and the eigenvalues $\lambda_j^{\pm}$ are given in (5.99) with (5.98). We use $\boldsymbol{d} := \mathrm{diag}_{N+1}(1/2, 1, \ldots, 1, 1/2)$.

Corollary 5.12. *For $\omega = 1/2$ the solution in Theorem 5.11 for $k \geq 1$ simplifies to*

$$\boldsymbol{R}^k = \sum_{j=1}^{N-1} \cos^k(y_j) \left(\boldsymbol{R}^0 \cdot \mathring{\boldsymbol{s}}^j\right) \mathring{\boldsymbol{s}}^j + \tau \sum_{l=0}^{k-1} \sum_{j=1}^{N-1} \cos^{k-l-1}(y_j) \left(\boldsymbol{f}^l \cdot \mathring{\boldsymbol{s}}^j\right) \mathring{\boldsymbol{s}}^j,$$

$$\boldsymbol{J}^k = - \sum_{j=1}^{N-1} \sin(y_j) \cos^{k-1}(y_j) \left(\boldsymbol{R}^0 \cdot \mathring{\boldsymbol{s}}^j\right) \mathring{\boldsymbol{t}}^j + \tau g^{k-1}$$

$$- \tau \sum_{l=0}^{k-2} \sum_{j=1}^{N-1} \sin(y_j) \cos^{k-l-2}(y_j) \left(\boldsymbol{f}^l \cdot \mathring{\boldsymbol{s}}^j\right) \mathring{\boldsymbol{t}}^j.$$

For $\omega \in (0, 1/2)$ there might be values of N such that the time evolution matrices are not diagonalizable. The comments and computations for the periodic case apply here with slight changes.

For the flux problem on VC grids we introduce the orthonormal basis

$$\mathring{\boldsymbol{S}}_c^0 := \frac{1}{\sqrt{2N}} \left[\begin{bmatrix} 1 \\ 1 \end{bmatrix}\right]_{i=1,\ldots,N}, \qquad \mathring{\boldsymbol{T}}_s^N := \frac{1}{\sqrt{2N}} \left[\begin{bmatrix} (-1)^{i+1} \\ -(-1)^{i+1} \end{bmatrix}\right]_{i=1,\ldots,N},$$

$$\mathring{\boldsymbol{S}}_c^l := \frac{1}{\sqrt{N}} \left[\begin{bmatrix} \cos(ly_{i-1/2}) \\ \cos(ly_{i-1/2}) \end{bmatrix}\right]_{i=1,\ldots,N}, \qquad \mathring{\boldsymbol{T}}_s^l := \frac{1}{\sqrt{N}} \left[\begin{bmatrix} \sin(ly_{i-1/2}) \\ -\sin(ly_{i-1/2}) \end{bmatrix}\right]_{i=1,\ldots,N}, \qquad (5.103)$$

$$\text{for } l = 1, \ldots, N-1.$$

We find the eigenvalues

$$\lambda_0^+ = 1, \qquad \lambda_N^- = -1,$$
$$\lambda_l^+ = \beta_l + V_l, \quad \lambda_l^- = \beta_l - V_l, \quad \text{for } l = 1, \ldots, N-1 \qquad (5.104)$$

and the eigenvectors of the matrix $\boldsymbol{M}$ (given in (5.18) and (5.20))

$$\mathring{\boldsymbol{V}}_+^0 = \mathring{\boldsymbol{S}}_c^0, \quad \mathring{\boldsymbol{V}}_-^N = \mathring{\boldsymbol{T}}_s^N,$$
$$\left.\begin{aligned} \mathring{\boldsymbol{V}}_+^l &= \alpha_l \mathring{\boldsymbol{S}}_c^l - (\omega - V_l)\mathring{\boldsymbol{T}}_s^l, \\ \mathring{\boldsymbol{V}}_-^l &= \alpha_l \mathring{\boldsymbol{S}}_c^l - (\omega + V_l)\mathring{\boldsymbol{T}}_s^l, \end{aligned}\right\} \quad \text{for } l = 1, \ldots, N-1.$$

The values α_l, β_l and V_l are defined in (5.98).

Hence, we get eigenvectors of $\boldsymbol{K}$ (given in (5.45)) via transformation by $\mathring{\boldsymbol{T}} := \boldsymbol{T}/\sqrt{2}$ with $\boldsymbol{T}$ given in (5.43)

$$\mathring{\boldsymbol{P}}_+^0 = \mathring{\boldsymbol{T}}\mathring{\boldsymbol{V}}_+^0 = \begin{bmatrix} \mathring{\boldsymbol{t}}^0 \\ \boldsymbol{0} \end{bmatrix}, \qquad \mathring{\boldsymbol{P}}_-^N = \mathring{\boldsymbol{T}}\mathring{\boldsymbol{V}}_-^N = \begin{bmatrix} \mathring{\boldsymbol{t}}^N \\ \boldsymbol{0} \end{bmatrix},$$

$$\mathring{\boldsymbol{P}}_+^l = \mathring{\boldsymbol{T}}\mathring{\boldsymbol{V}}_+^l = \begin{bmatrix} (\alpha_l \cos(y_l/2) - (\omega - V_l)\sin(y_l/2))\mathring{\boldsymbol{t}}^l \\ (\alpha_l \sin(y_l/2) + (\omega - V_l)\cos(y_l/2))\mathring{\boldsymbol{s}}^l \end{bmatrix},$$

$$\mathring{\boldsymbol{P}}_-^l = \mathring{\boldsymbol{T}}\mathring{\boldsymbol{V}}_-^l = \begin{bmatrix} (\alpha_l \cos(y_l/2) - (\omega + V_l)\sin(y_l/2))\mathring{\boldsymbol{t}}^l \\ (\alpha_l \sin(y_l/2) + (\omega + V_l)\cos(y_l/2))\mathring{\boldsymbol{s}}^l \end{bmatrix}.$$

Theorem 5.13. *Assume that $\omega \in (1/2, 1)$. Let the initial values be given by $\boldsymbol{P}^0 = [\boldsymbol{R}^0; \boldsymbol{J}^0]$ and let the right hand side be given by $\boldsymbol{G}^l = [\boldsymbol{f}^l; g^l]$ for $l = 0, \ldots, k-1$. Take $\boldsymbol{C}_0^l := c_0^l \mathring{\boldsymbol{P}}_+^0$ with $c_0^l := -(1-\omega)(j_R^k - j_L^k)/N$. Then the discrete Fourier solution $\boldsymbol{P}_{LR}^k = [\boldsymbol{R}^k; \boldsymbol{J}^k]$ of the system*

$$\boldsymbol{P}_{LR}^{l+1} = \boldsymbol{K}\boldsymbol{P}_{LR}^l + \tau \boldsymbol{G}_{LR}^l + \boldsymbol{C}_0^l \quad \text{for } l = 0, \ldots, k-1$$

for the lattice Boltzmann VCFD equations (5.1) with reduced flux boundary conditions (5.8) can be represented in terms of the bases (5.100) and (5.101) by

$$
\boldsymbol{R}^k = \left(\boldsymbol{R}^0 \cdot d\mathring{\boldsymbol{t}}^0\right)\mathring{\boldsymbol{t}}^0 + \sum_{l=0}^{k-1} c_0^l \mathring{\boldsymbol{t}}^0 + (-1)^k \left(\boldsymbol{R}^0 \cdot d\mathring{\boldsymbol{t}}^N\right)\mathring{\boldsymbol{t}}^N
$$

$$
+ \sum_{j=1}^{N-1} \left(\frac{1}{2}(1+A_j)(\lambda_j^+)^k + \frac{1}{2}(1-A_j)(\lambda_j^-)^k\right)\left(\boldsymbol{R}^0 \cdot d\mathring{\boldsymbol{t}}^j\right)\mathring{\boldsymbol{t}}^j
$$

$$
+ \frac{\theta}{2\omega}\sum_{j=1}^{N-1} B_j\left((\lambda_j^+)^k - (\lambda_j^-)^k\right)\left(\boldsymbol{J}^0 \cdot \mathring{\boldsymbol{s}}^j\right)\mathring{\boldsymbol{t}}^j
$$

$$
+ \tau\sum_{l=0}^{k-1}\left(\boldsymbol{f}^l \cdot d\mathring{\boldsymbol{t}}^0\right)\mathring{\boldsymbol{t}}^0 + \tau\sum_{l=0}^{k-1}(-1)^{k-l-1}\left(\boldsymbol{f}^l \cdot d\mathring{\boldsymbol{t}}^N\right)\mathring{\boldsymbol{t}}^N
$$

$$
+ \tau\sum_{l=0}^{k-1}\sum_{j=1}^{N-1}\left(\frac{1}{2}(1+A_j)(\lambda_j^+)^{k-l-1} + \frac{1}{2}(1-A_j)(\lambda_j^-)^{k-l-1}\right)\left(\boldsymbol{f}^l \cdot d\mathring{\boldsymbol{t}}^j\right)\mathring{\boldsymbol{t}}^j
$$

$$
+ \frac{\theta}{2\omega}\tau\sum_{l=0}^{k-2}\sum_{j=1}^{N-1} B_j\left((\lambda_j^+)^{k-l-1} - (\lambda_j^-)^{k-l-1}\right)\left(\boldsymbol{g}^l \cdot \mathring{\boldsymbol{s}}^j\right)\mathring{\boldsymbol{t}}^j,
$$

$$
\boldsymbol{J}^k = \sum_{j=1}^{N-1}\left(\frac{1}{2}(1-A_j)(\lambda_j^+)^k + \frac{1}{2}(1+A_j)(\lambda_j^-)^k\right)\left(\boldsymbol{J}^0 \cdot \mathring{\boldsymbol{s}}^j\right)\mathring{\boldsymbol{s}}^j
$$

$$
+ \frac{1}{2\omega}\sum_{j=1}^{N-1} B_j\left((\lambda_j^+)^k - (\lambda_j^-)^k\right)\left(\boldsymbol{R}^0 \cdot d\mathring{\boldsymbol{t}}^j\right)\mathring{\boldsymbol{s}}^j
$$

$$
+ \tau\sum_{l=0}^{k-1}\sum_{j=1}^{N-1}\left(\frac{1}{2}(1-A_j)(\lambda_j^+)^{k-l-1} + \frac{1}{2}(1+A_j)(\lambda_j^-)^{k-l-1}\right)\left(\boldsymbol{g}^l \cdot \mathring{\boldsymbol{s}}^j\right)\mathring{\boldsymbol{s}}^j
$$

$$
+ \frac{\tau}{2\omega}\sum_{l=0}^{k-2}\sum_{j=1}^{N-1} B_j\left((\lambda_j^+)^{k-l-1} - (\lambda_j^-)^{k-l-1}\right)\left(\boldsymbol{f}^l \cdot d\mathring{\boldsymbol{t}}^j\right)\mathring{\boldsymbol{s}}^j,
$$

where A_j and B_j are defined in (5.102) and the eigenvalues $\lambda_j^{\pm}$ are given in (5.104) with (5.98). We use $\boldsymbol{d} := \mathrm{diag}_{N+1}(1/2, 1, \ldots, 1, 1/2)$.

Corollary 5.14. *For $\omega = 1/2$ the solution in Theorem 5.13 for $k \geq 1$ simplifies to*

$$
\boldsymbol{R}^k = \sum_{j=0}^{N} \cos^k(y_j)\left(\boldsymbol{R}^0 \cdot d\mathring{\boldsymbol{t}}^j\right)\mathring{\boldsymbol{t}}^j + \tau\sum_{l=0}^{k-1}\sum_{j=0}^{N} \cos^{k-l-1}(y_j)\left(\boldsymbol{f}^l \cdot d\mathring{\boldsymbol{t}}^j\right)\mathring{\boldsymbol{t}}^j,
$$

$$
\boldsymbol{J}^k = \sum_{j=1}^{N-1} \sin(y_j)\cos^{k-1}(y_j)\left(\boldsymbol{R}^0 \cdot d\mathring{\boldsymbol{t}}^j\right)\mathring{\boldsymbol{s}}^j + \tau\boldsymbol{g}^{k-1}
$$

$$
+ \tau\sum_{l=0}^{k-2}\sum_{j=1}^{N-1} \sin(y_j)\cos^{k-l-2}(y_j)\left(\boldsymbol{f}^j \cdot d\mathring{\boldsymbol{t}}^j\right)\mathring{\boldsymbol{s}}^j.
$$

Now we examine solutions of the density boundary value problem on CCFD grids. To this aim, we choose the following orthonormal basis in $\mathbb{R}^{2N}$

$$\overline{T}_c^0 := \frac{1}{\sqrt{2N}} \left[\begin{bmatrix} -1 \\ 1 \end{bmatrix}\right]_{i=1,\ldots,N}, \qquad \overline{T}_c^N := \frac{1}{\sqrt{2N}} \left[\begin{bmatrix} (-1)^i \\ (-1)^i \end{bmatrix}\right]_{i=1,\ldots,N},$$

$$\overline{S}_s^l := \frac{1}{\sqrt{N}} \left[\begin{bmatrix} \sin(ly_{i-1}) \\ \sin(ly_i) \end{bmatrix}\right]_{i=1,\ldots,N}, \qquad \overline{T}_c^l := \frac{1}{\sqrt{N}} \left[\begin{bmatrix} -\cos(ly_{i-1}) \\ \cos(ly_i) \end{bmatrix}\right]_{i=1,\ldots,N}, \tag{5.105}$$

$$\text{for } l = 1, \ldots, N-1.$$

For the time evolution matrix M (given in (5.25) and (5.26)) we find the eigenvectors

$$\overline{W}_-^0 := \overline{T}_c^0, \qquad \overline{W}_-^N := \overline{T}_c^N,$$

$$\left.\begin{aligned} \overline{W}_+^l &:= \alpha_l \overline{S}_s^l + (\omega - V_l)\overline{T}_c^l, \\ \overline{W}_-^l &:= \alpha_l \overline{S}_s^l + (\omega + V_l)\overline{T}_c^l, \end{aligned}\right\} \qquad \text{for} \quad l = 1, \ldots, N-1,$$

with the eigenvalues given by

$$\lambda_0^- = -\theta, \qquad \lambda_N^- = -1,$$
$$\lambda_l^+ = \beta_l + V_l, \quad \lambda_l^- = \beta_l - V_l, \quad \text{for } l = 1, \ldots, N-1. \tag{5.106}$$

The definition of α_l, β_l and V_l is given in (5.98).

On CC grids we consider two sets of orthonormal basis vectors in $\mathbb{R}^N$. We choose the bases

$$\overline{s}^l := \sqrt{\frac{2}{N}} \left[\sin(ly_{i-1/2})\right]_{i=1,\ldots,N}, \qquad \overline{s}^N := \frac{1}{\sqrt{N}} \left[(-1)^{i+1}\right]_{i=1,\ldots,N}, \tag{5.107}$$
$$l = 1, \ldots, N-1,$$

and

$$\overline{t}^0 := \frac{1}{\sqrt{N}} \left[1\right]_{i=1,\ldots,N}, \qquad \overline{t}^l := \sqrt{\frac{2}{N}} \left[\cos(ly_{i-1/2})\right]_{i=1,\ldots,N}, \quad l = 1, \ldots, N-1. \tag{5.108}$$

Upon transformation via $\overline{T} := T/\sqrt{2}$ with T given in (5.53) we find the eigenvectors of K (given in (5.55)) by

$$\overline{Q}_-^0 := \overline{T}\overline{W}_-^0 = \begin{bmatrix} 0 \\ -\overline{t}^0 \end{bmatrix}, \qquad \overline{Q}_-^N := \overline{T}\overline{W}_-^N = \begin{bmatrix} -\overline{s}^N \\ 0 \end{bmatrix},$$

$$\overline{Q}_+^l := \overline{T}\overline{W}_+^l = \begin{bmatrix} (\alpha_l \cos(y_l/2) - (\omega - V_l)\sin(y_l/2))\,\overline{s}^l \\ -(\alpha_l \sin(y_l/2) + (\omega - V_l)\cos(y_l/2))\,\overline{t}^l \end{bmatrix},$$

$$\overline{Q}_-^l := \overline{T}\overline{W}_-^l = \begin{bmatrix} (\alpha_l \cos(y_l/2) - (\omega + V_l)\sin(y_l/2))\,\overline{s}^l \\ -(\alpha_l \sin(y_l/2) + (\omega + V_l)\cos(y_l/2))\,\overline{t}^l \end{bmatrix}.$$

We have

$$K\overline{Q}_-^0 = -\theta\overline{Q}_-^0, \qquad K\overline{Q}_-^N = -\overline{Q}_-^N,$$
$$K\overline{Q}_+^l = \lambda_l^+\overline{Q}_+^l, \qquad K\overline{Q}_-^l = \lambda_l^-\overline{Q}_-^l, \qquad \text{for } l = 1, \ldots, N-1.$$

Theorem 5.15. *Assume that $\omega \in (1/2, 1)$. Let the initial values be given by $P^0 = [R^0; J^0]$ and let the right hand be given by $G^l = [f^l; g^l]$ for $l = 0, \ldots, k-1$. Then the discrete Fourier solution $P_{LR}^k = [R^k; J^k]$ of the system*

$$P_{LR}^{l+1} = KP_{LR}^l + \tau G_{LR}^l \quad \text{for } l = 0, \ldots, k-1$$

for the CCFD lattice Boltzmann equations (5.2) with reduced density boundary conditions (5.12) can be represented in terms of the bases (5.107) and (5.108) as

$$\boldsymbol{R}^k = (-1)^k \left(\boldsymbol{R}^0 \cdot \overline{\boldsymbol{s}}^N\right) \overline{\boldsymbol{s}}^N + \tau \sum_{l=0}^{k-1} (-1)^{k-l-1} \left(\boldsymbol{f}^l \cdot \overline{\boldsymbol{s}}^N\right) \overline{\boldsymbol{s}}^N$$

$$+ \sum_{j=1}^{N-1} \left(\frac{1}{2}(1 + A_j)(\lambda_j^+)^k + \frac{1}{2}(1 - A_j)(\lambda_j^-)^k\right) \left(\boldsymbol{R}^0 \cdot \overline{\boldsymbol{s}}^j\right) \overline{\boldsymbol{s}}^j$$

$$- \frac{\theta}{2\omega} \sum_{j=1}^{N-1} B_j \left((\lambda_j^+)^k - (\lambda_j^-)^k\right) \left(\boldsymbol{J}^0 \cdot \overline{\boldsymbol{t}}^j\right) \overline{\boldsymbol{s}}^j$$

$$+ \tau \sum_{l=0}^{k-1} \sum_{j=1}^{N-1} \left(\frac{1}{2}(1 + A_j)(\lambda_j^+)^{k-l-1} + \frac{1}{2}(1 - A_j)(\lambda_j^-)^{k-l-1}\right) \left(\boldsymbol{f}^l \cdot \overline{\boldsymbol{s}}^j\right) \overline{\boldsymbol{s}}^j$$

$$- \frac{\theta}{2\omega}\tau \sum_{l=0}^{k-2} \sum_{j=1}^{N-1} B_j \left((\lambda_j^+)^{k-l-1} - (\lambda_j^-)^{k-l-1}\right) \left(\boldsymbol{g}^l \cdot \overline{\boldsymbol{t}}^j\right) \overline{\boldsymbol{s}}^j,$$

$$\boldsymbol{J}^k = (-\theta)^k \left(\boldsymbol{J}^0 \cdot \overline{\boldsymbol{t}}^0\right) \overline{\boldsymbol{t}}^0 + \tau \sum_{l=0}^{k-1} (-\theta)^{k-l-1} \left(\boldsymbol{g}^l \cdot \overline{\boldsymbol{t}}^0\right) \overline{\boldsymbol{t}}^0$$

$$+ \sum_{j=1}^{N-1} \left(\frac{1}{2}(1 - A_j)(\lambda_j^+)^k + \frac{1}{2}(1 + A_j)(\lambda_j^-)^k\right) \left(\boldsymbol{J}^0 \cdot \overline{\boldsymbol{t}}^j\right) \overline{\boldsymbol{t}}^j$$

$$- \frac{1}{2\omega} \sum_{j=1}^{N-1} B_j \left((\lambda_j^+)^k - (\lambda_j^-)^k\right) \left(\boldsymbol{R}^0 \cdot \overline{\boldsymbol{s}}^j\right) \overline{\boldsymbol{t}}^j$$

$$+ \tau \sum_{l=0}^{k-1} \sum_{j=1}^{N-1} \left(\frac{1}{2}(1 - A_j)(\lambda_j^+)^{k-l-1} + \frac{1}{2}(1 + A_j)(\lambda_j^-)^{k-l-1}\right) \left(\boldsymbol{g}^l \cdot \overline{\boldsymbol{t}}^j\right) \overline{\boldsymbol{t}}^j$$

$$- \frac{\tau}{2\omega} \sum_{l=0}^{k-2} \sum_{j=1}^{N-1} B_j \left((\lambda_j^+)^{k-l-1} - (\lambda_j^-)^{k-l-1}\right) \left(\boldsymbol{f}^l \cdot \overline{\boldsymbol{s}}^j\right) \overline{\boldsymbol{t}}^j,$$

where A_j and B_j are defined in (5.102) and the eigenvalues $\lambda_j^{\pm}$ are given in (5.106) with (5.98).

Corollary 5.16. *For $\omega = 1/2$ the solution in Theorem 5.15 for $k \geq 1$ simplifies to*

$$\boldsymbol{R}^k = \sum_{j=1}^{N} \cos^k(y_j) \left(\boldsymbol{R}^0 \cdot \overline{\boldsymbol{s}}^j\right) \overline{\boldsymbol{s}}^j + \tau \sum_{l=0}^{k-1} \sum_{j=1}^{N} \cos^{k-l-1}(y_j) \left(\boldsymbol{f}^l \cdot \overline{\boldsymbol{s}}^j\right) \overline{\boldsymbol{s}}^j,$$

$$\boldsymbol{J}^k = - \sum_{j=1}^{N-1} \sin(y_j) \cos^{k-1}(y_j) \left(\boldsymbol{R}^0 \cdot \overline{\boldsymbol{s}}^j\right) \overline{\boldsymbol{t}}^j + \tau g^{k-1}$$

$$- \tau \sum_{l=0}^{k-2} \sum_{j=1}^{N-1} \sin(y_j) \cos^{k-l-2}(y_j) \left(\boldsymbol{f}^l \cdot \overline{\boldsymbol{s}}^j\right) \overline{\boldsymbol{t}}^j.$$

For the solution of the flux boundary value problem on CCFD grids we define the orthonormal basis in $\mathbb{R}^{2N}$ by

$$\overline{\boldsymbol{S}}_c^0 := \frac{1}{\sqrt{2N}} \left[\begin{bmatrix} 1 \\ 1 \end{bmatrix} \right]_{i=1,\dots,N}, \qquad \overline{\boldsymbol{S}}_c^N := \frac{1}{\sqrt{2N}} \left[\begin{bmatrix} -(-1)^i \\ (-1)^i \end{bmatrix} \right]_{i=1,\dots,N},$$

$$\overline{\boldsymbol{S}}_c^l = \frac{1}{\sqrt{N}} \left[\begin{bmatrix} \cos(ly_{i-1}) \\ \cos(ly_i) \end{bmatrix} \right]_{i=1,\dots,N}, \qquad \boldsymbol{T}_s^l = \frac{1}{\sqrt{N}} \left[\begin{bmatrix} -\sin(ly_{i-1}) \\ \sin(ly_i) \end{bmatrix} \right]_{i=1,\dots,N},$$

$$\text{for } l = 1, \dots, N-1.$$

(5.109)

We find the eigenvalues

$$\lambda_0^+ = 1, \qquad \lambda_N^+ = \theta,$$

$$\lambda_l^+ = \beta_l + V_l, \quad \lambda_l^- = \beta_l - V_l, \quad \text{for } l = 1, \dots, N-1.$$

(5.110)

The eigenvectors of $\boldsymbol{M}$ (given in (5.25) and (5.27)) are given by

$$\overline{\boldsymbol{V}}_+^0 := \overline{\boldsymbol{S}}_c^0, \quad \overline{\boldsymbol{V}}_+^N := \overline{\boldsymbol{S}}_c^N,$$

$$\left.\begin{aligned} \overline{\boldsymbol{V}}_+^l &:= \alpha_l \overline{\boldsymbol{S}}_c^l - (\omega - V_l)\overline{\boldsymbol{T}}_s^l, \\ \overline{\boldsymbol{V}}_-^l &:= \alpha_l \overline{\boldsymbol{S}}_c^l - (\omega + V_l)\overline{\boldsymbol{T}}_s^l, \end{aligned}\right\} \quad \text{for } l = 1, \dots, N-1.$$

We get the eigenvectors of $\boldsymbol{K}$ (given in (5.58)) via transformation by $\bar{\boldsymbol{T}} := \boldsymbol{T}/\sqrt{2}$ with $\boldsymbol{T}$ given in (5.53)

$$\overline{\boldsymbol{P}}_+^0 := \bar{\boldsymbol{T}}\overline{\boldsymbol{V}}_+^0 = \begin{bmatrix} \bar{t}^0 \\ 0 \end{bmatrix}, \quad \overline{\boldsymbol{P}}_+^N := \bar{\boldsymbol{T}}\overline{\boldsymbol{V}}_+^N = \begin{bmatrix} 0 \\ \bar{s}^N \end{bmatrix},$$

$$\boldsymbol{P}_+^l := \boldsymbol{T}\overline{\boldsymbol{V}}_+^l = \begin{bmatrix} (\alpha_l \cos(y_l/2) - (\omega - V_l)\sin(y_l/2))\,\bar{t}^l \\ (\alpha_l \sin(y_l/2) + (\omega - V_l)\cos(y_l/2))\,\bar{s}^l \end{bmatrix},$$

$$\boldsymbol{P}_-^l := \boldsymbol{T}\overline{\boldsymbol{V}}_-^l = \begin{bmatrix} (\alpha_l \cos(y_l/2) - (\omega + V_l)\sin(y_l/2))\,\bar{t}^l \\ (\alpha_l \sin(y_l/2) + (\omega + V_l)\cos(y_l/2))\,\bar{s}^l \end{bmatrix}.$$

In the case $\omega = 1/2$ and $N = 2n$ the rank of the given eigenvectors is only $2N - 1$. We find

$$\boldsymbol{K} \begin{bmatrix} \bar{t}^n \\ 0 \end{bmatrix} = \begin{bmatrix} 0 \\ \bar{s}^n \end{bmatrix} = \sqrt{2}\overline{\boldsymbol{P}}_+^n = \sqrt{2}\overline{\boldsymbol{P}}_-^n, \quad \boldsymbol{K}^2 \begin{bmatrix} \bar{t}^n \\ 0 \end{bmatrix} = \boldsymbol{0}.$$

This renders a Jordan block of length two belonging to the eigenvalue 0. Hence, $\boldsymbol{K}$ is not diagonalizable in this case.

Theorem 5.17. *Assume that $\omega \in (1/2, 1)$. Let the initial values be given by $\boldsymbol{P}^0 = [\boldsymbol{R}^0; \boldsymbol{J}^0]$ and let the right hand be given by $\boldsymbol{G}^l = [\boldsymbol{f}^l; \boldsymbol{g}^l]$ for $l = 0, \dots, k-1$. Take $\boldsymbol{C}_0^l := c_0^l \overline{\boldsymbol{P}}_+^0$ with $c_0^l := -(1 - \omega)(j_R^l - j_L^l)/N$. Then the discrete Fourier solution $\boldsymbol{P}_{LR}^k = [\boldsymbol{R}^k; \boldsymbol{J}^k]$ of the system*

$$\boldsymbol{P}_{LR}^{l+1} = \boldsymbol{K}\boldsymbol{P}_{LR}^l + \tau\boldsymbol{G}_{LR}^l + \boldsymbol{C}_0^l \quad \text{for } l = 0, \dots, k-1$$

for the CCFD lattice Boltzmann equations (5.2) with reduced flux boundary conditions (5.13) can be represented in terms of the bases (5.107) and (5.108) by

$$\boldsymbol{R}^k = \left(\boldsymbol{R}^0 \cdot \bar{t}^0\right)\bar{t}^0 + \tau \sum_{l=0}^{k-1} \left(\boldsymbol{f}^l \cdot \bar{t}^0\right)\bar{t}^0 + \sum_{l=0}^{k-1} c_0^l \bar{t}^0$$

$$\sum_{j=1}^{N-1} \left(\frac{1}{2}(1 + A_j)(\lambda_j^+)^k + \frac{1}{2}(1 - A_j)(\lambda_j^-)^k\right)\left(\boldsymbol{R}^0 \cdot \bar{t}^j\right)\bar{t}^j$$

$$+ \frac{\theta}{2\omega} \sum_{j=1}^{N-1} B_j \left((\lambda_j^+)^k - (\lambda_j^-)^k\right) \left(\boldsymbol{J}^0 \cdot \overline{\boldsymbol{s}}^j\right) \overline{\boldsymbol{t}}^j$$

$$+ \tau \sum_{l=0}^{k-1} \sum_{j=1}^{N-1} \left(\frac{1}{2}(1 + A_j)(\lambda_j^+)^{k-l-1} + \frac{1}{2}(1 - A_j)(\lambda_j^-)^{k-l-1}\right) \left(\boldsymbol{f}^l \cdot \overline{\boldsymbol{t}}^j\right) \overline{\boldsymbol{t}}^j$$

$$+ \frac{\theta}{2\omega}\tau \sum_{l=0}^{k-2} \sum_{j=1}^{N-1} B_j \left((\lambda_j^+)^{k-l-1} - (\lambda_j^-)^{k-l-1}\right) \left(\boldsymbol{g}^l \cdot \overline{\boldsymbol{s}}^j\right) \overline{\boldsymbol{t}}^j,$$

$$\boldsymbol{J}^k = \theta^k \left(\boldsymbol{J}^0 \cdot \overline{\boldsymbol{s}}^N\right) \overline{\boldsymbol{s}}^N + \tau \sum_{l=0}^{k-1} \theta^{k-l-1} \left(\boldsymbol{g}^l \cdot \overline{\boldsymbol{s}}^N\right) \overline{\boldsymbol{s}}^N$$

$$+ \sum_{j=1}^{N-1} \left(\frac{1}{2}(1 - A_j)(\lambda_j^+)^k + \frac{1}{2}(1 + A_j)(\lambda_j^-)^k\right) \left(\boldsymbol{J}^0 \cdot \overline{\boldsymbol{s}}^j\right) \overline{\boldsymbol{s}}^j$$

$$+ \frac{1}{2\omega} \sum_{j=1}^{N-1} B_j \left((\lambda_j^+)^k - (\lambda_j^-)^k\right) \left(\boldsymbol{R}^0 \cdot \overline{\boldsymbol{t}}^j\right) \overline{\boldsymbol{s}}^j$$

$$+ \tau \sum_{l=0}^{k-1} \sum_{j=1}^{N-1} \left(\frac{1}{2}(1 - A_j)(\lambda_j^+)^{k-l-1} + \frac{1}{2}(1 + A_j)(\lambda_j^-)^{k-l-1}\right) \left(\boldsymbol{g}^l \cdot \overline{\boldsymbol{s}}^j\right) \overline{\boldsymbol{s}}^j$$

$$+ \frac{\tau}{2\omega} \sum_{l=0}^{k-2} \sum_{j=1}^{N-1} B_j \left((\lambda_j^+)^{k-l-1} - (\lambda_j^-)^{k-l-1}\right) \left(\boldsymbol{f}^l \cdot \overline{\boldsymbol{t}}^j\right) \overline{\boldsymbol{s}}^j,$$

where A_j and B_j are defined in (5.102) and the eigenvalues $\lambda_j^\pm$ are given in (5.110) with (5.98).

Corollary 5.18. *For $\omega = 1/2$ the solution in Theorem 5.17 for $k \geq 1$ simplifies to*

$$\boldsymbol{R}^k = \sum_{j=0}^{N-1} \cos^k(y_j) \left(\boldsymbol{R}^0 \cdot \overline{\boldsymbol{t}}^j\right) \overline{\boldsymbol{t}}^j + \tau \sum_{l=0}^{k-1} \sum_{j=0}^{N-1} \cos^{k-l-1}(y_j) \left(\boldsymbol{f}^l \cdot \overline{\boldsymbol{t}}^j\right) \overline{\boldsymbol{t}}^j,$$

$$\boldsymbol{J}^k = \sum_{j=1}^{N-1} \sin(y_j) \cos^{k-1}(y_j) \left(\boldsymbol{R}^0 \cdot \overline{\boldsymbol{t}}^j\right) \overline{\boldsymbol{s}}^j + \tau \boldsymbol{g}^{k-1}$$

$$+ \tau \sum_{l=0}^{k-2} \sum_{j=1}^{N-1} \sin(y_j) \cos^{k-l-2}(y_j) \left(\boldsymbol{f}^j \cdot \overline{\boldsymbol{t}}^j\right) \overline{\boldsymbol{s}}^j.$$

5.11. Discrete Fourier Solutions for the FV Lattice Boltzmann Schemes

In this section we consider the discrete Fourier solutions for the finite volume lattice Boltzmann equations (5.3) and (5.4) subject to the provided boundary conditions. First, we take a look at the periodic problem. We define for $\omega \in (0, 1)$

$$s_l := \sin(2y_l), \quad c_l := \cos(2y_l),$$

$$W_l := \left(\left(c_l - \frac{\omega}{2}(1 + c_l)\right)^2 - 1 + \omega(1 + c_l)\right)^{1/2}, \tag{5.111}$$

for $l = 1, \ldots, n-1$. We find that $\boldsymbol{S}^\dagger \boldsymbol{M} \boldsymbol{S}$, with the periodic time evolution matrix $\boldsymbol{M}$ given in (5.17) and $\boldsymbol{S}$ defined in (5.85), is a block-diagonal matrix with diagonal elements ± 1 and $-\theta$ ($\theta = 2\omega - 1$) and with the 2-by-2 blocks

$$\begin{bmatrix} c_l & s_l \\ -(1-\omega)s_l & (1-\omega)c_l - \omega \end{bmatrix}, \quad \begin{bmatrix} c_l & -s_l \\ (1-\omega)s_l & (1-\omega)c_l - \omega \end{bmatrix}$$

for $l = 1, \ldots, n-1$ on the diagonal. We derive the eigenvalues of $\boldsymbol{M}$

$$
\begin{aligned}
\lambda_0^+ &= 1, \quad \lambda_0^- = 1 - 2\omega, \\
\lambda_l^+ &= c_l - \frac{\omega}{2}(1 + c_l) + W_l, \\
\lambda_l^- &= c_l - \frac{\omega}{2}(1 + c_l) - W_l, \\
\lambda_n^- &= -1 \quad \text{if } N = 2n.
\end{aligned}
\left.\right\} \quad \text{for } l = 1, \ldots, n-1, \qquad (5.112)
$$

Furthermore, we get the eigenvectors of $\boldsymbol{M}$ in terms of the basis (5.84)

$$
\begin{aligned}
\widetilde{\boldsymbol{V}}_+^0 &:= \widetilde{\boldsymbol{S}}_c^0, \qquad \widetilde{\boldsymbol{W}}_-^0 := \widetilde{\boldsymbol{T}}_c^0, \\
\widetilde{\boldsymbol{V}}_+^l &:= -s_l \widetilde{\boldsymbol{S}}_c^l + \left(\frac{\omega}{2}(1 + c_l) - W_l\right) \widetilde{\boldsymbol{T}}_s^l, \\
\widetilde{\boldsymbol{V}}_-^l &:= -s_l \widetilde{\boldsymbol{S}}_c^l + \left(\frac{\omega}{2}(1 + c_l) + W_l\right) \widetilde{\boldsymbol{T}}_s^l, \\
\widetilde{\boldsymbol{W}}_+^l &:= s_l \widetilde{\boldsymbol{S}}_s^l + \left(\frac{\omega}{2}(1 + c_l) - W_l\right) \widetilde{\boldsymbol{T}}_c^l, \\
\widetilde{\boldsymbol{W}}_-^l &:= s_l \widetilde{\boldsymbol{S}}_s^l + \left(\frac{\omega}{2}(1 + c_l) + W_l\right) \widetilde{\boldsymbol{T}}_c^l, \\
\widetilde{\boldsymbol{V}}_-^n &:= \widetilde{\boldsymbol{T}}_s^n, \qquad \widetilde{\boldsymbol{W}}_-^n := \widetilde{\boldsymbol{S}}_s^n, \qquad \text{if } N = 2n.
\end{aligned}
\left.\right\} \quad \text{for } l = 1, \ldots, n-1,
$$

Using the monotone decreasing function $f : (0,1) \to (-1,1)$

$$
f(\omega) := \frac{8 - (2 + \omega)^2}{(2 - \omega)^2},
$$

we find

$$
W_l = \left(1 - \frac{\omega}{2}\right) \sqrt{(1 + c_l)(c_l - f(\omega))}
$$

Hence, we see that W_l is a real number for $l/N \leq \arccos(f(w))/2\pi$ and imaginary otherwise. In contrast to the FD schemes we get complex eigenvalues and complex eigenvectors for all values of ω. The complex eigenvalues are located on a circle with midpoint $-\omega/(2-\omega) < 0$ and radius $2(1-\omega)/(2-\omega)$; see Figure 5.7. In this figure we use $l^* := \lfloor N \arccos(f(\omega))/2\pi \rfloor$. Since the modulus of the eigenvalues is equal or less then 1 for all l and $\omega \in (0,1)$, the solutions for the underlying lattice Boltzmann schemes are stable. Increasing ω raises the number of real eigenvalues, which leads to an improvement of the convergence.

The eigenvectors are not orthogonal, especially we see

$$
\widetilde{\boldsymbol{V}}_+^l \cdot \widetilde{\boldsymbol{V}}_-^l = (2-w)s_l^2, \quad \widetilde{\boldsymbol{W}}_+^l \cdot \widetilde{\boldsymbol{W}}_-^l = (2-w)s_l^2
$$

for $l/N \leq \arccos(f(w))/2\pi$. By employing the orthogonal transformation via $\widetilde{\boldsymbol{T}} := \boldsymbol{T}/\sqrt{2}$ with $\boldsymbol{T}$ as in (5.33), we get the eigenvectors of $\boldsymbol{K}$ (given in (5.60)) in terms

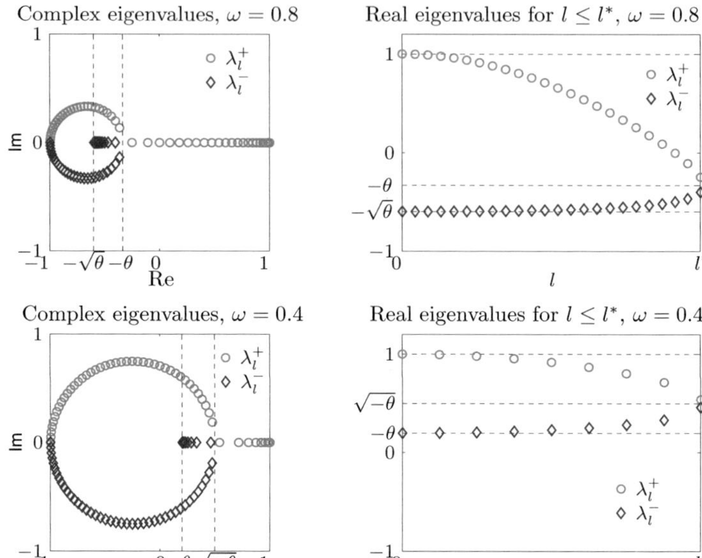

FIGURE 5.7. Complex eigenvalues of $\boldsymbol{M}$ and $\boldsymbol{K}$ in the periodic case for $\omega = 0.8$, $\omega = 0.4$ and $N = 100$.

of the basis (5.88) in the form

$$
\widetilde{\boldsymbol{P}}_+^0 := \widetilde{\boldsymbol{T}}\widetilde{\boldsymbol{V}}_+^0 = \begin{bmatrix} \widetilde{\boldsymbol{t}}^0 \\ \boldsymbol{0} \end{bmatrix}, \qquad \widetilde{\boldsymbol{Q}}_-^0 := \widetilde{\boldsymbol{T}}\widetilde{\boldsymbol{W}}_-^0 = \begin{bmatrix} \boldsymbol{0} \\ -\widetilde{\boldsymbol{t}}^0 \end{bmatrix},
$$

$$
\widetilde{\boldsymbol{P}}_+^l := \widetilde{\boldsymbol{T}}\widetilde{\boldsymbol{V}}_+^l = \begin{bmatrix} -\left(\sin(2y_l)\cos(y_l) - \left(\frac{\omega}{2}(1+c_l) - W_l\right)\sin(y_l)\right)\widetilde{\boldsymbol{t}}^l \\ -\left(\sin(2y_l)\sin(y_l) + \left(\frac{\omega}{2}(1+c_l) - W_l\right)\cos(y_l)\right)\widetilde{\boldsymbol{s}}^l \end{bmatrix},
$$

$$
\widetilde{\boldsymbol{P}}_-^l := \widetilde{\boldsymbol{T}}\widetilde{\boldsymbol{V}}_-^l = \begin{bmatrix} -\left(\sin(2y_l)\cos(y_l) - \left(\frac{\omega}{2}(1+c_l) + W_l\right)\sin(y_l)\right)\widetilde{\boldsymbol{t}}^l \\ -\left(\sin(2y_l)\sin(y_l) + \left(\frac{\omega}{2}(1+c_l) + W_l\right)\cos(y_l)\right)\widetilde{\boldsymbol{s}}^l \end{bmatrix},
$$

$$
\widetilde{\boldsymbol{Q}}_+^l := \widetilde{\boldsymbol{T}}\widetilde{\boldsymbol{W}}_+^l = \begin{bmatrix} \left(\sin(2y_l)\cos(y_l) - \left(\frac{\omega}{2}(1+c_l) - W_l\right)\sin(y_l)\right)\widetilde{\boldsymbol{s}}^l \\ -\left(\sin(2y_l)\sin(y_l) + \left(\frac{\omega}{2}(1+c_l) - W_l\right)\cos(y_l)\right)\widetilde{\boldsymbol{t}}^l \end{bmatrix},
$$

$$
\widetilde{\boldsymbol{Q}}_-^l := \widetilde{\boldsymbol{T}}\widetilde{\boldsymbol{W}}_-^l = \begin{bmatrix} \left(\sin(2y_l)\cos(y_l) - \left(\frac{\omega}{2}(1+c_l) + W_l\right)\sin(y_l)\right)\widetilde{\boldsymbol{s}}^l \\ -\left(\sin(2y_l)\sin(y_l) + \left(\frac{\omega}{2}(1+c_l) + W_l\right)\cos(y_l)\right)\widetilde{\boldsymbol{t}}^l \end{bmatrix},
$$

$$
\widetilde{\boldsymbol{P}}_-^n := \widetilde{\boldsymbol{T}}\widetilde{\boldsymbol{V}}_-^n = \begin{bmatrix} \widetilde{\boldsymbol{t}}^n \\ \boldsymbol{0} \end{bmatrix}, \qquad \widetilde{\boldsymbol{Q}}_-^n := \widetilde{\boldsymbol{T}}\widetilde{\boldsymbol{W}}_-^n = \begin{bmatrix} \boldsymbol{0} \\ -\widetilde{\boldsymbol{t}}^n \end{bmatrix}, \quad \text{if } N = 2n.
$$

If there is a l such that $2\pi l/N = \arccos(f(w))$, we have $W_l = 0$ and hence, the rank of the presented eigenvectors is only $2N - 2$. In this case we find

$$
\widetilde{\boldsymbol{P}}_J^l := \frac{1}{2-\omega}\begin{bmatrix} -\left(\sqrt{1-\omega}\cos(y_l) + \sin(y_l)\right)\widetilde{\boldsymbol{t}}^l \\ -\left(\sqrt{1-\omega}\sin(y_l) - \cos(y_l)\right)\widetilde{\boldsymbol{s}}^l \end{bmatrix},
$$

$$
\widetilde{\boldsymbol{Q}}_J^l := \frac{1}{2-\omega}\begin{bmatrix} \left(\sqrt{1-\omega}\cos(y_l) + \sin(y_l)\right)\widetilde{\boldsymbol{s}}^l \\ -\left(\sqrt{1-\omega}\sin(y_l) - \cos(y_l)\right)\widetilde{\boldsymbol{t}}^l \end{bmatrix}
$$

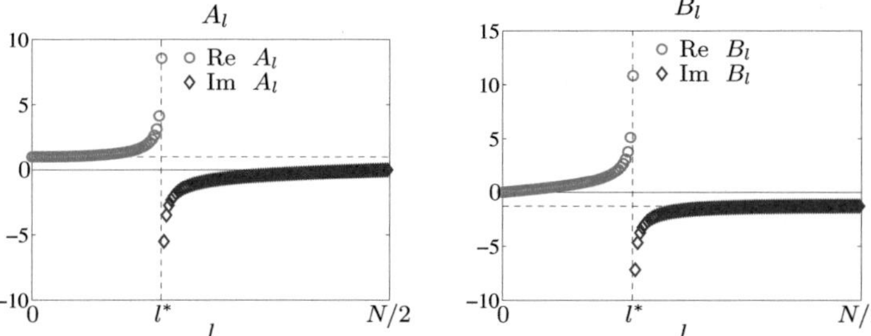

FIGURE 5.8. Behavior of A_l and B_l for $l = 0, \ldots, N/2$.

with $(\boldsymbol{K} - \lambda_l \boldsymbol{Id})\widetilde{\boldsymbol{P}}_J^l = \widetilde{\boldsymbol{P}}_+^l = \widetilde{\boldsymbol{P}}_-^l$ and $(\boldsymbol{K} - \lambda_l \boldsymbol{Id})\widetilde{\boldsymbol{Q}}_J^l = \widetilde{\boldsymbol{Q}}_+^l = \widetilde{\boldsymbol{Q}}_-^l$ with $\lambda_l :=$ $c_l - \omega(1 + c_l)/2$. This renders two Jordan blocks of length two belonging to the eigenvalue λ_l.

For $l = 1, \ldots, n - 1$ and $W_l \neq 0$ we define

$$\gamma_l^\pm := \sin(2y_l)\cos(y_l) - \left(\frac{\omega}{2}(1 + c_l) \mp W_l\right)\sin(y_l)$$

$$= \left(2\cos^2(y_l) - \left(\frac{\omega}{2}(1 + c_l) \mp W_l\right)\right)\sin(y_l),$$

$$\delta_l^\pm := \sin(2y_l)\sin(y_l) + \left(\frac{\omega}{2}(1 + c_l) \mp W_l\right)\cos(y_l)$$

$$= \left(2\sin^2(y_l) + \left(\frac{\omega}{2}(1 + c_l) \mp W_l\right)\right)\cos(y_l).$$

We use (5.90) in order to define the basis transformations

$$\widetilde{\boldsymbol{P}}_+^l = -\gamma_l^+ \widetilde{\boldsymbol{T}}_N^l - \delta_l^+ \widetilde{\boldsymbol{S}}_N^l, \qquad \widetilde{\boldsymbol{P}}_-^l = -\gamma_l^- \widetilde{\boldsymbol{T}}_N^l - \delta_l^- \widetilde{\boldsymbol{S}}_N^l,$$

$$\widetilde{\boldsymbol{Q}}_+^l = \gamma_l^+ \widetilde{\boldsymbol{S}}_D^l - \delta_l^+ \widetilde{\boldsymbol{T}}_D^l, \qquad \widetilde{\boldsymbol{Q}}_-^l = \gamma_l^- \widetilde{\boldsymbol{S}}_D^l - \delta_l^- \widetilde{\boldsymbol{T}}_D^l,$$

and

$$\widetilde{\boldsymbol{S}}_N^l = \frac{\gamma_l^-}{\gamma_l^+ \delta_l^- - \gamma_l^- \delta_l^+}\widetilde{\boldsymbol{P}}_+^l - \frac{\gamma_l^+}{\gamma_l^+ \delta_l^- - \gamma_l^- \delta_l^+}\widetilde{\boldsymbol{P}}_-^l,$$

$$\widetilde{\boldsymbol{T}}_N^l = \frac{-\delta_l^-}{\gamma_l^+ \delta_l^- - \gamma_l^- \delta_l^+}\widetilde{\boldsymbol{P}}_+^l + \frac{\delta_l^+}{\gamma_l^+ \delta_l^- - \gamma_l^- \delta_l^+}\widetilde{\boldsymbol{P}}_-^l,$$

$$\widetilde{\boldsymbol{S}}_D^l = \frac{\delta_l^-}{\gamma_l^+ \delta_l^- - \gamma_l^- \delta_l^+}\widetilde{\boldsymbol{Q}}_+^l - \frac{\delta_l^+}{\gamma_l^+ \delta_l^- - \gamma_l^- \delta_l^+}\widetilde{\boldsymbol{Q}}_-^l,$$

$$\widetilde{\boldsymbol{T}}_D^l = \frac{\gamma_l^-}{\gamma_l^+ \delta_l^- - \gamma_l^- \delta_l^+}\widetilde{\boldsymbol{Q}}_+^l - \frac{\gamma_l^+}{\gamma_l^+ \delta_l^- - \gamma_l^- \delta_l^+}\widetilde{\boldsymbol{Q}}_-^l$$

for $l = 1, \ldots, n - 1$. Furthermore, we find

$$\frac{\gamma_l^+ \delta_l^-}{\gamma_l^+ \delta_l^- - \gamma_l^- \delta_l^+} = \frac{1}{2}(1 + A_l), \qquad \frac{\gamma_l^+ \gamma_l^-}{\gamma_l^+ \delta_l^- - \gamma_l^- \delta_l^+} = \frac{1 - \omega}{2\omega}B_l,$$

$$\frac{-\gamma_l^- \delta_l^+}{\gamma_l^+ \delta_l^- - \gamma_l^- \delta_l^+} = \frac{1}{2}(1 - A_l), \qquad \frac{\delta_l^+ \delta_l^-}{\gamma_l^+ \delta_l^- - \gamma_l^- \delta_l^+} = \frac{1}{2\omega}B_l,$$

where we use the definitions

$$A_l := \frac{\omega}{2W_l}(1 + \cos(2y_l)), \quad B_l := \frac{\omega}{W_l}\sin(2y_l) \quad \text{for } l = 1, \ldots, n - 1. \tag{5.113}$$

This transformation fails in the case $W_l = 0$, that is, there is one l such that

$c_l = f(\omega)$. For this reason, we assume

$$\frac{N}{2\pi}\arccos(f(\omega)) \notin \mathbb{N}. \tag{A}$$

If the assumption (A) is not fulfilled, then the matrices $\boldsymbol{M}$ and $\boldsymbol{K}$ are not diagonalizable anymore, since the rank of the given eigenvectors is then only $2N - 2$. Thus we get Jordan blocks of length two.

The behavior of A_l and B_l depending on ω and l is plotted in Figure 5.8. The values of A_l and B_l become imaginary for $l > l^* := \lfloor N \arccos(f(\omega))/(2\pi) \rfloor$. There are poles at l^*.

Theorem 5.19. *Assume that $\omega \in (0,1)$, $N = 2n$ and (A). Let the initial values be given by $\boldsymbol{P}^0 = [\boldsymbol{R}^0; \boldsymbol{J}^0]$ and let the right hand side be given by $\boldsymbol{G}^l = [\boldsymbol{f}^l; \boldsymbol{g}^l]$ for $l = 0, \ldots, k - 1$. Then the discrete Fourier solution $\boldsymbol{P}^k = [\boldsymbol{R}^k; \boldsymbol{J}^k]$ of the system*

$$\boldsymbol{P}^{l+1} = \boldsymbol{K}\boldsymbol{P}^l + \tau\boldsymbol{G}^l \quad \text{for } l = 0, \ldots, k-1$$

for the VCFV lattice Boltzmann equations (5.3) with periodic boundary conditions (5.5) can be represented in terms of the orthonormal basis (5.88) for $k \geq 0$ by

$$\boldsymbol{R}^k = \left(\boldsymbol{R}^0 \cdot \widetilde{\boldsymbol{t}}^0\right)\widetilde{\boldsymbol{t}}^0 + (-1)^k\left(\boldsymbol{R}^0 \cdot \widetilde{\boldsymbol{t}}^n\right)\widetilde{\boldsymbol{t}}^n$$

$$+ \sum_{j=1}^{n-1}\left(\frac{1}{2}(1 + A_j)(\lambda_j^+)^k + \frac{1}{2}(1 - A_j)(\lambda_j^-)^k\right)\left(\left(\boldsymbol{R}^0 \cdot \widetilde{\boldsymbol{t}}^j\right)\widetilde{\boldsymbol{t}}^j + \left(\boldsymbol{R}^0 \cdot \widetilde{\boldsymbol{s}}^j\right)\widetilde{\boldsymbol{s}}^j\right)$$

$$- \frac{1-\omega}{2\omega}\sum_{j=1}^{n-1}B_j\left((\lambda_j^+)^k - (\lambda_j^-)^k\right)\left(\left(\boldsymbol{J}^0 \cdot \widetilde{\boldsymbol{s}}^j\right)\widetilde{\boldsymbol{t}}^j - \left(\boldsymbol{J}^0 \cdot \widetilde{\boldsymbol{t}}^j\right)\widetilde{\boldsymbol{s}}^j\right)$$

$$+ \tau\sum_{l=0}^{k-1}\left(\boldsymbol{f}^l \cdot \widetilde{\boldsymbol{t}}^0\right)\widetilde{\boldsymbol{t}}^0 + \tau\sum_{l=0}^{k-1}(-1)^{k-l-1}\left(\boldsymbol{f}^l \cdot \widetilde{\boldsymbol{t}}^n\right)\widetilde{\boldsymbol{t}}^n$$

$$+ \tau\sum_{l=0}^{k-1}\sum_{j=1}^{n-1}\frac{1}{2}(1 + A_j)(\lambda_j^+)^{k-l-1}\left(\left(\boldsymbol{f}^l \cdot \widetilde{\boldsymbol{t}}^j\right)\widetilde{\boldsymbol{t}}^j + \left(\boldsymbol{f}^l \cdot \widetilde{\boldsymbol{s}}^j\right)\widetilde{\boldsymbol{s}}^j\right)$$

$$+ \tau\sum_{l=0}^{k-1}\sum_{j=1}^{n-1}\frac{1}{2}(1 - A_j)(\lambda_j^-)^{k-l-1}\left(\left(\boldsymbol{f}^l \cdot \widetilde{\boldsymbol{t}}^j\right)\widetilde{\boldsymbol{t}}^j + \left(\boldsymbol{f}^l \cdot \widetilde{\boldsymbol{s}}^j\right)\widetilde{\boldsymbol{s}}^j\right)$$

$$- \frac{1-\omega}{2\omega}\tau\sum_{l=0}^{k-2}\sum_{j=1}^{n-1}B_j\left((\lambda_j^+)^{k-l-1} - (\lambda_j^-)^{k-l-1}\right)\left(\left(\boldsymbol{g}^l \cdot \widetilde{\boldsymbol{s}}^j\right)\widetilde{\boldsymbol{t}}^j - \left(\boldsymbol{g}^l \cdot \widetilde{\boldsymbol{t}}^j\right)\widetilde{\boldsymbol{s}}^j\right),$$

$$\boldsymbol{J}^k = (-\theta)^k\left(\boldsymbol{J}^0 \cdot \widetilde{\boldsymbol{t}}^0\right)\widetilde{\boldsymbol{t}}^0 + (-1)^k\left(\boldsymbol{J}^0 \cdot \widetilde{\boldsymbol{t}}^n\right)\widetilde{\boldsymbol{t}}^n$$

$$+ \sum_{j=1}^{n-1}\left(\frac{1}{2}(1 - A_j)(\lambda_j^+)^k + \frac{1}{2}(1 + A_j)(\lambda_j^-)^k\right)\left(\left(\boldsymbol{J}^0 \cdot \widetilde{\boldsymbol{t}}^j\right)\widetilde{\boldsymbol{t}}^j + \left(\boldsymbol{J}^0 \cdot \widetilde{\boldsymbol{s}}^j\right)\widetilde{\boldsymbol{s}}^j\right)$$

$$- \frac{1}{2\omega}\sum_{j=1}^{n-1}B_j\left((\lambda_j^+)^k - (\lambda_j^-)^k\right)\left(\left(\boldsymbol{R}^0 \cdot \widetilde{\boldsymbol{s}}^j\right)\widetilde{\boldsymbol{t}}^j - \left(\boldsymbol{R}^0 \cdot \widetilde{\boldsymbol{t}}^j\right)\widetilde{\boldsymbol{s}}^j\right)$$

$$+ \tau\sum_{l=0}^{k-1}(-\theta)^{k-l-1}\left(\boldsymbol{g}^l \cdot \widetilde{\boldsymbol{t}}^0\right)\widetilde{\boldsymbol{t}}^0 + \tau\sum_{l=0}^{k-1}(-1)^{k-l-1}\left(\boldsymbol{g}^l \cdot \widetilde{\boldsymbol{t}}^n\right)\widetilde{\boldsymbol{t}}^n$$

$$+ \tau\sum_{l=0}^{k-1}\sum_{j=1}^{n-1}\frac{1}{2}(1 - A_j)(\lambda_j^+)^{k-l-1}\left(\left(\boldsymbol{g}^l \cdot \widetilde{\boldsymbol{t}}^j\right)\widetilde{\boldsymbol{t}}^j + \left(\boldsymbol{g}^l \cdot \widetilde{\boldsymbol{s}}^j\right)\widetilde{\boldsymbol{s}}^j\right)$$

$$+ \tau \sum_{l=0}^{k-1} \sum_{j=1}^{n-1} \frac{1}{2}(1 + A_j)(\lambda_j^-)^{k-l-1} \left(\left(\boldsymbol{g}^l \cdot \widetilde{\boldsymbol{t}}^j \right) \widetilde{\boldsymbol{t}}^j + \left(\boldsymbol{g}^l \cdot \widetilde{\boldsymbol{s}}^j \right) \widetilde{\boldsymbol{s}}^j \right)$$

$$- \frac{\tau}{2\omega} \sum_{l=0}^{k-2} \sum_{j=1}^{n-1} B_j \left((\lambda_j^+)^{k-l-1} - (\lambda_j^-)^{k-l-1} \right) \left(\left(\boldsymbol{f}^l \cdot \widetilde{\boldsymbol{s}}^j \right) \widetilde{\boldsymbol{t}}^j - \left(\boldsymbol{f}^l \cdot \widetilde{\boldsymbol{t}}^j \right) \widetilde{\boldsymbol{s}}^j \right),$$

where the eigenvalues $\lambda_j^\pm$ are given in (5.112) with (5.111) and A_j and B_j are defined in (5.113). For odd $N = 2n - 1$ the parts with $\widetilde{\boldsymbol{t}}^n$ disappear.

Although the eigenvalues and the corresponding eigenvectors become complex for large values of j, the above representations are real.

If there is a j with $\cos(2y_j) = f(\omega)$, then we have the jth summand

$$\left(\lambda_j^k + \frac{\omega}{2}(1 + c_j)k\lambda_j^{k-1} \right) \left(\left(\boldsymbol{R}^0 \cdot \widetilde{\boldsymbol{t}}^j \right) \widetilde{\boldsymbol{t}}^j + \left(\boldsymbol{R}^0 \cdot \widetilde{\boldsymbol{s}}^j \right) \widetilde{\boldsymbol{s}}^j \right)$$

$$+ (1 - \omega)s_j k\lambda_j^{k-1} \left(\left(\boldsymbol{J}^0 \cdot \widetilde{\boldsymbol{s}}^j \right) \widetilde{\boldsymbol{t}}^j - \left(\boldsymbol{J}^0 \cdot \widetilde{\boldsymbol{t}}^j \right) \widetilde{\boldsymbol{s}}^j \right)$$

for the $\boldsymbol{R}$-equation and

$$\left(\lambda_j^k - \frac{\omega}{2}(1 + c_j)k\lambda_j^{k-1} \right) \left(\left(\boldsymbol{J}^0 \cdot \widetilde{\boldsymbol{t}}^j \right) \widetilde{\boldsymbol{t}}^j + \left(\boldsymbol{J}^0 \cdot \widetilde{\boldsymbol{s}}^j \right) \widetilde{\boldsymbol{s}}^j \right)$$

$$- s_j k\lambda_j^{k-1} \left(\left(\boldsymbol{R}^0 \cdot \widetilde{\boldsymbol{s}}^j \right) \widetilde{\boldsymbol{t}}^j - \left(\boldsymbol{R}^0 \cdot \widetilde{\boldsymbol{t}}^j \right) \widetilde{\boldsymbol{s}}^j \right)$$

for the $\boldsymbol{J}$-equation. Here, we use $\lambda_j := c_j - \omega(1 + c_j)/2$. Accordingly, we have to add corresponding terms for the right hand sides.

For the low-frequency coefficients in the solution representation we find

$$A_j = 1 + \frac{1 - \omega}{2\omega^2}(2y_j)^2 + \mathcal{O}(y_j^4) \qquad \text{for } j \ll N,$$

$$B_j = 2y_j + \frac{\omega^2 - 6\omega + 6}{12\omega^2}(2y_j)^3 + \mathcal{O}(y_j^5) \quad \text{for } j \ll N.$$

For the corresponding eigenvalues we find

$$\lambda_j^+ = 1 - \frac{1}{2\omega}(2y_j)^2 + \mathcal{O}(y_j^4) \quad \text{for } j \ll N,$$

$$(\lambda_j^+)^k \approx e^{-4\nu j^2 \pi^2 t_k / |\Omega|^2} \qquad \text{for } j \ll N,$$

where we take $\nu := \gamma/(2\omega)$ and $t_k := k\tau = k|\Omega|^2/(\gamma N^2)$. Hence, the leading order terms

$$\frac{1}{2}(1 + A_j)(\lambda_j^+)^k \approx e^{-4\nu j^2 \pi^2 t_k / |\Omega|^2} \quad \text{for } j \ll N$$

show the correct approximation property. In contrast we have $1 - A_j = \mathcal{O}(y_j^2)$ and $(\lambda_j^-)^k$ is oscillating and it decays very fast. Since B_j and $\boldsymbol{J}^0$ are of order $\mathcal{O}(h)$, the terms induced by $\boldsymbol{J}^0$ in the $\boldsymbol{R}$-representation are only of order $\mathcal{O}(h^2)$.

The decisive term in the $\boldsymbol{J}$-representation is $B_j(\lambda_j^+)^k$. With the background of partial integration we find

$$- \frac{B_j}{2\omega} \left(\left(\boldsymbol{R}^0 \cdot \widetilde{\boldsymbol{s}}^j \right) - \left(\boldsymbol{R}^0 \cdot \widetilde{\boldsymbol{t}}^j \right) \right) \approx - \frac{h}{2\omega} \left(\left(D_h \boldsymbol{R}^0 \cdot \widetilde{\boldsymbol{t}}^j \right) + \left(D_h \boldsymbol{R}^0 \cdot \widetilde{\boldsymbol{s}}^j \right) \right),$$

where we define $D_h \boldsymbol{R}^0 := [\partial_x r_0(x_i)]_i$. Hence, $\boldsymbol{J}^k$ is an approximation of $-h\partial_x r/(2\omega)$. The choice of $\boldsymbol{J}^0$ plays a subordinate role.

In the $\boldsymbol{R}$-representation the oscillating parts $(\lambda_j^-)^k$ are hidden in an order $\mathcal{O}(h^2)$-term. In the $\boldsymbol{J}$-representation the oscillating terms $B_j(\lambda_j^-)^k$ are apparent. Hence, we get the wrong viscosity in the first steps. After several iterations the oscillations have faded away and the correct viscosity is reached. This can be seen in numerical experiments.

The behavior of the time-dependent coefficients

$$R_j^k := \frac{1}{2}(1 + A_j)(\lambda_j^+)^k + \frac{1}{2}(1 - A_j)(\lambda_j^-)^k$$

is examined in Figure 5.9. The poles in A_j are deleted. For large time steps only the coefficients with j close to 0 and $N/2$ are of significant size. Oscillations are observed for large values of j.

By switching to the basis (5.95), we find

Theorem 5.20. *Assume that $\omega \in (0,1)$, $N = 2n$ and (A). Let the initial values be given by $\boldsymbol{P}^0 = [\boldsymbol{R}^0; \boldsymbol{J}^0]$ and let the right hand side be given by $\boldsymbol{G}^l = [\boldsymbol{f}^l; \boldsymbol{g}^l]$ for $l = 0, \ldots, k-1$. Then the discrete Fourier solution $\boldsymbol{P}^k = [\boldsymbol{R}^k; \boldsymbol{J}^k]$ of the system*

$$\boldsymbol{P}^{l+1} = \boldsymbol{K}\boldsymbol{P}^l + \tau\boldsymbol{G}^l \quad \text{for } l = 0, \ldots, k-1$$

for the VCFV lattice Boltzmann equation (5.3) with periodic boundary conditions (5.5) can be represented in terms of the orthonormal basis (5.96). The representation is the same as in Theorem 5.19 with $\widetilde{\boldsymbol{s}}^l$ replaced by $\widehat{\boldsymbol{s}}^l$ for $l = 1, \ldots n-1$, $\widetilde{\boldsymbol{t}}^l$ replaced by $\widehat{\boldsymbol{t}}^l$ for $l = 0, \ldots, n-1$ and $\widetilde{\boldsymbol{t}}^n$ replaced by $\widehat{\boldsymbol{s}}^n$.

For the consideration of the boundary value problems we define for $\omega \in (0,1)$

$$s_l := \sin(y_l), \quad c_l := \cos(y_l),$$

$$W_l := \left(\left(\left(c_l - \frac{\omega}{2}(1 + c_l)\right)^2 - 1 + \omega(1 + c_l) \right) \right)^{1/2} \tag{5.114}$$

for $l = 0, \ldots, N$.

The eigenvectors of the time evolution matrix $\boldsymbol{M}$ (given in (5.21) and (5.22)) can be written in terms of the basis (5.97)

$$\mathring{\boldsymbol{W}}_-^0 = \mathring{\boldsymbol{T}}_c^0, \quad \mathring{\boldsymbol{W}}_-^N = \mathring{\boldsymbol{S}}_s^N,$$

$$\left.\begin{aligned}
\mathring{\boldsymbol{W}}_+^l &= s_l \mathring{\boldsymbol{S}}_s^l + \left(\frac{\omega}{2}(1 + c_l) - W_l\right)\mathring{\boldsymbol{T}}_c^l, \\
\mathring{\boldsymbol{W}}_-^l &= s_l \mathring{\boldsymbol{S}}_s^l + \left(\frac{\omega}{2}(1 + c_l) + W_l\right)\mathring{\boldsymbol{T}}_c^l,
\end{aligned}\right\} \quad \text{for} \quad l = 1, \ldots, N-1.$$

We find the corresponding eigenvalues

$$\lambda_0^- = -\theta, \quad \lambda_N^- = -1,$$

$$\lambda_l^+ = c_l - \frac{\omega}{2}(1 + c_l) + W_l, \quad \lambda_l^- = c_l - \frac{\omega}{2}(1 + c_l) - W_l, \tag{5.115}$$

$$\text{for } l = 1, \ldots, N-1.$$

The eigenvectors of $\boldsymbol{K}$ (given in (5.61)) are found via the transformation $\mathring{\boldsymbol{T}} := \boldsymbol{T}/\sqrt{2}$ with $\boldsymbol{T}$ given in (5.38) in the form

$$\mathring{\boldsymbol{Q}}_-^0 = \mathring{\boldsymbol{T}}\mathring{\boldsymbol{W}}_-^0 = \begin{bmatrix} \boldsymbol{0} \\ -\mathring{\boldsymbol{t}}^0 \end{bmatrix}, \quad \mathring{\boldsymbol{Q}}_-^N = \mathring{\boldsymbol{T}}\mathring{\boldsymbol{W}}_-^N = \begin{bmatrix} \boldsymbol{0} \\ -\mathring{\boldsymbol{t}}^N \end{bmatrix},$$

$$\mathring{\boldsymbol{Q}}_+^l = \mathring{\boldsymbol{T}}\mathring{\boldsymbol{W}}_+^l = \begin{bmatrix} \left(\sin(y_l)\cos(y_l/2) - \left(\frac{\omega}{2}(1 + c_l) - W_l\right)\sin(y_l/2)\right)\mathring{\boldsymbol{s}}^l \\ -\left(\sin(y_l)\sin(y_l/2) + \left(\frac{\omega}{2}(1 + c_l) - W_l\right)\cos(y_l/2)\right)\mathring{\boldsymbol{t}}^l \end{bmatrix},$$

$$\mathring{\boldsymbol{Q}}_-^l = \mathring{\boldsymbol{T}}\mathring{\boldsymbol{W}}_-^l = \begin{bmatrix} \left(\sin(y_l)\cos(y_l/2) - \left(\frac{\omega}{2}(1 + c_l) + W_l\right)\sin(y_l/2)\right)\mathring{\boldsymbol{s}}^l \\ -\left(\sin(y_l)\sin(y_l/2) + \left(\frac{\omega}{2}(1 + c_l) + W_l\right)\cos(y_l/2)\right)\mathring{\boldsymbol{t}}^l \end{bmatrix}.$$

For the representation of the discrete Fourier solutions we define

$$A_l := \frac{\omega}{2W_l}(1 + \cos(y_l)), \quad B_l := \frac{\omega}{W_l}\sin(y_l) \quad \text{for } l = 1, \ldots, N-1. \tag{5.116}$$

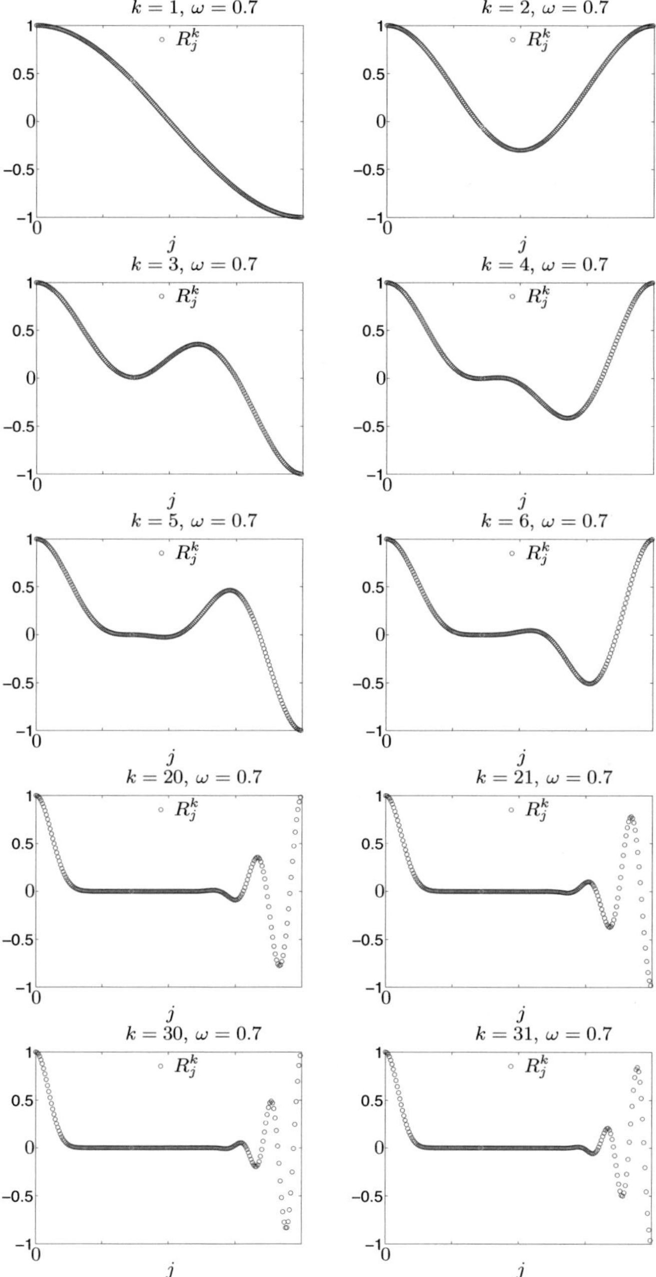

FIGURE 5.9. Behavior of R_j^k for $N = 200$ and selected time steps k.

Furthermore, we assume

$$\frac{N}{\pi}\arccos(f(\omega)) \notin \mathbb{N}. \tag{A'}$$

Theorem 5.21. *Assume that $\omega \in (0,1)$ and (A'). Let the initial values be given by $\boldsymbol{P}^0 = [\boldsymbol{R}^0; \boldsymbol{J}^0]$ and let the right hand side be given by $\boldsymbol{G}^l = [\boldsymbol{f}^l; \boldsymbol{g}^l]$ for $l = 0, \ldots, k-1$. Then the discrete Fourier solution $\boldsymbol{P}^k_{LR} = [\boldsymbol{R}^k; \boldsymbol{J}^k]$ of the system*

$$\boldsymbol{P}^{l+1}_{LR} = \boldsymbol{K}\boldsymbol{P}^l_{LR} + \tau \boldsymbol{G}^l_{LR} \quad \text{for } l = 0, \ldots, k-1$$

for the VCFV lattice Boltzmann equations (5.3) with reduced density boundary conditions (5.6) can be represented in terms of the bases (5.100) and (5.101) by

$$\boldsymbol{R}^k = \sum_{j=1}^{N-1} \left(\frac{1}{2}(1 + A_j)(\lambda_j^+)^k + \frac{1}{2}(1 - A_j)(\lambda_j^-)^k \right) \left(\boldsymbol{R}^0 \cdot \mathring{\boldsymbol{s}}^j \right) \mathring{\boldsymbol{s}}^j$$

$$+ \frac{1-\omega}{2\omega} \sum_{j=1}^{N-1} B_j \left((\lambda_j^+)^k - (\lambda_j^-)^k \right) \left(\boldsymbol{J}^0 \cdot \mathring{\boldsymbol{dt}}^j \right) \mathring{\boldsymbol{s}}^j$$

$$+ \tau \sum_{l=0}^{k-1} \sum_{j=1}^{N-1} \left(\frac{1}{2}(1 + A_j)(\lambda_j^+)^{k-l-1} + \frac{1}{2}(1 - A_j)(\lambda_j^-)^{k-l-1} \right) \left(\boldsymbol{f}^l \cdot \mathring{\boldsymbol{s}}^j \right) \mathring{\boldsymbol{s}}^j$$

$$+ \frac{1-\omega}{2\omega} \tau \sum_{l=0}^{k-2} \sum_{j=1}^{N-1} B_j \left((\lambda_j^+)^{k-l-1} - (\lambda_j^-)^{k-l-1} \right) \left(\boldsymbol{g}^l \cdot \mathring{\boldsymbol{dt}}^j \right) \mathring{\boldsymbol{s}}^j,$$

$$\boldsymbol{J}^k = (-\theta)^k \left(\boldsymbol{J}^0 \cdot \mathring{\boldsymbol{dt}}^0 \right) \mathring{\boldsymbol{t}}^0 + (-1)^k \left(\boldsymbol{J}^0 \cdot \mathring{\boldsymbol{dt}}^N \right) \mathring{\boldsymbol{t}}^N$$

$$+ \sum_{j=1}^{N-1} \left(\frac{1}{2}(1 - A_j)(\lambda_j^+)^k + \frac{1}{2}(1 + A_j)(\lambda_j^-)^k \right) \left(\boldsymbol{J}^0 \cdot \mathring{\boldsymbol{dt}}^j \right) \mathring{\boldsymbol{t}}^j$$

$$- \frac{1}{2\omega} \sum_{j=1}^{N-1} B_j \left((\lambda_j^+)^k - (\lambda_j^-)^k \right) \left(\boldsymbol{R}^0 \cdot \mathring{\boldsymbol{s}}^j \right) \mathring{\boldsymbol{t}}^j$$

$$+ \tau \sum_{l=0}^{k-1} (-\theta)^{k-l-1} \left(\boldsymbol{g}^l \cdot \mathring{\boldsymbol{dt}}^0 \right) \mathring{\boldsymbol{t}}^0 + \tau \sum_{l=0}^{k-1} (-1)^{k-l-1} \left(\boldsymbol{g}^l \cdot \mathring{\boldsymbol{dt}}^N \right) \mathring{\boldsymbol{t}}^N$$

$$+ \tau \sum_{l=0}^{k-1} \sum_{j=1}^{N-1} \left(\frac{1}{2}(1 - A_j)(\lambda_j^+)^{k-l-1} + \frac{1}{2}(1 + A_j)(\lambda_j^-)^{k-l-1} \right) \left(\boldsymbol{g}^l \cdot \mathring{\boldsymbol{dt}}^j \right) \mathring{\boldsymbol{t}}^j$$

$$- \frac{\tau}{2\omega} \sum_{l=0}^{k-2} \sum_{j=1}^{N-1} B_j \left((\lambda_j^+)^{k-l-1} - (\lambda_j^-)^{k-l-1} \right) \left(\boldsymbol{f}^l \cdot \mathring{\boldsymbol{s}}^j \right) \mathring{\boldsymbol{t}}^j,$$

where the eigenvalues $\lambda_j^{\pm}$ are given in (5.115) with (5.114) and A_j and B_j are defined in (5.116). We use $\boldsymbol{d} := \operatorname{diag}_{N+1}(1/2, 1, \ldots, 1, 1/2)$.

In the flux case the eigenvectors of the matrix $\boldsymbol{M}$ (given in (5.21) and (5.23)) are

$$\mathring{\boldsymbol{V}}^0_+ = \mathring{\boldsymbol{S}}^0_c, \quad \mathring{\boldsymbol{V}}^N_- = \mathring{\boldsymbol{T}}^N_s,$$

$$\left. \begin{aligned} \mathring{\boldsymbol{V}}^l_+ &= -s_l \mathring{\boldsymbol{S}}^l_c - \left(\frac{\omega}{2}(1 + c_l) - W_l \right) \mathring{\boldsymbol{T}}^l_s, \\ \mathring{\boldsymbol{V}}^l_- &= -s_l \mathring{\boldsymbol{S}}^l_c - \left(\frac{\omega}{2}(1 + c_l) - W_l \right) \mathring{\boldsymbol{T}}^l_s, \end{aligned} \right\} \quad \text{for } l = 1, \ldots, N-1.$$

The values s_l, c_l and W_l are defined in (5.114). We find the eigenvalues

$$\lambda_0^+ = 1, \quad \lambda_N^- = -1,$$
$$\lambda_l^+ = c_l - \frac{\omega}{2}(1 + c_l) + W_l, \quad \lambda_l^- = c_l - \frac{\omega}{2}(1 + c_l) - W_l, \tag{5.117}$$
$$\text{for } l = 1, \ldots, N - 1.$$

Hence, we get the eigenvectors of $\boldsymbol{K}$ (given in (5.62)) via transformation by $\mathring{\boldsymbol{T}} := \boldsymbol{T}/\sqrt{2}$ with $\boldsymbol{T}$ given in (5.43) in the form

$$\mathring{\boldsymbol{P}}_+^0 = \mathring{\boldsymbol{T}}\mathring{\boldsymbol{V}}_+^0 = \begin{bmatrix} \mathring{\boldsymbol{t}}^0 \\ \boldsymbol{0} \end{bmatrix}, \quad \mathring{\boldsymbol{P}}_-^N = \mathring{\boldsymbol{T}}\mathring{\boldsymbol{V}}_-^N = \begin{bmatrix} \mathring{\boldsymbol{t}}^N \\ \boldsymbol{0} \end{bmatrix},$$

$$\mathring{\boldsymbol{P}}_+^l = \mathring{\boldsymbol{T}}\mathring{\boldsymbol{V}}_+^l = \begin{bmatrix} -\left(\sin(y_l)\cos(y_l/2) - \left(\frac{\omega}{2}(1 + c_l) - W_l\right)\sin(y_l/2)\right)\mathring{\boldsymbol{t}}^l \\ -\left(\sin(y_l)\sin(y_l/2) + \left(\frac{\omega}{2}(1 + c_l) - W_l\right)\cos(y_l/2)\right)\mathring{\boldsymbol{s}}^l \end{bmatrix},$$

$$\mathring{\boldsymbol{P}}_-^l = \mathring{\boldsymbol{T}}\mathring{\boldsymbol{V}}_-^l = \begin{bmatrix} -\left(\sin(y_l)\cos(y_l/2) - \left(\frac{\omega}{2}(1 + c_l) + W_l\right)\sin(y_l/2)\right)\mathring{\boldsymbol{t}}^l \\ -\left(\sin(y_l)\sin(y_l/2) + \left(\frac{\omega}{2}(1 + c_l) + W_l\right)\cos(y_l/2)\right)\mathring{\boldsymbol{s}}^l \end{bmatrix}.$$

We now arrive at

Theorem 5.22. *Assume that $\omega \in (0,1)$ and (A'). Let the initial values be given by $\boldsymbol{P}^0 = [\boldsymbol{R}^0; \boldsymbol{J}^0]$ and let the right hand side be given by $\boldsymbol{G}^l = [\boldsymbol{f}^l; \boldsymbol{g}^l]$ for $l = 0, \ldots, k - 1$. Take $\boldsymbol{C}_0^l := c_0^l\mathring{\boldsymbol{P}}_+^0$ with $c_0^l := -(j_R^l - j_L^l)/N$. Then the discrete Fourier solution $\boldsymbol{P}_{LR}^k = [\boldsymbol{R}^k; \boldsymbol{J}^k]$ of the system*

$$\boldsymbol{P}_{LR}^{l+1} = \boldsymbol{K}\boldsymbol{P}_{LR}^l + \tau\boldsymbol{G}_{LR}^l + \boldsymbol{C}_0^l \quad \text{for } l = 0, \ldots, k - 1$$

for the lattice Boltzmann VCFV equations (5.3) with reduced flux boundary conditions (5.8) can be represented in terms of the bases (5.100) and (5.101) by

$$\boldsymbol{R}^k = \left(\boldsymbol{R}^0 \cdot d\mathring{\boldsymbol{t}}^0\right)\mathring{\boldsymbol{t}}^0 + \sum_{l=0}^{k-1} c_0^l\mathring{\boldsymbol{t}}^0 + (-1)^k\left(\boldsymbol{R}^0 \cdot d\mathring{\boldsymbol{t}}^N\right)\mathring{\boldsymbol{t}}^N$$

$$+ \sum_{j=1}^{N-1}\left(\frac{1}{2}(1 + A_j)(\lambda_j^+)^k + \frac{1}{2}(1 - A_j)(\lambda_j^-)^k\right)\left(\boldsymbol{R}^0 \cdot d\mathring{\boldsymbol{t}}^j\right)\mathring{\boldsymbol{t}}^j$$

$$+ \frac{\theta}{2\omega}\sum_{j=1}^{N-1} B_j\left((\lambda_j^+)^k - (\lambda_j^-)^k\right)\left(\boldsymbol{J}^0 \cdot \mathring{\boldsymbol{s}}^j\right)\mathring{\boldsymbol{t}}^j$$

$$+ \tau\sum_{l=0}^{k-1}\left(\boldsymbol{f}^l \cdot d\mathring{\boldsymbol{t}}^0\right)\mathring{\boldsymbol{t}}^0 + \tau\sum_{l=0}^{k-1}(-1)^{k-l-1}\left(\boldsymbol{f}^l \cdot d\mathring{\boldsymbol{t}}^N\right)\mathring{\boldsymbol{t}}^N$$

$$+ \tau\sum_{l=0}^{k-1}\sum_{j=1}^{N-1}\left(\frac{1}{2}(1 + A_j)(\lambda_j^+)^{k-l-1} + \frac{1}{2}(1 - A_j)(\lambda_j^-)^{k-l-1}\right)\left(\boldsymbol{f}^l \cdot d\mathring{\boldsymbol{t}}^j\right)\mathring{\boldsymbol{t}}^j$$

$$+ \frac{\theta}{2\omega}\tau\sum_{l=0}^{k-2}\sum_{j=1}^{N-1} B_j\left((\lambda_j^+)^{k-l-1} - (\lambda_j^-)^{k-l-1}\right)\left(\boldsymbol{g}^l \cdot \mathring{\boldsymbol{s}}^j\right)\mathring{\boldsymbol{t}}^j,$$

$$\boldsymbol{J}^k = \sum_{j=1}^{N-1}\left(\frac{1}{2}(1 - A_j)(\lambda_j^+)^k + \frac{1}{2}(1 + A_j)(\lambda_j^-)^k\right)\left(\boldsymbol{J}^0 \cdot \mathring{\boldsymbol{s}}^j\right)\mathring{\boldsymbol{s}}^j$$

$$+ \frac{1}{2\omega}\sum_{j=1}^{N-1} B_j\left((\lambda_j^+)^k - (\lambda_j^-)^k\right)\left(\boldsymbol{R}^0 \cdot d\mathring{\boldsymbol{t}}^j\right)\mathring{\boldsymbol{s}}^j$$

$$+ \tau \sum_{l=0}^{k-1} \sum_{j=1}^{N-1} \left(\frac{1}{2}(1 - A_j)(\lambda_j^+)^{k-l-1} + \frac{1}{2}(1 + A_j)(\lambda_j^-)^{k-l-1} \right) \left(\boldsymbol{g}^l \cdot \mathring{\boldsymbol{s}}^j \right) \mathring{\boldsymbol{s}}^j$$

$$+ \frac{\tau}{2\omega} \sum_{l=0}^{k-2} \sum_{j=1}^{N-1} B_j \left((\lambda_j^+)^{k-l-1} - (\lambda_j^-)^{k-l-1} \right) \left(\boldsymbol{f}^l \cdot \boldsymbol{d}\mathring{\boldsymbol{t}}^j \right) \mathring{\boldsymbol{s}}^j,$$

where the eigenvalues $\lambda_j^\pm$ are given in (5.117) with (5.114) and A_j and B_j are defined in (5.116). We use $\boldsymbol{d} := \mathrm{diag}_{N+1}(1/2, 1, \ldots, 1, 1/2)$.

Now we examine solutions of the density boundary value problem on CCFV grids. For the time evolution matrix $\boldsymbol{M}$ (given in (5.28) and (5.29)) we find the eigenvectors in terms of the basis (5.105)

$$\overline{\boldsymbol{W}}_-^0 := \overline{\boldsymbol{T}}_c^0, \qquad \overline{\boldsymbol{W}}_-^N := \overline{\boldsymbol{T}}_c^N,$$

$$\left. \begin{aligned} \overline{\boldsymbol{W}}_+^l &:= s_l \overline{\boldsymbol{S}}_s^l + \left(\frac{\omega}{2}(1 + c_l) - W_l \right) \overline{\boldsymbol{T}}_c^l, \\ \overline{\boldsymbol{W}}_-^l &:= s_l \overline{\boldsymbol{S}}_s^l + \left(\frac{\omega}{2}(1 + c_l) + W_l \right) \overline{\boldsymbol{T}}_c^l, \end{aligned} \right\} \quad \text{for} \quad l = 1, \ldots, N-1$$

with the eigenvalues given by

$$\begin{aligned} \lambda_0^- &= -\theta, \quad \lambda_N^- = -1, \\ \lambda_l^+ &= c_l - \frac{\omega}{2}(1 + c_l) + W_l, \quad \lambda_l^- = c_l - \frac{\omega}{2}(1 + c_l) - W_l, \\ &\quad \text{for } l = 1, \ldots, N-1. \end{aligned} \tag{5.118}$$

The definitions of s_l, c_l and W_l are given in (5.114).

By transformation via $\bar{\boldsymbol{T}} := \boldsymbol{T}/\sqrt{2}$ with $\boldsymbol{T}$ given in (5.53) we find the eigenvectors of $\boldsymbol{K}$ (given in (5.64)) by

$$\overline{\boldsymbol{Q}}_-^0 := \bar{\boldsymbol{T}}\overline{\boldsymbol{W}}_-^0 = \begin{bmatrix} \boldsymbol{0} \\ -\bar{\boldsymbol{t}}^0 \end{bmatrix}, \qquad \overline{\boldsymbol{Q}}_-^N := \bar{\boldsymbol{T}}\overline{\boldsymbol{W}}_-^N = \begin{bmatrix} -\bar{\boldsymbol{s}}^N \\ \boldsymbol{0} \end{bmatrix},$$

$$\overline{\boldsymbol{Q}}_+^l := \bar{\boldsymbol{T}}\overline{\boldsymbol{W}}_+^l = \begin{bmatrix} (\alpha_l \cos(y_l/2) - (\omega - V_l) \sin(y_l/2)) \bar{\boldsymbol{s}}^l \\ -(\alpha_l \sin(y_l/2) + (\omega - V_l) \cos(y_l/2)) \bar{\boldsymbol{t}}^l \end{bmatrix},$$

$$\overline{\boldsymbol{Q}}_-^l := \bar{\boldsymbol{T}}\overline{\boldsymbol{W}}_-^l = \begin{bmatrix} (\alpha_l \cos(y_l/2) - (\omega + V_l) \sin(y_l/2)) \bar{\boldsymbol{s}}^l \\ -(\alpha_l \sin(y_l/2) + (\omega + V_l) \cos(y_l/2)) \bar{\boldsymbol{t}}^l \end{bmatrix}.$$

We have

$$\boldsymbol{K}\overline{\boldsymbol{Q}}_-^0 = -\theta\overline{\boldsymbol{Q}}_-^0, \quad \boldsymbol{K}\overline{\boldsymbol{Q}}_-^N = -\overline{\boldsymbol{Q}}_-^N,$$

$$\boldsymbol{K}\overline{\boldsymbol{Q}}_+^l = \lambda_l^+ \overline{\boldsymbol{Q}}_+^l, \quad \boldsymbol{K}\overline{\boldsymbol{Q}}_-^l = \lambda_l^- \overline{\boldsymbol{Q}}_-^l, \quad \text{for } l = 1, \ldots, N-1.$$

Theorem 5.23. *Assume that $\omega \in (0,1)$ and (A'). Let the initial values be given by $\boldsymbol{P}^0 = [\boldsymbol{R}^0; \boldsymbol{J}^0]$ and let the right hand be given by $\boldsymbol{G}^l = [\boldsymbol{f}^l; \boldsymbol{g}^l]$ for $l = 0, \ldots, k-1$. Then the discrete Fourier solution $\boldsymbol{P}_{LR}^k = [\boldsymbol{R}^k; \boldsymbol{J}^k]$ of the system*

$$\boldsymbol{P}_{LR}^{l+1} = \boldsymbol{K}\boldsymbol{P}_{LR}^l + \tau\boldsymbol{G}_{LR}^l \quad \text{for } l = 0, \ldots, k-1$$

for the CCFV lattice Boltzmann equations (5.4) with reduced density boundary conditions (5.12) can be represented in terms of the bases (5.107) and (5.108) by

$$\boldsymbol{R}^k = (-1)^k \left(\boldsymbol{R}^0 \cdot \bar{\boldsymbol{s}}^N \right) \bar{\boldsymbol{s}}^N + \tau \sum_{l=0}^{k-1} (-1)^{k-l-1} \left(\boldsymbol{f}^l \cdot \bar{\boldsymbol{s}}^N \right) \bar{\boldsymbol{s}}^N$$

$$+ \sum_{j=1}^{N-1} \left(\frac{1}{2}(1 + A_j)(\lambda_j^+)^k + \frac{1}{2}(1 - A_j)(\lambda_j^-)^k \right) \left(\boldsymbol{R}^0 \cdot \bar{\boldsymbol{s}}^j \right) \bar{\boldsymbol{s}}^j$$

$$- \frac{\theta}{2\omega} \sum_{j=1}^{N-1} B_j \left((\lambda_j^+)^k - (\lambda_j^-)^k \right) \left(\boldsymbol{J}^0 \cdot \bar{\boldsymbol{t}}^j \right) \bar{\boldsymbol{s}}^j$$

$$+ \tau \sum_{l=0}^{k-1} \sum_{j=1}^{N-1} \left(\frac{1}{2}(1 + A_j)(\lambda_j^+)^{k-l-1} + \frac{1}{2}(1 - A_j)(\lambda_j^-)^{k-l-1} \right) \left(\boldsymbol{f}^l \cdot \bar{\boldsymbol{s}}^j \right) \bar{\boldsymbol{s}}^j$$

$$- \frac{\theta}{2\omega} \tau \sum_{l=0}^{k-2} \sum_{j=1}^{N-1} B_j \left((\lambda_j^+)^{k-l-1} - (\lambda_j^-)^{k-l-1} \right) \left(\boldsymbol{g}^l \cdot \bar{\boldsymbol{t}}^j \right) \bar{\boldsymbol{s}}^j,$$

$$\boldsymbol{J}^k = (-\theta)^k \left(\boldsymbol{J}^0 \cdot \bar{\boldsymbol{t}}^0 \right) \bar{\boldsymbol{t}}^0 + \tau \sum_{l=0}^{k-1} (-\theta)^{k-l-1} \left(\boldsymbol{g}^l \cdot \bar{\boldsymbol{t}}^0 \right) \bar{\boldsymbol{t}}^0$$

$$+ \sum_{j=1}^{N-1} \left(\frac{1}{2}(1 - A_j)(\lambda_j^+)^k + \frac{1}{2}(1 + A_j)(\lambda_j^-)^k \right) \left(\boldsymbol{J}^0 \cdot \bar{\boldsymbol{t}}^j \right) \bar{\boldsymbol{t}}^j$$

$$- \frac{1}{2\omega} \sum_{j=1}^{N-1} B_j \left((\lambda_j^+)^k - (\lambda_j^-)^k \right) \left(\boldsymbol{R}^0 \cdot \bar{\boldsymbol{s}}^j \right) \bar{\boldsymbol{t}}^j$$

$$+ \tau \sum_{l=0}^{k-1} \sum_{j=1}^{N-1} \left(\frac{1}{2}(1 - A_j)(\lambda_j^+)^{k-l-1} + \frac{1}{2}(1 + A_j)(\lambda_j^-)^{k-l-1} \right) \left(\boldsymbol{g}^l \cdot \bar{\boldsymbol{t}}^j \right) \bar{\boldsymbol{t}}^j$$

$$- \frac{\tau}{2\omega} \sum_{l=0}^{k-2} \sum_{j=1}^{N-1} B_j \left((\lambda_j^+)^{k-l-1} - (\lambda_j^-)^{k-l-1} \right) \left(\boldsymbol{f}^l \cdot \bar{\boldsymbol{s}}^j \right) \bar{\boldsymbol{t}}^j,$$

where the eigenvalues $\lambda_j^\pm$ are given in (5.118) with (5.114) and A_j and B_j are defined in (5.116).

The eigenvectors of $\boldsymbol{M}$ (given in (5.28) and (5.30)) are given in terms of the basis (5.109) by

$$\bar{\boldsymbol{V}}_+^0 := \bar{\boldsymbol{S}}_c^0, \quad \bar{\boldsymbol{V}}_-^N := \bar{\boldsymbol{S}}_c^N,$$

$$\left.\begin{aligned}
\bar{\boldsymbol{V}}_+^l &:= s_l \bar{\boldsymbol{S}}_c^l - \left(\frac{\omega}{2}(1 + c_l - W_l) \bar{\boldsymbol{T}}_s^l, \right. \\
\bar{\boldsymbol{V}}_-^l &:= s_l \bar{\boldsymbol{S}}_c^l - \left(\frac{\omega}{2}(1 + c_l - W_l) \bar{\boldsymbol{T}}_s^l, \right.
\end{aligned}\right\} \quad \text{for } l = 1, \ldots, N-1.$$

We find the eigenvalues

$$\begin{aligned}
\lambda_0^+ &= 1, \quad \lambda_N^- = -1, \\
\lambda_l^+ &= c_l - \frac{\omega}{2}(1 + c_l) + W_l, \quad \lambda_l^- = c_l - \frac{\omega}{2}(1 + c_l) - W_l, \\
&\quad \text{for } l = 1, \ldots, N-1.
\end{aligned} \tag{5.119}$$

We get eigenvectors of $\boldsymbol{K}$ (given in (5.58)) via transformation by $\bar{\boldsymbol{T}} := \boldsymbol{T}/\sqrt{2}$ with $\boldsymbol{T}$ given in (5.53)

$$\bar{\boldsymbol{P}}_+^0 := \bar{\boldsymbol{T}} \bar{\boldsymbol{V}}_+^0 = \begin{bmatrix} \bar{\boldsymbol{t}}^0 \\ \boldsymbol{0} \end{bmatrix}, \quad \bar{\boldsymbol{P}}_-^N := \bar{\boldsymbol{T}} \bar{\boldsymbol{V}}_-^N = \begin{bmatrix} \boldsymbol{0} \\ \bar{\boldsymbol{s}}^N \end{bmatrix},$$

$$\bar{\boldsymbol{P}}_+^l := \bar{\boldsymbol{T}} \bar{\boldsymbol{V}}_+^l = \begin{bmatrix} \left(\sin(y_l) \cos(y_l/2) - \left(\frac{\omega}{2}(1 + c_l) - W_l \right) \sin(y_l/2) \right) \bar{\boldsymbol{t}}^l \\ - \left(\sin(y_l) \sin(y_l/2) + \left(\frac{\omega}{2}(1 + c_l) - W_l \right) \cos(y_l/2) \right) \bar{\boldsymbol{s}}^l \end{bmatrix},$$

$$\bar{\boldsymbol{P}}_-^l := \bar{\boldsymbol{T}} \bar{\boldsymbol{V}}_-^l = \begin{bmatrix} - \left(\sin(y_l) \cos(y_l/2) - \left(\frac{\omega}{2}(1 + c_l) + W_l \right) \sin(y_l/2) \right) \bar{\boldsymbol{t}}^l \\ - \left(\sin(y_l) \sin(y_l/2) + \left(\frac{\omega}{2}(1 + c_l) + W_l \right) \cos(y_l/2) \right) \bar{\boldsymbol{s}}^l \end{bmatrix}.$$

Theorem 5.24. *Assume that $\omega \in (0, 1)$ and (A'). Let the initial values be given by $\boldsymbol{P}^0 = [\boldsymbol{R}^0; \boldsymbol{J}^0]$ and let the right hand be given by $\boldsymbol{G}^l = [\boldsymbol{f}^l; \boldsymbol{g}^l]$ for $l = 0, \ldots, k-1$.*

Take $\boldsymbol{C}_0^l := c_0^l \overline{\boldsymbol{P}}_+^0$ with $c_0^l := -(j_R^l - j_L^l)/N$. Then the discrete Fourier solution $\boldsymbol{P}_{LR}^k = [\boldsymbol{R}^k; \boldsymbol{J}^k]$ of the system

$$\boldsymbol{P}_{LR}^{l+1} = \boldsymbol{K}\boldsymbol{P}_{LR}^l + \tau \boldsymbol{G}_{LR}^l + \boldsymbol{C}_0^l \quad \text{for } l = 0, \dots, k-1$$

for the CCFV lattice Boltzmann equations (5.4) with reduced flux boundary conditions (5.14) can be represented in terms of the bases (5.107) and (5.108) by

$$\boldsymbol{R}^k = \left(\boldsymbol{R}^0 \cdot \overline{\boldsymbol{t}}^0\right) \overline{\boldsymbol{t}}^0 + \tau \sum_{l=0}^{k-1} \left(\boldsymbol{f}^l \cdot \overline{\boldsymbol{t}}^0\right) \overline{\boldsymbol{t}}^0 + \sum_{l=0}^{k-1} c_0^l \overline{\boldsymbol{t}}^0$$

$$\sum_{j=1}^{N-1} \left(\frac{1}{2}(1 + A_j)(\lambda_j^+)^k + \frac{1}{2}(1 - A_j)(\lambda_j^-)^k\right) \left(\boldsymbol{R}^0 \cdot \overline{\boldsymbol{t}}^j\right) \overline{\boldsymbol{t}}^j$$

$$-\frac{1-\omega}{2\omega} \sum_{j=1}^{N-1} B_j \left((\lambda_j^+)^k - (\lambda_j^-)^k\right) \left(\boldsymbol{J}^0 \cdot \overline{\boldsymbol{s}}^j\right) \overline{\boldsymbol{t}}^j$$

$$+\tau \sum_{l=0}^{k-1} \sum_{j=1}^{N-1} \left(\frac{1}{2}(1 + A_j)(\lambda_j^+)^{k-l-1} + \frac{1}{2}(1 - A_j)(\lambda_j^-)^{k-l-1}\right) \left(\boldsymbol{f}^l \cdot \overline{\boldsymbol{t}}^j\right) \overline{\boldsymbol{t}}^j$$

$$-\frac{1-\omega}{2\omega}\tau \sum_{l=0}^{k-2} \sum_{j=1}^{N-1} B_j \left((\lambda_j^+)^{k-l-1} - (\lambda_j^-)^{k-l-1}\right) \left(\boldsymbol{g}^l \cdot \overline{\boldsymbol{s}}^j\right) \overline{\boldsymbol{t}}^j,$$

$$\boldsymbol{J}^k = (-1)^k \left(\boldsymbol{J}^0 \cdot \overline{\boldsymbol{s}}^N\right) \overline{\boldsymbol{s}}^N + \tau \sum_{l=0}^{k-1} (-1)^{k-l-1} \left(\boldsymbol{g}^l \cdot \overline{\boldsymbol{s}}^N\right) \overline{\boldsymbol{s}}^N$$

$$+\sum_{j=1}^{N-1} \left(\frac{1}{2}(1 - A_j)(\lambda_j^+)^k + \frac{1}{2}(1 + A_j)(\lambda_j^-)^k\right) \left(\boldsymbol{J}^0 \cdot \overline{\boldsymbol{s}}^j\right) \overline{\boldsymbol{s}}^j$$

$$+\frac{1}{2\omega} \sum_{j=1}^{N-1} B_j \left((\lambda_j^+)^k - (\lambda_j^-)^k\right) \left(\boldsymbol{R}^0 \cdot \overline{\boldsymbol{t}}^j\right) \overline{\boldsymbol{s}}^j$$

$$+\tau \sum_{l=0}^{k-1} \sum_{j=1}^{N-1} \left(\frac{1}{2}(1 - A_j)(\lambda_j^+)^{k-l-1} + \frac{1}{2}(1 + A_j)(\lambda_j^-)^{k-l-1}\right) \left(\boldsymbol{g}^l \cdot \overline{\boldsymbol{s}}^j\right) \overline{\boldsymbol{s}}^j$$

$$+\frac{\tau}{2\omega} \sum_{l=0}^{k-2} \sum_{j=1}^{N-1} B_j \left((\lambda_j^+)^{k-l-1} - (\lambda_j^-)^{k-l-1}\right) \left(\boldsymbol{f}^l \cdot \overline{\boldsymbol{t}}^j\right) \overline{\boldsymbol{s}}^j,$$

where the eigenvalues $\lambda_j^\pm$ are given in (5.119) with (5.114) and A_j and B_j are defined in (5.116).

CHAPTER 6

Convergence of the Lattice Boltzmann Solutions

In this chapter we prove the convergence of the FD lattice Boltzmann solutions by following the concept of stability and consistency. Since stability estimates are only available for the lattice Boltzmann schemes (5.1) on VCFD grids and (5.2) on CCFD grids, we restrict our attention to these cases. For the convergence of the FV schemes (5.3) and (5.4), one has to invoke the discrete Fourier solutions.

For a fixed end time, we prove second order convergence for the density and the h-scaled flux in the discrete $L^\infty(L^2)$-norm and third order convergence for the h-scaled flux in the discrete $L^2(L^2)$-norm towards solutions of the heat equation with respect to the grid size h.

Special care has to be taken at the boundaries, where the residuals do not show the right order. This problem can be overcome, however, by methods similar to those employed when we reduced the boundary values.

We begin with formal asymptotic expansions of the problem.

6.1. A Formal Approach

The question of the numerical viscosity of the lattice Boltzmann schemes has first been answered by the computation of the discrete Fourier solutions. We now wish to confirm these results with the aid of a more formal approach.

By using Taylor expansions up to second order in h for the terms u_{l+1}^{k+1} and v_{l-1}^{k+1} in the FD lattice Boltzmann scheme (5.1) with the coupling $\gamma\tau = h^2$, and disregarding the indices in $u_l^k := u(t_k, x_l)$, $v_l^k := v(t_k, x_l)$ and $f_l^k := f(t_k, x_l)$, we gain

$$h^2 \partial_t u + \gamma h \partial_x u + \gamma \frac{h^2}{2} \partial_x^2 u + \gamma \omega (u - v) = \frac{h^2}{2} f,$$

$$h^2 \partial_t v - \gamma h \partial_x v + \gamma \frac{h^2}{2} \partial_x^2 v - \gamma \omega (u - v) = \frac{h^2}{2} f.$$

Next, we plug in expansions of the form

$$u = u^0 + h u^1 + h^2 u^2 + \ldots,$$

$$v = v^0 + h v^1 + h^2 v^2 + \ldots.$$

We take $r := u + v$ and $j := u - v$. In zeroth order, we find

$$u^0 = v^0 = \frac{1}{2} r^0 := \frac{1}{2}(u^0 + v^0) \quad j^0 := u^0 - v^0 = 0.$$

In first order, we have

$$\partial_x u^0 = -\omega(u^1 - v^1),$$

$$-\partial_x v^0 = \omega(u^1 - v^1)$$

and hence we get

$$j^1 := u^1 - v^1 = -\frac{1}{\omega} \partial_x u^0 = -\frac{1}{\omega} \partial_x v^0 = -\frac{1}{2\omega} \partial_x r^0.$$

107

Observe that the first order density term $r^1 := u^1 + v^1$ is yet undetermined. Later on the numerical convergence results imply that $r^1 = 0$. In second order, we get

$$\partial_t u^0 + \gamma \partial_x u^1 + \frac{\gamma}{2} \partial_x^2 u^0 + \gamma \omega (u^2 - v^2) = \frac{1}{2} f,$$

$$\partial_t v^0 - \gamma \partial_x v^1 + \frac{\gamma}{2} \partial_x^2 v^0 - \gamma \omega (u^2 - v^2) = \frac{1}{2} f.$$

Taking the sum of the previous equations yields

$$\partial_t r^0 + \gamma \partial_x j^1 + \frac{\gamma}{2} \partial_x^2 r^0 = \partial_t r^0 - \frac{1 - \omega}{2\omega} \gamma \partial_x^2 r^0 = f.$$

Hence, we find the numerical viscosity

$$\nu^\cdot = \frac{1 - \omega}{2\omega} \gamma$$

for the FD lattice Boltzmann schemes. From $r^1 = u^1 + v^1 = 0$ we get $j^2 := u^2 - v^2 = 0$. We formally have

$$u = \frac{1}{2} r^0 - \frac{h}{4\omega} \partial_x r^0 + \frac{h^2}{2} r^2 + \mathcal{O}(h^3),$$

$$v = \frac{1}{2} r^0 + \frac{h}{4\omega} \partial_x r^0 + \frac{h^2}{2} r^2 + \mathcal{O}(h^3),$$

which implies the approximation of the density and the flux of the orders

$$r = r^0 + h^2 r^2 + \mathcal{O}(h^3),$$

$$j = -\frac{h}{2\omega} \partial_x r^0 + \mathcal{O}(h^3).$$

Here, r^0 is the solution of the heat equation and r^2 is an unknown function. This result is quantified in the ensuing sections.

Using Taylor expansions up to second order in h for the terms u_{l+1}^{k+1}, v_{l-1}^{k+1} in the FV lattice Boltzmann scheme (5.3) with the coupling $\gamma \tau = h^2$ and omitting the indices for u_l^k, v_l^k and f_l^k yields

$$h^2 \partial_t u + \gamma h \partial_x u + \gamma \frac{h^2}{2} \partial_x^2 u + \gamma \frac{\omega}{2} (2j + h \partial_x j + \frac{h^2}{2} \partial_x^2 j) = \frac{h^2}{2} f,$$

$$h^2 \partial_t v - \gamma h \partial_x v + \gamma \frac{h^2}{2} \partial_x^2 v - \gamma \frac{\omega}{2} (2j - h \partial_x j + \frac{h^2}{2} \partial_x^2 j) = \frac{h^2}{2} f.$$

Using the expansions

$$u = u^0 + h u^1 + h^2 u^2 + \dots,$$

$$v = v^0 + h v^1 + h^2 v^2 + \dots$$

yields in zeroth order

$$u^0 = v^0 = \frac{1}{2} r^0, \quad j^0 = 0.$$

In first order, we have

$$\partial_x u^0 = -\omega j^1 - \frac{\omega}{2} \partial_x j^0 = -\omega j^1,$$

$$-\partial_x v^0 = \omega j^1 - \frac{\omega}{2} \partial_x j^0 = \omega j^1$$

and hence we gain

$$j^1 = -\frac{1}{\omega} \partial_x u^0 = -\frac{1}{\omega} \partial_x v^0 = -\frac{1}{2\omega} \partial_x r^0.$$

In second order, we get

$$\partial_t u^0 + \gamma \partial_x u^1 + \frac{\gamma}{2}\partial_x^2 u^0 + \gamma\omega j^2 + \gamma\frac{\omega}{2}\partial_x j^1 + \gamma\frac{\omega}{4}\partial_x^2 j^0 = \frac{1}{2}f,$$

$$\partial_t v^0 - \gamma \partial_x v^1 + \frac{\gamma}{2}\partial_x^2 v^0 - \gamma\omega j^2 + \gamma\frac{\omega}{2}\partial_x j^1 - \gamma\frac{\omega}{4}\partial_x^2 j^0 = \frac{1}{2}f.$$

Taking the sum of the previous equations yields

$$\partial_t r^0 + \gamma \partial_x j^1 + \frac{\gamma}{2}\partial_x^2 r^0 + \gamma\omega\partial_x j^1 = \partial_t r^0 - \frac{\gamma}{2\omega}\partial_x^2 r^0 = f.$$

Hence, we obtain the numerical viscosity as

$$\nu = \frac{1}{2\omega}\gamma$$

for the FV lattice Boltzmann schemes. The approximation properties are the same as in the FD case. As a consequence, the time steps for the FV schemes can be chosen a factor $1/(1-\omega)$ larger, so that in total fewer time steps have to be performed.

Transformation of the heat equation (2.1) to an advection system by defining $u := (r+j)/2$, $v := (r-j)/2$ with $j := -h\partial_x r/(2\omega)$ leads to

$$\partial_t u + \frac{1}{\epsilon}\partial_x u + \frac{1-\omega}{2\nu\epsilon^2}(u-v) = \frac{1}{2}f + \frac{1}{2}\partial_t j - \frac{\omega}{2(1-\omega)}\nu\partial_x^2 r,$$

$$\partial_t v - \frac{1}{\epsilon}\partial_x v - \frac{1-\omega}{2\nu\epsilon^2}(u-v) = \frac{1}{2}f - \frac{1}{2}\partial_t j + \frac{\omega}{2(1-\omega)}\nu\partial_x^2 r,$$

where we applied the couplings $\epsilon := h/\gamma$, $\gamma\tau = h^2$ and $\gamma = 2\omega\nu/(1-\omega)$ on FD grids. In comparison to (3.14), we see that the solution of the FD lattice Boltzmann schemes that is consistent to the heat equation, cannot be consistent to the solutions of the advection system (3.1).

For the FV couplings $\epsilon := h/\gamma$, $\gamma\tau = h^2$ and $\gamma = 2\omega\nu$, we gain

$$\partial_t u + \frac{1}{\epsilon}\partial_x u + \frac{1}{2\nu\epsilon^2}(u-v) = \frac{1}{2}f + \frac{1}{2}\partial_t j,$$

$$\partial_t v - \frac{1}{\epsilon}\partial_x v - \frac{1}{2\nu\epsilon^2}(u-v) = \frac{1}{2}f - \frac{1}{2}\partial_t j,$$

which is exactly (3.14). Hence, solutions of the FV lattice Boltzmann schemes are consistent with the heat equation and with the advection system simultaneously.

From these formal results we now step back to a rigorous study of the convergence. Let us now first consider the consistency of the lattice Boltzmann schemes.

6.2. Consistency of the FD Lattice Boltzmann Schemes

Let r be a solution of the heat equation (2.1) with viscosity ν and source term f. For a given $\omega \in (0,1)$ and grid size $h := |\Omega|/N$ we define $j := -h\partial_x r/(2\omega)$. Furthermore, we define $r_l^k := r(t_k, x_l)$, $j_l^k := j(t_k, x_l)$ and $f_l^k := f(t_k, x_l)$. In the inner points we consider the residuals for the FD lattice Boltzmann schemes (5.1) and (5.2) in the RJ-formulation implied by (5.32). The residuals read

$$\begin{aligned}
\xi_l^k &:= r_l^{k+1} - r_l^k - \frac{1}{2}\left(r_{l+1}^k - 2r_l^k + r_{l-1}^k\right) \\
&\quad - \frac{2\omega-1}{2}\left(j_{l+1}^k - j_{l-1}^k\right) - \frac{\tau}{2}\left(f_{l+1}^k + f_{l-1}^k\right), \\
\eta_l^k &:= j_l^{k+1} - j_l^k + \frac{1}{2}\left(r_{l+1}^k - r_{l-1}^k\right) \\
&\quad + \frac{2\omega-1}{2}\left(j_{l+1}^k + j_{l-1}^k\right) + j_l^k + \frac{\tau}{2}\left(f_{l+1}^k - f_{l-1}^k\right).
\end{aligned} \tag{6.1}$$

In the periodic case the spatial index l takes the values $l = 0, \ldots, N - 1$. In the boundary case we choose $l = 2, \ldots, N - 2$ on VC grids and $l = 3/2, \ldots, N - 3/2$ on CC grids.

By using the relations $\gamma\tau = h^2$ and $\nu = (1 - \omega)\gamma/(2\omega)$ and Taylor series expansions, we find

$$
\begin{aligned}
\xi_l^k &= h^4 \left(\frac{6\omega^2 - 8\omega + 3}{24\omega^2} \partial_x^4 r_l^k + \frac{1}{2\gamma^2} \partial_t f_l^k + \frac{1 - 3\omega}{4\omega\gamma} \partial_x^2 f_l^k \right) + \mathcal{O}(h^6), \\
\eta_l^k &= h^3 \left(\frac{-4\omega^2 + 6\omega - 3}{12\omega^2} \partial_x^3 r_l^k + \frac{2\omega - 1}{2\omega\gamma} \partial_x f_l^k \right) + \mathcal{O}(h^5).
\end{aligned}
\tag{6.2}
$$

Here, we use the definitions $\partial_x^j r_l^k := \partial_x^j r(t_k, x_l)$ for $j \in \mathbb{N}$ and $\partial_t f_l^k := \partial_t f(t_k, x_l)$.

For the boundary value problems, we have to consider in addition the residuals at the boundary. For the VCFD lattice Boltzmann equations (5.1) with density boundary conditions (5.6) we find

$$
\begin{aligned}
\xi_1^k &:= r_1^{k+1} - r_1^k - \frac{1}{2} \left(r_2^k - 2r_1^k + r_L^k \right) \\
&\quad - \frac{2\omega - 1}{2} \left(j_2^k - j_0^k \right) - \frac{\tau}{2} \left(f_2^k + f_0^k \right), \\
\xi_{N-1}^k &:= r_{N-1}^{k+1} - r_{N-1}^k - \frac{1}{2} \left(r_R^k - 2r_{N-1}^k + r_{N-2}^k \right) \\
&\quad - \frac{2\omega - 1}{2} \left(j_N^k - j_{N-2}^k \right) - \frac{\tau}{2} \left(f_N^k + f_{N-2}^k \right), \\
\eta_0^k &:= j_0^{k+1} - j_0^k + r_1^k - r_L^{k+1} + (2\omega - 1)j_1^k + j_0^k + \tau f_1^k, \\
\eta_1^k &:= j_1^{k+1} - j_1^k + \frac{1}{2} \left(r_2^k - r_L^k \right) \\
&\quad + \frac{2\omega - 1}{2} \left(j_2^k + j_0^k \right) + j_1^k + \frac{\tau}{2} \left(f_2^k - f_0^k \right), \\
\eta_{N-1}^k &:= j_{N-1}^{k+1} - j_{N-1}^k + \frac{1}{2} \left(r_R^k - r_{N-2}^k \right) \\
&\quad + \frac{2\omega - 1}{2} \left(j_N^k + j_{N-2}^k \right) + j_{N-1}^k + \frac{\tau}{2} \left(f_N^k - f_{N-2}^k \right), \\
\eta_N^k &:= j_N^{k+1} - j_N^k + r_R^{k+1} - r_{N-1}^k + (2\omega - 1)j_{N-1}^k + j_N^k - \tau f_{N-1}^k,
\end{aligned}
\tag{6.3}
$$

with r_L^k and r_R^k defined in (5.7). While the residuals ξ_1^k, ξ_{N-1}^k, η_1^k and η_{N-1}^k have the same behavior as in the inner points, we see

$$
\begin{aligned}
\eta_0^k &= h^3 \left(\frac{-4\omega^2 + 6\omega - 3}{12\omega^2} \partial_x^3 r_0^k + \frac{2\omega - 1}{2\omega\gamma} \partial_x f_0^k \right) + \mathcal{O}(h^4), \\
\eta_N^k &= h^3 \left(\frac{-4\omega^2 + 6\omega - 3}{12\omega^2} \partial_x^3 r_N^k + \frac{2\omega - 1}{2\omega\gamma} \partial_x f_N^k \right) + \mathcal{O}(h^4).
\end{aligned}
\tag{6.4}
$$

For the VCFD lattice Boltzmann equations (5.1) with flux boundary conditions (5.8), we find

$$\xi_0^k := r_0^{k+1} - r_0^k - r_1^k + r_0^k - (2\omega - 1)j_1^k - j_L^{k+1} - \tau f_1^k,$$

$$\xi_1^k := r_1^{k+1} - r_1^k - \frac{1}{2}\left(r_2^k - 2r_1^k + r_0^k\right)$$
$$- \frac{2\omega - 1}{2}\left(j_2^k - j_L^k\right) - \frac{\tau}{2}\left(f_2^k + f_0^k\right),$$

$$\xi_{N-1}^k := r_{N-1}^{k+1} - r_{N-1}^k - \frac{1}{2}\left(r_N^k - 2r_{N-1}^k + r_{N-2}^k\right)$$
$$- \frac{2\omega - 1}{2}\left(j_R^k - j_{N-2}^k\right) - \frac{\tau}{2}\left(f_N^k + f_{N-2}^k\right),$$

$$\xi_N^k := r_N^{k+1} - r_N^k + r_N^k - r_{N-1}^k + (2\omega - 1)j_{N-1}^k + j_R^{k+1} - \tau f_{N-1}^k,$$

$$\eta_1^k := j_1^{k+1} - j_1^k + \frac{1}{2}\left(r_2^k - r_0^k\right)$$
$$+ \frac{2\omega - 1}{2}\left(j_L^k + j_2^k\right) + j_1^k + \frac{\tau}{2}\left(f_2^k - f_0^k\right),$$

$$\eta_{N-1}^k := j_{N-1}^{k+1} - j_{N-1}^k + \frac{1}{2}\left(r_N^k - r_{N-2}^k\right)$$
$$+ \frac{2\omega - 1}{2}\left(j_R^k + j_{N-2}^k\right) + j_{N-1}^k + \frac{\tau}{2}\left(f_N^k - f_{N-2}^k\right),$$

$$(6.5)$$

with j_L^k and j_R^k defined in (5.9). While the residuals ξ_1^k, ξ_{N-1}^k, η_1^k and η_{N-1}^k have the same behavior as in the inner points, we gain

$$\xi_0^k = -h^3\left(-\frac{4\omega^2 + 6\omega - 3}{12\omega^2}\partial_x^3 r_0^k + \frac{2\omega - 1}{2\omega\gamma}\partial_x f_0^k\right) + \mathcal{O}(h^4),$$
$$\xi_N^k = \ h^3\left(-\frac{4\omega^2 + 6\omega - 3}{12\omega^2}\partial_x^3 r_N^k + \frac{2\omega - 1}{2\omega\gamma}\partial_x f_N^k\right) + \mathcal{O}(h^4).$$

$$(6.6)$$

For the VCFD lattice Boltzmann equations (5.1) with inflow boundary conditions (5.10), we find

$$\xi_0^k := r_0^{k+1} - r_0^k - \frac{1}{2}r_1^k + r_0^k - \frac{2\omega - 1}{2}j_1^k - u_L^{k+1} - \frac{\tau}{2}f_1^k,$$

$$\xi_1^k := r_1^{k+1} - r_1^k - \frac{1}{2}r_2^k + r_1^k - \omega r_0^k$$
$$- \frac{2\omega - 1}{2}j_2^k + (2\omega - 1)u_L^k - \frac{\tau}{2}\left(f_2^k + f_0^k\right),$$

$$\xi_{N-1}^k := r_{N-1}^{k+1} - r_{N-1}^k - \frac{1}{2}r_{N-2}^k + r_{N-1}^k - \omega r_N^k$$
$$+ \frac{2\omega - 1}{2}j_{N-2}^k + (2\omega - 1)v_R^k - \frac{\tau}{2}\left(f_N^k + f_{N-2}^k\right),$$

$$\xi_N^k := r_N^{k+1} - r_N^k - \frac{1}{2}r_{N-1}^k + r_N^k + \frac{2\omega - 1}{2}j_{N-1}^k - v_R^{k+1} - \frac{\tau}{2}f_{N-1}^k,$$

$$\eta_1^k := j_1^{k+1} - j_1^k + \frac{1}{2}r_2^k - \omega r_0^k$$
$$+ \frac{2\omega - 1}{2}j_2^k + j_1^k + (2\omega - 1)u_L^k + \frac{\tau}{2}\left(f_2^k - f_0^k\right),$$

$$\eta_{N-1}^k := j_{N-1}^{k+1} - j_{N-1}^k - \frac{1}{2}r_{N-2}^k + \omega r_N^k$$
$$+ \frac{2\omega - 1}{2}j_{N-2}^k + j_{N-1}^k - (2\omega - 1)v_R^k + \frac{\tau}{2}\left(f_N^k - f_{N-2}^k\right),$$

$$(6.7)$$

with u_L^k and v_R^k defined in (5.11). The parameter ϑ for the Robin boundary condition (2.5) for the heat equation has to be chosen as $\vartheta := h/(2\omega)$. For the residuals, we gain

$$
\begin{aligned}
\xi_0^k &= -\frac{1}{2}h^3\left(\frac{-4\omega^2+6\omega-3}{12\omega^2}\partial_x^3 r_0^k + \frac{2\omega-1}{2\omega\gamma}\partial_x f_0^k\right) + \mathcal{O}(h^4),\\[4pt]
\xi_1^k &= \mathcal{O}(h^4),\quad \xi_{N-1}^k = \mathcal{O}(h^4),\\[4pt]
\xi_N^k &= \frac{1}{2}h^3\left(\frac{-4\omega^2+6\omega-3}{12\omega^2}\partial_x^3 r_N^k + \frac{2\omega-1}{2\omega\gamma}\partial_x f_N^k\right) + \mathcal{O}(h^4),\\[4pt]
\eta_1^k &= \frac{1}{2}h^3\left(\frac{-4\omega^2+6\omega-3}{12\omega^2}\partial_x^3 r_1^k + \frac{2\omega-1}{2\omega\gamma}\partial_x f_1^k\right) + \mathcal{O}(h^4),\\[4pt]
\eta_{N-1}^k &= \frac{1}{2}h^3\left(\frac{-4\omega^2+6\omega-3}{12\omega^2}\partial_x^3 r_{N-1}^k + \frac{2\omega-1}{2\omega\gamma}\partial_x f_{N-1}^k\right) + \mathcal{O}(h^4).
\end{aligned}
\tag{6.8}
$$

For the CCFD lattice Boltzmann equations (5.2) with density boundary conditions (5.12), we find

$$
\begin{aligned}
\xi_{1/2}^k &:= r_{1/2}^{k+1} - r_{1/2}^k - \frac{1}{2}\left(r_{3/2}^k - 3r_{1/2}^k + 2r_L^k\right) - r_L^{k+\delta} + r_L^k\\
&\quad - \frac{2\omega-1}{2}\left(j_{3/2}^k - j_{1/2}^k\right) - \frac{\tau}{2}\left(f_{3/2}^k - f_{1/2}^k\right),\\[4pt]
\xi_{N-1/2}^k &:= r_{N-1/2}^{k+1} - r_{N-1/2}^k - \frac{1}{2}\left(2r_R^k - 3r_{N-1/2}^k + r_{N-3/2}^k\right) - r_R^{k+\delta} + r_R^k\\
&\quad - \frac{2\omega-1}{2}\left(j_{N-1/2}^k - j_{N-3/2}^k\right) + \frac{\tau}{2}\left(f_{N-1/2}^k - f_{N-3/2}^k\right),\\[4pt]
\eta_{1/2}^k &:= j_{1/2}^{k+1} - j_{1/2}^k + \frac{1}{2}\left(r_{1/2}^k + r_{3/2}^k\right)\\
&\quad + \frac{2\omega-1}{2}\left(j_{1/2}^k + j_{3/2}^k\right) + j_{1/2}^k - r_L^{k+\delta} + \frac{\tau}{2}\left(f_{1/2}^k + f_{3/2}^k\right),\\[4pt]
\eta_{N-1/2}^k &:= j_{N-1/2}^{k+1} - j_{N-1/2}^k - \frac{1}{2}\left(r_{N-1/2}^k + r_{N-3/2}^k\right)\\
&\quad + \frac{2\omega-1}{2}\left(j_{N-1/2}^k + j_{N-3/2}^k\right) + j_{N-1/2}^k + r_R^{k+\delta} - \frac{\tau}{2}\left(f_{N-1/2}^k + f_{N-3/2}^k\right),
\end{aligned}
\tag{6.9}
$$

with $r_L^{k+\delta}$ and $r_R^{k+\delta}$ defined in (5.7) and evaluated in $t_{k+\delta} = t_k + \delta\tau$ for a given $\delta \in [0,1]$. Taylor expansions render

$$
\begin{aligned}
\xi_{1/2}^k &= h^2 s_L^k + \mathcal{O}(h^4),\\[4pt]
\xi_{N-1/2}^k &= h^2 s_R^k + \mathcal{O}(h^4),\\[4pt]
\eta_{1/2}^k &= h^2 s_L^k + h^3\left(\frac{-4\omega^2+6\omega-3}{12\omega^2}\partial_x^3 r_{1/2}^k + \frac{2\omega-1}{2\omega\gamma}\partial_x f_{1/2}^k\right) + \mathcal{O}(h^4),\\[4pt]
\eta_{N-1/2}^k &= -h^2 s_R^k + h^3\left(\frac{-4\omega^2+6\omega-3}{12\omega^2}\partial_x^3 r_{N-1/2}^k + \frac{2\omega-1}{2\omega\gamma}\partial_x f_{N-1/2}^k\right) + \mathcal{O}(h^4),
\end{aligned}
\tag{6.10}
$$

where we use the definitions

$$
\begin{aligned}
s_L^k &:= \frac{2 - 3\omega - 4\delta(1-\omega)}{8\omega}\left(\partial_x^2 r_{1/2}^k - \frac{h}{2}\partial_x^3 r_{1/2}^k\right) \\
&\quad + \frac{1-\delta}{\gamma}\left(f_{1/2}^k - \frac{h}{2}\partial_x f_{1/2}^k\right), \\
s_R^k &:= \frac{2 - 3\omega - 4\delta(1-\omega)}{8\omega}\left(\partial_x^2 r_{N-1/2}^k + \frac{h}{2}\partial_x^3 r_{N-1/2}^k\right) \\
&\quad + \frac{1-\delta}{\gamma}\left(f_{N-1/2}^k + \frac{h}{2}\partial_x f_{N-1/2}^k\right).
\end{aligned}
\tag{6.11}
$$

For the CCFD lattice Boltzmann equations (5.2) with flux boundary conditions (5.13), we find

$$
\begin{aligned}
\xi_{1/2}^k &:= r_{1/2}^{k+1} - r_{1/2}^k - \frac{1}{2}\left(r_{3/2}^k + r_{1/2}^k\right) + r_{1/2}^k \\
&\quad - \frac{2\omega - 1}{2}\left(j_{3/2}^k + j_{1/2}^k\right) - (1-\omega)j_L^{k+\delta} - \frac{\tau}{2}\left(f_{3/2}^k + f_{1/2}^k\right), \\
\xi_{N-1/2}^k &:= r_{N-1/2}^{k+1} - r_{N-1/2}^k - \frac{1}{2}\left(r_{N-3/2}^k + r_{N-1/2}^k\right) + r_{N-1/2}^k \\
&\quad + \frac{2\omega - 1}{2}\left(j_{N-1/2}^k + j_{N-3/2}^k\right) + (1-\omega)j_R^{k+\delta} - \frac{\tau}{2}\left(f_{N-3/2}^k + f_{N-1/2}^k\right), \\
\eta_{1/2}^k &:= j_{1/2}^{k+1} - j_{1/2}^k + \frac{1}{2}\left(r_{3/2}^k - r_{1/2}^k\right) + \frac{\tau}{2}\left(f_{3/2}^k - f_{1/2}^k\right) \\
&\quad + \frac{2\omega - 1}{2}\left(j_{3/2}^k - j_{1/2}^k\right) + j_{1/2}^k - (1-\omega)j_L^{k+\delta}, \\
\eta_{N-1/2}^k &:= j_{N-1/2}^{k+1} - j_{N-1/2}^k + \frac{1}{2}\left(r_{N-1/2}^k - r_{N-3/2}^k\right) + \frac{\tau}{2}\left(f_{N-1/2}^k - f_{N-3/2}^k\right) \\
&\quad - \frac{2\omega - 1}{2}\left(j_{N-1/2}^k - j_{N-3/2}^k\right) + j_{N-1/2}^k - (1-\omega)j_R^{k+\delta},
\end{aligned}
\tag{6.12}
$$

with $j_L^{k+\delta}$ and $j_R^{k+\delta}$ defined in (5.9) and evaluated in $t_{k+\delta} = t_k + \delta\tau$ for a given $\delta \in [0,1]$. Taylor expansions render

$$
\begin{aligned}
\xi_{1/2}^k &= h^3 s_L^k + h^3 t_L^k \\
&\quad - h^3\left(\frac{-4\omega^2 + 6\omega - 3}{12\omega^2}\partial_x^3 r_{1/2}^k + \frac{2\omega - 1}{2\omega\gamma}\partial_x f_{1/2}^k\right) + \mathcal{O}(h^4), \\
\xi_{N-1/2}^k &= -h^3 s_R^k - h^3 t_R^k \\
&\quad + h^3\left(\frac{-4\omega^2 + 6\omega - 3}{12\omega^2}\partial_x^3 r_{N-1/2}^k + \frac{2\omega - 1}{2\omega\gamma}\partial_x f_{N-1/2}^k\right) + \mathcal{O}(h^4), \\
\eta_{1/2}^k &= h^3 s_L^k - h^3 t_L^k + \mathcal{O}(h^4), \\
\eta_{N-1/2}^k &= h^3 s_R^k - h^3 t_R^k + \mathcal{O}(h^4),
\end{aligned}
\tag{6.13}
$$

where we use the definitions

$$
\begin{aligned}
s_L^k &:= \frac{-5\omega^2 + 15\omega - 12 + 12\delta(1-\omega)^2}{48\omega^2}\partial_x^3 r_{1/2}^k - \frac{(1-\delta)(1-\omega)}{2\omega\gamma}\partial_x f_{1/2}^k, \\
t_L^k &:= \frac{1-\omega}{8\omega}\partial_x^3 r_{1/2}^k, \\
s_R^k &:= \frac{-5\omega^2 + 15\omega - 12 + 12\delta(1-\omega)^2}{48\omega^2}\partial_x^3 r_{N-1/2}^k - \frac{(1-\delta)(1-\omega)}{2\omega\gamma}\partial_x f_{N-1/2}^k, \\
t_R^k &:= \frac{1-\omega}{8\omega}\partial_x^3 r_{N-1/2}^k.
\end{aligned}
\tag{6.14}
$$

The residuals appear as right hand sides of the error equations. With the stability estimates we can now prove convergence of the FD lattice Boltzmann solutions.

6.3. Convergence of the FD Lattice Boltzmann Solutions

Let r be a solution of the heat equation (2.1) in $(0, t_M) \times \Omega$ with viscosity ν, source term f and initial value r_0. Define $j := -h\partial_x r/(2\omega)$. Let $\boldsymbol{P}^k := [\boldsymbol{R}^k; \boldsymbol{J}^k]$, $k = 0, \ldots, M$, be the discrete solutions of the matrix equations (5.37) and (5.52) belonging to the FD lattice Boltzmann schemes (5.1) on VC grids and (5.2) on CC grids with grid size $h := |\Omega|/N$ and time step $\tau := h^2/\gamma$ and the relation $\nu = (1 - \omega)\gamma/(2\omega)$.

We define the errors

$$(E_R)_l^k := r(t_k, x_l) - R_l^k,$$
$$(E_J)_l^k := j(t_k, x_l) - J_l^k.$$

Then the error $\boldsymbol{E}^k := \left[[(E_R)_l^k]_l; [(E_J)_l^k]_l \right]$ satisfies the equation

$$\boldsymbol{E}^{k+1} = \boldsymbol{K}\boldsymbol{E}^k + \mathbf{Res}^k$$

with the residuals

$$\mathbf{Res}^k := \left[\boldsymbol{\xi}^k; \boldsymbol{\eta}^k \right],$$

with $\boldsymbol{\xi}^k = [\xi_l^k]_l$ and $\boldsymbol{\eta}^k = [\eta_l^k]_l$. In the periodic case we gain the following convergence result.

Theorem 6.1. *Let $\boldsymbol{P}^k := [\boldsymbol{R}^k; \boldsymbol{J}^k]$, $k = 0, \ldots, M$, be the discrete solutions belonging to the VCFD lattice Boltzmann equations (5.1) with relaxation parameter ω and with periodic boundary conditions (5.5) and initial values $R_l^0 = r_0(x_l)$, $J_l^0 = -h\partial_x r_0(x_l)/(2\omega)$ for $l = 0, \ldots, N - 1$. The right hand sides are evaluated at the origin of the characteristics, that is, we use $F_{l,+}^k = F_{l,-}^k = f(t_k, x_l)$. Then we get the error estimate*

$$\|\boldsymbol{E}_R^M\|_2^2 + \|\boldsymbol{E}_J^M\|_2^2 + 2\omega(1 - \omega)\frac{1}{\tau}\sum_{i=0}^{M-1}\tau\|\boldsymbol{E}_J^i\|_2^2 \leq Kh^4,$$

with the discrete norms defined in (5.75). The constant K only depends on the relaxation parameter ω, the end time t_M, higher order derivatives of the solution of the heat equation and hence on the viscosity.

PROOF. According to Theorem 5.1 and Corollary 5.2, the errors obey the estimate

$$\|\boldsymbol{E}_R^k\|_2^2 + \|\boldsymbol{E}_J^k\|_2^2 + 2\omega(1 - \omega)\frac{1}{\tau}\sum_{i=0}^{k-1}\tau\|\boldsymbol{E}_J^i\|_2^2$$

$$\leq 2\|\boldsymbol{E}_R^0\|_2^2 + 2\|\boldsymbol{E}_J^0\|_2^2 + 2\sum_{i=0}^{k-1}\|\boldsymbol{\xi}^i\|_2^2 + 2\sum_{i=0}^{k-1}\|\boldsymbol{\eta}^i\|_2^2$$

$$+ \frac{5t_k}{\tau}\sum_{i=0}^{k-1}\|\frac{1}{2}\boldsymbol{A}^\dagger\boldsymbol{\xi}^i - \frac{1}{2}\boldsymbol{D}^\dagger\boldsymbol{\eta}^i\|_2^2 + \frac{\theta^2}{\omega(1 - \omega)}\sum_{i=0}^{k-1}\|\frac{1}{2}\boldsymbol{D}^\dagger\boldsymbol{\xi}^i - \frac{1}{2}\boldsymbol{A}^\dagger\boldsymbol{\eta}^i\|_2^2,$$

with $\boldsymbol{A}$ and $\boldsymbol{D}$ given in (5.35). The residuals $\boldsymbol{\xi}^k = [\xi_l^k]_{l=0,\ldots,N-1}$ and $\boldsymbol{\eta}^k = [\eta_l^k]_{l=0,\ldots,N-1}$ are defined in (6.1). We find

$$\frac{1}{2}\boldsymbol{A}^\dagger\boldsymbol{\xi}^k - \frac{1}{2}\boldsymbol{D}^\dagger\boldsymbol{\eta}^k = \frac{1}{2}\big[\xi_{l-1}^k + \xi_{l+1}^k - \eta_{l-1}^k + \eta_{l+1}^k\big]_{l=0,\ldots,N-1},$$

$$\frac{1}{2}\boldsymbol{D}^\dagger\boldsymbol{\xi}^k - \frac{1}{2}\boldsymbol{A}^\dagger\boldsymbol{\eta}^k = \frac{1}{2}\big[\xi_{l-1}^k - \xi_{l+1}^k - \eta_{l-1}^k - \eta_{l+1}^k\big]_{l=0,\ldots,N-1}.$$

With (6.2) we find

$$\xi_l^k = \mathcal{O}(h^4), \quad \eta_l^k = \mathcal{O}(h^3),$$

$$\xi_{l-1}^k + \xi_{l+1}^k - \eta_{l-1}^k + \eta_{l+1}^k = \mathcal{O}(h^4),$$

$$\xi_{l-1}^k - \xi_{l+1}^k - \eta_{l-1}^k - \eta_{l+1}^k = \mathcal{O}(h^3)$$

for $l = 0, \ldots, N - 1$ and periodic continuations. By keeping a factor $\tau = h^2/\gamma$ for the summation over the time intervals, we find that the first sum in the estimate is of order $\mathcal{O}(h^6)$ and the last three sums are of order $\mathcal{O}(h^4)$. With the choice of the initial data, we have $\boldsymbol{E}_R^0 = \boldsymbol{E}_J^0 = \boldsymbol{0}$. $\square$

An evaluation of the source terms along the characteristics, that is, the choice

$$F_{l,+}^k = f(t_k + s\tau, x_l + sh), \quad F_{l,-}^k = f(t_k + s\tau, x_l - sh) \quad \text{for a given } s \in (0,1],$$

does not lead to an improvement of the convergence.

Theorem 6.2. *Let $\boldsymbol{P}^k := [\boldsymbol{R}^k; \boldsymbol{J}^k]$, $k = 0, \ldots, M$, be the discrete solutions belonging to the VCFD lattice Boltzmann equations (5.1) with relaxation parameter ω and with density boundary conditions (5.6) and initial values $R_l^0 = r_0(x_l)$ for $l = 1, \ldots, N - 1$ and $J_l^0 = -h\partial_x r_0(x_l)/(2\omega)$ for $l = 0, \ldots, N$. The right hand sides are evaluated at the origin of the characteristics, that is, we use $F_{l,+}^k = F_{l,-}^k = f(t_k, x_l)$. Then we get the error estimate*

$$\|\boldsymbol{E}_R^M\|_2^2 + \|\boldsymbol{E}_J^M\|_2^2 + 2\omega(1-\omega)\frac{1}{\tau}\sum_{i=0}^{M-1}\tau\|\boldsymbol{E}_J^i\|_2^2 \leq Kh^4,$$

with the discrete norms defined in (5.77). The constant K only depends on the relaxation parameter ω, the end time t_M, higher order derivatives of the solution of the heat equation and hence on the viscosity.

Proof. According to Theorem 5.1 and Corollary 5.3, the errors obey the estimate

$$\|\boldsymbol{E}_R^k\|_2^2 + \|\boldsymbol{E}_J^k\|_2^2 + 2\omega(1-\omega)\frac{1}{\tau}\sum_{i=0}^{k-1}\tau\|\boldsymbol{E}_J^i\|_2^2$$

$$\leq 2\|\boldsymbol{E}_R^0\|_2^2 + 2\|\boldsymbol{E}_J^0\|_2^2 + 2\sum_{i=0}^{k-1}\|\boldsymbol{\xi}^i\|_2^2 + 2\sum_{i=0}^{k-1}\|\boldsymbol{\eta}^i\|_2^2$$

$$+ \frac{5t_k}{\tau}\sum_{i=0}^{k-1}\|\frac{1}{2}\boldsymbol{A}^\dagger\boldsymbol{\xi}^i - \frac{1}{2}\boldsymbol{E}^\dagger\boldsymbol{d}\boldsymbol{\eta}^i\|_2^2 + \frac{4\theta^2}{\omega(1-\omega)}\sum_{i=0}^{k-1}\|\frac{1}{2}\boldsymbol{D}^\dagger\boldsymbol{\xi}^i - \frac{1}{2}\boldsymbol{B}^\dagger\boldsymbol{d}\boldsymbol{\eta}^i\|_2^2,$$

with $\boldsymbol{A}$, $\boldsymbol{B}$, $\boldsymbol{D}$ and $\boldsymbol{E}$ given in (5.41) and $\boldsymbol{d}$ defined in (5.76). The residuals $\boldsymbol{\xi}^k = [\xi_l^k]_{l=1,\ldots,N-1}$ and $\boldsymbol{\eta}^k = [\eta_l^k]_{l=0,\ldots,N}$ are defined in (6.1) for $l = 2, \ldots, N - 2$ and in

(6.3). We find

$$\frac{1}{2}\boldsymbol{A}^\dagger \boldsymbol{\xi}^k - \frac{1}{2}\boldsymbol{E}^\dagger \boldsymbol{d\eta}^k = \frac{1}{2}\left[\begin{array}{c} \xi_2^k - \eta_0^k + \eta_2^k \\ [\xi_{l-1}^k + \xi_{l+1}^k - \eta_{l-1}^k + \eta_{l+1}^k]_{l=2,\ldots,N-2} \\ \xi_{N-2}^k - \eta_{N-2}^k + \eta_N^k \end{array}\right],$$

$$\frac{1}{2}\boldsymbol{D}^\dagger \boldsymbol{\xi}^k - \frac{1}{2}\boldsymbol{B}^\dagger \boldsymbol{d\eta}^k = \frac{1}{2}\left[\begin{array}{c} -\xi_1^k - \eta_1^k \\ -\xi_2^k - \eta_0^k - \eta_2^k \\ [\xi_{l-1}^k - \xi_{l+1}^k - \eta_{l-1}^k - \eta_{l+1}^k]_{l=2,\ldots,N-2} \\ \xi_{N-2}^k - \eta_{N-2}^k - \eta_N^k \\ \xi_{N-1}^k - \eta_{N-1}^k \end{array}\right].$$

With (6.2) and (6.4) we see

$$\xi_l^k = \mathcal{O}(h^4), \quad \eta_l^k = \mathcal{O}(h^3) \quad \text{for } l = 1,\ldots,N-1 \text{ and } l = 0,\ldots,N.$$

Furthermore, we obtain

$$\xi_2^k - \eta_0^k + \eta_2^k = \mathcal{O}(h^4),$$
$$\xi_{l-1}^k + \xi_{l+1}^k - \eta_{l-1}^k + \eta_{l+1}^k = \mathcal{O}(h^4) \quad \text{for } l = 2,\ldots,N-2,$$
$$\xi_{N-2}^k - \eta_{N-2}^k + \eta_N^k = \mathcal{O}(h^4),$$

and

$$-\xi_1^k - \eta_1^k = \mathcal{O}(h^3), \quad -\xi_2^k - \eta_0^k - \eta_2^k = \mathcal{O}(h^3),$$
$$\xi_{l-1}^k - \xi_{l+1}^k - \eta_{l-1}^k - \eta_{l+1}^k = \mathcal{O}(h^3) \quad \text{for } l = 2,\ldots,N-2,$$
$$\xi_{N-2}^k - \eta_{N-2}^k - \eta_N^k = \mathcal{O}(h^3), \quad \xi_{N-1}^k - \eta_{N-1}^k = \mathcal{O}(h^3).$$

By keeping a factor $\tau = h^2/\gamma$ for the summation over the time intervals, we find that the first sum in the estimate is of order $\mathcal{O}(h^6)$ and the last three sums are of order $\mathcal{O}(h^4)$. With the choice of the initial data, we have $\boldsymbol{E}_R^0 = \boldsymbol{E}_J^0 = \boldsymbol{0}$. $\qquad\square$

Theorem 6.3. *Let $\boldsymbol{P}^k := [\boldsymbol{R}^k; \boldsymbol{J}^k]$, $k = 0,\ldots,M$, be the discrete solutions belonging to the VCFD lattice Boltzmann equations (5.1) with relaxation parameter ω and with flux boundary conditions (5.8) and initial values $R_l^0 = r_0(x_l)$ for $l = 0,\ldots,N$ and $J_l^0 = -h\partial_x r_0(x_l)/(2\omega)$ for $l = 1,\ldots,N-1$. The right hand sides are evaluated at the origin of the characteristics, that is, we use $F_{l,+}^k = F_{l,-}^k = f(t_k,x_l)$. Then we get the error estimate*

$$\|\boldsymbol{E}_R^M\|_2^2 + \|\boldsymbol{E}_J^M\|_2^2 + 2\omega(1-\omega)\frac{1}{\tau}\sum_{i=0}^{M-1}\tau\|\boldsymbol{E}_J^i\|_2^2 \leq Kh^4,$$

with the discrete norms defined in (5.78). The constant K only depends on the relaxation parameter ω, the end time t_M, higher order derivatives of the solution of the heat equation and hence on the viscosity.

PROOF. According to Theorem 5.1 and Corollary 5.4, the errors obey the estimate

$$\|\boldsymbol{E}_R^k\|_2^2 + \|\boldsymbol{E}_J^k\|_2^2 + 2\omega(1-\omega)\frac{1}{\tau}\sum_{i=0}^{k-1}\tau\|\boldsymbol{E}_J^i\|_2^2$$

$$\leq 2\|\boldsymbol{E}_R^0\|_2^2 + 2\|\boldsymbol{E}_J^0\|_2^2 + 2\sum_{i=0}^{k-1}\|\boldsymbol{\xi}^i\|_2^2 + 2\sum_{i=0}^{k-1}\|\boldsymbol{\eta}^i\|_2^2$$

$$+ \frac{20t_k}{\tau}\sum_{i=0}^{k-1}\|\frac{1}{2}\boldsymbol{B}^\dagger \boldsymbol{d\xi}^i - \frac{1}{2}\boldsymbol{D}^\dagger \boldsymbol{\eta}^i\|_2^2 + \frac{\theta^2}{\omega(1-\omega)}\sum_{i=0}^{k-1}\|\frac{1}{2}\boldsymbol{E}^\dagger \boldsymbol{d\xi}^i - \frac{1}{2}\boldsymbol{A}^\dagger \boldsymbol{\eta}^i\|_2^2$$

with $\boldsymbol{A}$, $\boldsymbol{B}$, $\boldsymbol{D}$ and $\boldsymbol{E}$ given in (5.41) and $\boldsymbol{d}$ defined in (5.76). The residuals $\boldsymbol{\xi}^k = [\xi_l^k]_{l=0,\dots,N}$ and $\boldsymbol{\eta}^k = [\eta_l^k]_{l=1,\dots,N-1}$ are defined in (6.1) for $l = 2,\dots,N-2$ and in (6.5). We find

$$\frac{1}{2}\boldsymbol{B}^\dagger \boldsymbol{d}\boldsymbol{\xi}^k - \frac{1}{2}\boldsymbol{D}^\dagger \boldsymbol{\eta}^k = \frac{1}{2}\begin{bmatrix} \xi_1^k + \eta_1^k \\ \xi_0^k + \xi_2^k + \eta_2^k \\ [\xi_{l-1}^k + \xi_{l+1}^k - \eta_{l-1}^k + \eta_{l+1}^k]_{l=2,\dots,N-2} \\ \xi_{N-2}^k + \xi_N^k - \eta_{N-2}^k \\ \xi_{N-1}^k - \eta_{N-1}^k \end{bmatrix},$$

$$\frac{1}{2}\boldsymbol{E}^\dagger \boldsymbol{d}\boldsymbol{\xi}^k - \frac{1}{2}\boldsymbol{A}^\dagger \boldsymbol{\eta}^k = \frac{1}{2}\begin{bmatrix} \xi_0^k - \xi_2^k - \eta_2^k \\ [\xi_{l-1}^k - \xi_{l+1}^k - \eta_{l-1}^k - \eta_{l+1}^k]_{l=2,\dots,N-2} \\ \xi_{N-2}^k - \xi_N^k - \eta_{N-2}^k \end{bmatrix}.$$

With (6.2) and (6.6) we find

$$\xi_0^k = \mathcal{O}(h^3), \quad \xi_N^k = \mathcal{O}(h^3),$$
$$\xi_l^k = \mathcal{O}(h^4), \quad \eta_l^k = \mathcal{O}(h^3) \quad \text{for } l = 1,\dots,N-1.$$

Furthermore, we get

$$\xi_1^k + \eta_1^k = \mathcal{O}(h^3), \quad \xi_0^k + \xi_2^k + \eta_2^k = \mathcal{O}(h^4),$$
$$\xi_{l-1}^k + \xi_{l+1}^k - \eta_{l-1}^k + \eta_{l+1}^k = \mathcal{O}(h^4) \quad \text{for } l = 2,\dots,N-2,$$
$$\xi_{N-2}^k + \xi_N^k - \eta_{N-2}^k = \mathcal{O}(h^4), \quad \xi_{N-1}^k - \eta_{N-1}^k = \mathcal{O}(h^3),$$

and

$$\xi_0^k - \xi_2^k - \eta_2^k = \mathcal{O}(h^4),$$
$$\xi_{l-1}^k - \xi_{l+1}^k - \eta_{l-1}^k - \eta_{l+1}^k = \mathcal{O}(h^3) \quad \text{for } l = 2,\dots,N-2,$$
$$\xi_{N-2}^k - \xi_N^k - \eta_{N-2}^k = \mathcal{O}(h^3).$$

The decreased order of ξ_0^k and ξ_N^k only renders the first sum to be of order $\mathcal{O}(h^5)$, which is of subordinate importance. Since $\xi_1^k + \eta_1^k$ and $\xi_{N-1}^k - \eta_{N-1}^k$ do not show the right behavior, we have to reduce the residuals η_1^k and η_{N-1}^k by solving

$$-(\boldsymbol{K} - \boldsymbol{Id}_{2N})\boldsymbol{S}_L = [\boldsymbol{0}; \boldsymbol{e}_1],$$
$$-(\boldsymbol{K} - \boldsymbol{Id}_{2N})\boldsymbol{S}_R = [\boldsymbol{0}; \boldsymbol{e}_{N-1}].$$

The solutions are obtained in the form

$$\boldsymbol{S}_L = \frac{1}{2(1-\omega)}\left[2\theta\boldsymbol{e}_1 + \theta\boldsymbol{e}_2; \boldsymbol{e}_1\right],$$

$$\boldsymbol{S}_R = \frac{1}{2(1-\omega)}\left[-\theta\boldsymbol{e}_N - 2\theta\boldsymbol{e}_{N+1}; \boldsymbol{e}_{N-1}\right].$$

Next, we redefine the errors according to

$$\widetilde{\boldsymbol{E}}^k := \begin{bmatrix} \widetilde{\boldsymbol{E}}_R^k \\ \widetilde{\boldsymbol{E}}_J^k \end{bmatrix} := \begin{bmatrix} [r_l^k - R_l^k]_l \\ [j_l^k - J_l^k]_l \end{bmatrix} - \eta_1^k \boldsymbol{S}_L - \eta_{N-1}^k \boldsymbol{S}_R.$$

Then the error estimate holds for $\widetilde{\boldsymbol{E}}_R^k$ and $\widetilde{\boldsymbol{E}}_J^k$ with residuals of expected orders and with the additional higher order source term $-(\eta_1^{k+1} - \eta_1^k)\boldsymbol{S}_L - (\eta_{N-1}^{k+1} - \eta_{N-1}^k)\boldsymbol{S}_R$. The last three sums in the estimate are then of order $\mathcal{O}(h^4)$. The initial errors vanish due to the choice of the initial data. Since the perturbations of the errors are of order $\mathcal{O}(h^3)$, the estimate in Theorem 6.3 remains true for $\boldsymbol{E}_R$ and $\boldsymbol{E}_J$. $\square$

Theorem 6.4. *Let $\boldsymbol{P}^k := [\boldsymbol{R}^k; \boldsymbol{J}^k]$, $k = 0, \ldots, M$, be the discrete solutions belonging to the VCFD lattice Boltzmann equations (5.1) with relaxation parameter ω and with inflow boundary conditions (5.10) and initial values $R_l^0 = r_0(x_l)$ for $l = 0, \ldots, N$ and $J_l^0 = -h\partial_x r_0(x_l)/(2\omega)$ for $l = 1, \ldots, N-1$. The right hand sides are evaluated at the origin of the characteristics, that is, we use $F_{l,+}^k = F_{l,-}^k = f(t_k, x_l)$. Then we get the error estimate*

$$\|\boldsymbol{E}_R^M\|_2^2 + \|\boldsymbol{E}_J^M\|_2^2 + 2\omega(1-\omega)\frac{1}{\tau}\sum_{i=0}^{M-1}\tau\|\boldsymbol{E}_J^i\|_2^2 \le Kh^4,$$

with the discrete norms defined in (5.80). The constant K only depends on the relaxation parameter ω, the end time t_M, higher order derivatives of the solution of the heat equation and hence on the viscosity.

PROOF. According to Theorem 5.1 and Corollary 5.5, the errors obey the estimate

$$\|\boldsymbol{E}_R^k\|_2^2 + \|\boldsymbol{E}_J^k\|_2^2 + 2\omega(1-\omega)\frac{1}{\tau}\sum_{i=0}^{k-1}\tau\|\boldsymbol{E}_J^i\|_2^2$$

$$\le 2\|\boldsymbol{E}_R^0\|_2^2 + 2\|\boldsymbol{E}_J^0\|_2^2 + 2\sum_{i=0}^{k-1}\|\boldsymbol{\xi}^i\|_2^2 + 2\sum_{i=0}^{k-1}\|\boldsymbol{\eta}^i\|_2^2$$

$$+ \frac{5t_k}{\tau}\sum_{i=0}^{k-1}\|\frac{1}{2}\boldsymbol{A}^\dagger e\boldsymbol{\xi}^i - \frac{1}{2}\boldsymbol{E}^\dagger\boldsymbol{\eta}^i\|_2^2 + \frac{\theta^2}{\omega(1-\omega)}\sum_{i=0}^{k-1}\|\frac{1}{2}\boldsymbol{D}^\dagger e\boldsymbol{\xi}^i - \frac{1}{2}\boldsymbol{B}^\dagger\boldsymbol{\eta}^i\|_2^2,$$

with $\boldsymbol{A}$, $\boldsymbol{B}$, $\boldsymbol{D}$ and $\boldsymbol{E}$ given in (5.50) and e defined in (5.79). The residuals $\boldsymbol{\xi}^k = [\xi_l^k]_{l=0,\ldots,N}$ and $\boldsymbol{\eta}^k = [\eta_l^k]_{l=1,\ldots,N-1}$ are defined in (6.1) for $l = 2, \ldots, N-2$ and in (6.7). We find

$$\frac{1}{2}\boldsymbol{A}^\dagger e\boldsymbol{\xi}^k - \frac{1}{2}\boldsymbol{E}^\dagger\boldsymbol{\eta}^k = \frac{1}{2}\begin{bmatrix} 2\omega\xi_1^k + 2\omega\eta_1^k \\ 2\xi_0^k + \xi_2^k + \eta_2^k \\ [\xi_{l-1}^k + \xi_{l+1}^k - \eta_{l-1}^k + \eta_{l+1}^k]_{l=2,\ldots,N-2} \\ \xi_{N-2}^k + 2\xi_N^k - \eta_{N-2}^k \\ 2\omega\xi_{N-1}^k - 2\omega\eta_{N-1}^k \end{bmatrix},$$

$$\frac{1}{2}\boldsymbol{D}^\dagger e\boldsymbol{\xi}^k - \frac{1}{2}\boldsymbol{B}^\dagger\boldsymbol{\eta}^k = \frac{1}{2}\begin{bmatrix} 2\xi_0^k - \xi_2^k - \eta_2^k \\ [\xi_{l-1}^k - \xi_{l+1}^k - \eta_{l-1}^k - \eta_{l+1}^k]_{l=2,\ldots,N-2} \\ \xi_{N-2}^k - 2\xi_N^k - \eta_{N-2}^k \end{bmatrix}.$$

With (6.2) and (6.8) we find

$$\xi_0^k = \mathcal{O}(h^3), \quad \xi_N^k = \mathcal{O}(h^3),$$
$$\xi_l^k = \mathcal{O}(h^4), \quad \eta_l^k = \mathcal{O}(h^3) \quad \text{for } l = 1, \ldots, N-1.$$

Furthermore, we obtain

$$2\omega\xi_1^k + 2\omega\eta_1^k = \mathcal{O}(h^3), \quad 2\xi_0^k + \xi_2^k + \eta_2^k = \mathcal{O}(h^4),$$
$$\xi_{l-1}^k + \xi_{l+1}^k - \eta_{l-1}^k + \eta_{l+1}^k = \mathcal{O}(h^4) \quad \text{for } l = 2, \ldots, N-2,$$
$$\xi_{N-2}^k + 2\xi_N^k - \eta_{N-2}^k = \mathcal{O}(h^4), \quad 2\omega\xi_{N-1}^k - 2\omega\eta_{N-1}^k = \mathcal{O}(h^3)$$

and

$$2\xi_0^k - \xi_2^k - \eta_2^k = \mathcal{O}(h^3),$$
$$\xi_{l-1}^k - \xi_{l+1}^k - \eta_{l-1}^k - \eta_{l+1}^k = \mathcal{O}(h^3) \quad \text{for } l = 2, \ldots, N-2,$$
$$\xi_{N-2}^k - 2\xi_N^k - \eta_{N-2}^k = \mathcal{O}(h^3).$$

Since $2\omega\xi_1^k + 2\omega\eta_1^k$ and $2\omega\xi_{N-1}^k - 2\omega\eta_{N-1}^k$ do not show the right behavior, we have to reduce the residuals η_1^k and η_{N-1}^k by solving

$$-(\boldsymbol{K} - \boldsymbol{Id}_{2N})\boldsymbol{S}_L = [0; e_1],$$
$$-(\boldsymbol{K} - \boldsymbol{Id}_{2N})\boldsymbol{S}_R = [0; e_{N-1}].$$

The solutions are obtained by

$$\boldsymbol{S}_L = \frac{1}{2(1-\omega)}\left[\begin{array}{l} -\theta\left[\dfrac{1+2\omega(N+1-i)}{1+N\omega}\right]_{i=1,\ldots,N+1} + \theta(2e_1+e_2) \\[2ex] -\theta\left[\dfrac{1}{1+N\omega}\right]_{i=1,\ldots,N-1} + e_1 \end{array}\right],$$

$$\boldsymbol{S}_R = \frac{1}{2(1-\omega)}\left[\begin{array}{l} \theta\left[\dfrac{1+2\omega(i-1)}{1+N\omega}\right]_{i=1,\ldots,N+1} - \theta(e_N+2e_{N+1}) \\[2ex] -\theta\left[\dfrac{1}{1+N\omega}\right]_{i=1,\ldots,N-1} + e_{N-1} \end{array}\right].$$

We redefine the errors

$$\widetilde{\boldsymbol{E}}^k := \begin{bmatrix} \widetilde{\boldsymbol{E}}_R^k \\[1ex] \widetilde{\boldsymbol{E}}_J^k \end{bmatrix} := \begin{bmatrix} [r_l^k - R_l^k]_l \\[1ex] [j_l^k - J_l^k]_l \end{bmatrix} - 2\omega\eta_1^k \boldsymbol{S}_L - 2\omega\eta_{N-1}^k \boldsymbol{S}_R.$$

Then the error estimate holds for $\widetilde{\boldsymbol{E}}_R^k$ and $\widetilde{\boldsymbol{E}}_J^k$ with residuals of expected orders and with the additional higher order source term $-2\omega(\eta_1^{k+1} - \eta_1^k)\boldsymbol{S}_L - 2\omega(\eta_{N-1}^{k+1} - \eta_{N-1}^k)\boldsymbol{S}_R$. The first sum in the estimate is of order $\mathcal{O}(h^5)$, the last three sums are of order $\mathcal{O}(h^4)$. The initial errors vanish due to the choice of the initial data. Since the perturbations of the errors are of order $\mathcal{O}(h^3)$, the estimate in Theorem 6.4 remains true for $\boldsymbol{E}_R$ and $\boldsymbol{E}_J$. $\qquad\square$

Theorem 6.5. *Let $\boldsymbol{P}^k := [\boldsymbol{R}^k; \boldsymbol{J}^k]$, $k = 0, \ldots, M$, be the discrete solutions belonging to the CCFD lattice Boltzmann equations (5.2) with relaxation parameter ω and with density boundary conditions (5.12) and initial values $R_{l-1/2}^0 = r_0(x_{l-1/2})$ and $J_{l-1/2}^0 = -h\partial_x r_0(x_{l-1/2})/(2\omega)$ for $l = 1, \ldots, N$. The right hand sides are evaluated at the origin of the characteristics, that is, we use $F_{l-1/2,+}^k = F_{l-1/2,-}^k = f(t_k, x_{l-1/2})$. Then we get the error estimate*

$$\|\boldsymbol{E}_R^M\|_2^2 + \|\boldsymbol{E}_J^M\|_2^2 + 2\omega(1-\omega)\frac{1}{\tau}\sum_{i=0}^{M-1}\tau\|\boldsymbol{E}_J^i\|_2^2 \leq Kh^4,$$

with the discrete norms defined in (5.75). The constant K only depends on the relaxation parameter ω, the end time t_M, higher order derivatives of the solution of the heat equation and hence on the viscosity.

PROOF. According to Theorem 5.1 and Corollary 5.6, the errors obey the estimate

$$\|\boldsymbol{E}_R^k\|_2^2 + \|\boldsymbol{E}_J^k\|_2^2 + 2\omega(1-\omega)\frac{1}{\tau}\sum_{i=0}^{k-1}\tau\|\boldsymbol{E}_J^i\|_2^2$$

$$\leq 2\|\boldsymbol{E}_R^0\|_2^2 + 2\|\boldsymbol{E}_J^0\|_2^2 + 2\sum_{i=0}^{k-1}\|\boldsymbol{\xi}^i\|_2^2 + 2\sum_{i=0}^{k-1}\|\boldsymbol{\eta}^i\|_2^2$$

$$+ \frac{5t_k}{\tau}\sum_{i=0}^{k-1}\|\tfrac{1}{2}\boldsymbol{A}^\dagger\boldsymbol{\xi}^i - \tfrac{1}{2}\boldsymbol{E}^\dagger\boldsymbol{\eta}^i\|_2^2 + \frac{\theta^2}{\omega(1-\omega)}\sum_{i=0}^{k-1}\|\tfrac{1}{2}\boldsymbol{D}^\dagger\boldsymbol{\xi}^i - \tfrac{1}{2}\boldsymbol{B}^\dagger\boldsymbol{\eta}^i\|_2^2,$$

with $\boldsymbol{A}$, $\boldsymbol{B}$, $\boldsymbol{D}$ and $\boldsymbol{E}$ given in (5.56). The residuals $\boldsymbol{\xi}^k = [\xi^k_{l-1/2}]_{l=1,\ldots,N}$ and $\boldsymbol{\eta}^k = [\eta^k_{l-1/2}]_{l=1,\ldots,N}$ are defined in (6.1) for $l = 3/2,\ldots,N-3/2$ and in (6.9). We find

$$\frac{1}{2}\boldsymbol{A}^\dagger\boldsymbol{\xi}^k - \frac{1}{2}\boldsymbol{E}^\dagger\boldsymbol{\eta}^k = \frac{1}{2}\begin{bmatrix} -\xi^k_{1/2} + \xi^k_{3/2} - \eta^k_{1/2} + \eta^k_{3/2} \\ \left[\xi^k_{l-3/2} + \xi^k_{l+1/2} - \eta^k_{l-3/2} + \eta^k_{l+1/2}\right]_{l=2,\ldots,N-1} \\ \xi^k_{N-3/2} - \xi^k_{N-1/2} - \eta^k_{N-3/2} + \eta^k_{N-1/2} \end{bmatrix},$$

$$\frac{1}{2}\boldsymbol{D}^\dagger\boldsymbol{\xi}^k - \frac{1}{2}\boldsymbol{B}^\dagger\boldsymbol{\eta}^k = \frac{1}{2}\begin{bmatrix} -\xi^k_{1/2} - \xi^k_{3/2} - \eta^k_{1/2} - \eta^k_{3/2} \\ \left[\xi^k_{l-3/2} - \xi^k_{l+1/2} - \eta^k_{l-3/2} - \eta^k_{l+1/2}\right]_{l=2,\ldots,N-1} \\ \xi^k_{N-3/2} + \xi^k_{N-1/2} - \eta^k_{N-3/2} - \eta^k_{N-1/2} \end{bmatrix}.$$

With (6.2) and (6.10) we find

$$\xi^k_{1/2} = \mathcal{O}(h^2), \quad \eta^k_{1/2} = \mathcal{O}(h^2),$$
$$\xi^k_{l-1/2} = \mathcal{O}(h^4), \quad \eta^k_{l-1/2} = \mathcal{O}(h^3) \quad \text{for } l = 2,\ldots,N-1,$$
$$\xi^k_{N-1/2} = \mathcal{O}(h^2), \quad \eta^k_{N-1/2} = \mathcal{O}(h^2).$$

Furthermore, we gain

$$-\xi^k_{1/2} + \xi^k_{3/2} - \eta^k_{1/2} + \eta^k_{3/2} = \mathcal{O}(h^2),$$
$$\xi^k_{1/2} + \xi^k_{5/2} - \eta^k_{1/2} + \eta^k_{5/2} = \mathcal{O}(h^2),$$
$$\xi^k_{l-3/2} + \xi^k_{l+1/2} - \eta^k_{l-3/2} + \eta^k_{l+1/2} = \mathcal{O}(h^4) \quad \text{for } l = 3,\ldots,N-2,$$
$$\xi^k_{N-5/2} + \xi^k_{N-1/2} - \eta^k_{N-5/2} + \eta^k_{N-1/2} = \mathcal{O}(h^2),$$
$$\xi^k_{N-3/2} - \xi^k_{N-1/2} - \eta^k_{N-3/2} + \eta^k_{N-1/2} = \mathcal{O}(h^2),$$

and

$$-\xi^k_{1/2} - \xi^k_{3/2} - \eta^k_{1/2} - \eta^k_{3/2} = \mathcal{O}(h^2),$$
$$\xi^k_{1/2} - \xi^k_{5/2} - \eta^k_{1/2} - \eta^k_{5/2} = \mathcal{O}(h^2),$$
$$\xi^k_{l-3/2} - \xi^k_{l+1/2} - \eta^k_{l-3/2} - \eta^k_{l+1/2} = \mathcal{O}(h^3) \quad \text{for } l = 3,\ldots,N-2,$$
$$\xi^k_{N-5/2} - \xi^k_{N-1/2} - \eta^k_{N-5/2} - \eta^k_{N-1/2} = \mathcal{O}(h^2),$$
$$\xi^k_{N-3/2} + \xi^k_{N-1/2} - \eta^k_{N-3/2} - \eta^k_{N-1/2} = \mathcal{O}(h^2).$$

Since the residuals at the boundaries do not show the right behavior, we have to reduce the leading orders of $\xi^k_{1/2}$, $\eta^k_{1/2}$, $\xi^k_{N-1/2}$ and $\eta^k_{N-1/2}$ by solving

$$-(\boldsymbol{K} - \boldsymbol{Id}_{2N})\boldsymbol{S}_L = [\boldsymbol{e}_1; \boldsymbol{e}_1],$$
$$-(\boldsymbol{K} - \boldsymbol{Id}_{2N})\boldsymbol{S}_R = [\boldsymbol{e}_N; -\boldsymbol{e}_N].$$

The solutions are obtained in the form

$$\boldsymbol{S}_L = \left[\left[\frac{2(N-i)+1}{2N}\right]_{i=1,\ldots,N} ; \frac{1}{2\omega}\left[\frac{1}{N}\right]_{i=1,\ldots,N}\right],$$

$$\boldsymbol{S}_R = \left[\left[\frac{2i-1}{2N}\right]_{i=1,\ldots,N} ; -\frac{1}{2\omega}\left[\frac{1}{N}\right]_{i=1,\ldots,N}\right].$$

We redefine the errors

$$\widetilde{\boldsymbol{E}}^k := \begin{bmatrix} \widetilde{\boldsymbol{E}}^k_R \\ \widetilde{\boldsymbol{E}}^k_J \end{bmatrix} := \begin{bmatrix} [r^k_l - R^k_l]_l \\ [j^k_l - J^k_l]_l \end{bmatrix} - h^2 s^k_L \boldsymbol{S}_L - h^2 s^k_R \boldsymbol{S}_R,$$

with s_L^k and s_R^k defined in (6.11). Then the error estimate holds for $\widetilde{\boldsymbol{E}}_R^k$ and $\widetilde{\boldsymbol{E}}_J^k$ with residuals of expected orders and with additional higher order source terms. Since the perturbations of the errors are of order $\mathcal{O}(h^2)$ and the $\boldsymbol{J}$-component of the corrections is of order $\mathcal{O}(h)$, the estimate in Theorem 6.5 remains true for $\boldsymbol{E}_R$ and $\boldsymbol{E}_J$. $\qquad\square$

Theorem 6.6. *Let $\boldsymbol{P}^k := [\boldsymbol{R}^k; \boldsymbol{J}^k]$, $k = 0,\ldots,M$, be the discrete solutions belonging to the CCFD lattice Boltzmann equations (5.2) with relaxation parameter ω and with flux boundary conditions (5.13) and initial values $R_{l-1/2}^0 = r_0(x_{l-1/2})$ and $J_{l-1/2}^0 = -h\partial_x r_0(x_{l-1/2})/(2\omega)$ for $l = 1,\ldots,N$. The right hand sides are evaluated at the origin of the characteristics, that is, we use $F_{l-1/2,+}^k = F_{l-1/2,-}^k = f(t_k, x_{l-1/2})$. Then we get the error estimate*

$$\|\boldsymbol{E}_R^M\|_2^2 + \|\boldsymbol{E}_J^M\|_2^2 + 2\omega(1-\omega)\frac{1}{\tau}\sum_{i=0}^{M-1}\tau\|\boldsymbol{E}_J^i\|_2^2 \le Kh^4,$$

with the discrete norms defined in (5.75). The constant K only depends on the relaxation parameter ω, the end time t_M, higher order derivatives of the solution of the heat equation and hence on the viscosity.

PROOF. According to Theorem 5.1 and Corollary 5.7, the errors obey the estimate

$$\|\boldsymbol{E}_R^k\|_2^2 + \|\boldsymbol{E}_J^k\|_2^2 + 2\omega(1-\omega)\frac{1}{\tau}\sum_{i=0}^{k-1}\tau\|\boldsymbol{E}_J^i\|_2^2$$

$$\le 2\|\boldsymbol{E}_R^0\|_2^2 + 2\|\boldsymbol{E}_J^0\|_2^2 + 2\sum_{i=0}^{k-1}\|\boldsymbol{\xi}^i\|_2^2 + 2\sum_{i=0}^{k-1}\|\boldsymbol{\eta}^i\|_2^2$$

$$+ \frac{5t_k}{\tau}\sum_{i=0}^{k-1}\left\|\frac{1}{2}\boldsymbol{B}^\dagger\boldsymbol{\xi}^i - \frac{1}{2}\boldsymbol{D}^\dagger\boldsymbol{\eta}^i\right\|_2^2 + \frac{\theta^2}{\omega(1-\omega)}\sum_{i=0}^{k-1}\left\|\frac{1}{2}\boldsymbol{E}^\dagger\boldsymbol{\xi}^i - \frac{1}{2}\boldsymbol{A}^\dagger\boldsymbol{\eta}^i\right\|_2^2,$$

with $\boldsymbol{A}$, $\boldsymbol{B}$, $\boldsymbol{D}$ and $\boldsymbol{E}$ given in (5.56). The residuals $\boldsymbol{\xi}^k = [\xi_{l-1/2}^k]_{l=1,\ldots,N}$ and $\boldsymbol{\eta}^k = [\eta_{l-1/2}^k]_{l=1,\ldots,N}$ are defined in (6.1) for $l = 3/2,\ldots,N-3/2$ and in (6.12). We find

$$\frac{1}{2}\boldsymbol{B}^\dagger\boldsymbol{\xi}^k - \frac{1}{2}\boldsymbol{D}^\dagger\boldsymbol{\eta}^k = \frac{1}{2}\begin{bmatrix} \xi_{1/2}^k + \xi_{3/2}^k + \eta_{1/2}^k + \eta_{3/2}^k \\ \left[\xi_{l-3/2}^k + \xi_{l+1/2}^k - \eta_{l-3/2}^k + \eta_{l+1/2}^k\right]_{l=2,\ldots,N-1} \\ \xi_{N-3/2}^k + \xi_{N-1/2}^k - \eta_{N-3/2}^k - \eta_{N-1/2}^k \end{bmatrix},$$

$$\frac{1}{2}\boldsymbol{E}^\dagger\boldsymbol{\xi}^k - \frac{1}{2}\boldsymbol{A}^\dagger\boldsymbol{\eta}^k = \frac{1}{2}\begin{bmatrix} \xi_{1/2}^k - \xi_{3/2}^k + \eta_{1/2}^k - \eta_{3/2}^k \\ \left[\xi_{l-3/2}^k - \xi_{l+1/2}^k - \eta_{l-3/2}^k - \eta_{l+1/2}^k\right]_{l=2,\ldots,N-1} \\ \xi_{N-3/2}^k - \xi_{N-1/2}^k - \eta_{N-3/2}^k + \eta_{N-1/2}^k \end{bmatrix}.$$

With (6.2) and (6.13) we find

$$\xi_{1/2}^k = \mathcal{O}(h^3), \quad \eta_{1/2}^k = \mathcal{O}(h^3),$$
$$\xi_{l-1/2}^k = \mathcal{O}(h^4), \quad \eta_{l-1/2}^k = \mathcal{O}(h^3) \quad \text{for } l = 2,\ldots,N-1,$$
$$\xi_{N-1/2}^k = \mathcal{O}(h^3), \quad \eta_{N-1/2}^k = \mathcal{O}(h^3).$$

Furthermore, we gain

$$\xi^k_{1/2} + \xi^k_{3/2} + \eta^k_{1/2} + \eta^k_{3/2} = \mathcal{O}(h^3),$$

$$\xi^k_{1/2} + \xi^k_{5/2} - \eta^k_{1/2} + \eta^k_{5/2} = \mathcal{O}(h^3),$$

$$\xi^k_{l-3/2} + \xi^k_{l+1/2} - \eta^k_{l-3/2} + \eta^k_{l+1/2} = \mathcal{O}(h^4) \quad \text{for } l = 3, \dots, N-2,$$

$$\xi^k_{N-5/2} + \xi^k_{N-1/2} - \eta^k_{N-5/2} + \eta^k_{N-1/2} = \mathcal{O}(h^3),$$

$$\xi^k_{N-3/2} + \xi^k_{N-1/2} - \eta^k_{N-3/2} - \eta^k_{N-1/2} = \mathcal{O}(h^3),$$

and

$$\xi^k_{1/2} - \xi^k_{3/2} + \eta^k_{1/2} - \eta^k_{3/2} = \mathcal{O}(h^3),$$

$$\xi^k_{1/2} - \xi^k_{5/2} - \eta^k_{1/2} - \eta^k_{5/2} = \mathcal{O}(h^3),$$

$$\xi^k_{l-3/2} - \xi^k_{l+1/2} - \eta^k_{l-3/2} - \eta^k_{l+1/2} = \mathcal{O}(h^3) \quad \text{for } l = 3, \dots, N-2,$$

$$\xi^k_{N-5/2} - \xi^k_{N-1/2} - \eta^k_{N-5/2} - \eta^k_{N-1/2} = \mathcal{O}(h^3),$$

$$\xi^k_{N-3/2} - \xi^k_{N-1/2} - \eta^k_{N-3/2} + \eta^k_{N-1/2} = \mathcal{O}(h^3).$$

Since the residuals at the boundaries do not show the right behavior, we have to reduce the leading orders of $\xi^k_{1/2}$, $\eta^k_{1/2}$, $\xi^k_{N-1/2}$ and $\eta^k_{N-1/2}$ by solving

$$-(\boldsymbol{K} - \boldsymbol{Id}_{2N})\boldsymbol{S}_L = [\boldsymbol{e}_1; \boldsymbol{e}_1],$$

$$-(\boldsymbol{K} - \boldsymbol{Id}_{2N})\boldsymbol{S}_R = [-\boldsymbol{e}_N; \boldsymbol{e}_N],$$

$$-(\boldsymbol{K} - \boldsymbol{Id}_{2N})\boldsymbol{T}_L = [\boldsymbol{0}; \boldsymbol{e}_1],$$

$$-(\boldsymbol{K} - \boldsymbol{Id}_{2N})\boldsymbol{T}_R = [\boldsymbol{0}; \boldsymbol{e}_N].$$

The solutions are obtained by

$$\boldsymbol{S}_L = \frac{1}{1-\omega}\left[\omega\left[\frac{(2(N-i)+1)^2}{4N}\right]_{i=1,\dots,N} ; \left[\frac{2(N-i)+1}{2N}\right]_{i=1,\dots,N}\right],$$

$$\boldsymbol{S}_R = \frac{1}{1-\omega}\left[-\omega\left[\frac{(2i-1)^2}{4N}\right]_{i=1,\dots,N} ; \left[\frac{2i-1}{2N}\right]_{i=1,\dots,N}\right],$$

$$\boldsymbol{T}_L = \frac{1}{2(1-\omega)}\left[\theta\boldsymbol{e}_1; \boldsymbol{e}_1\right],$$

$$\boldsymbol{T}_R = \frac{1}{2(1-\omega)}\left[-\theta\boldsymbol{e}_N; \boldsymbol{e}_N\right].$$

We redefine the errors

$$\begin{bmatrix}\widetilde{\boldsymbol{E}}^k_R \\ \widetilde{\boldsymbol{E}}^k_J\end{bmatrix} := \begin{bmatrix}\boldsymbol{E}^k_R \\ \boldsymbol{E}^k_J\end{bmatrix} - h^3(s^k_L + t^k_L)\boldsymbol{S}_L - h^3(s^k_R + t^k_R)\boldsymbol{S}_R + 2h^3 t^k_L \boldsymbol{T}_L + 2h^3 t^k_R \boldsymbol{T}_R,$$

with s^k_L, s^k_R, t^k_L, t^k_R defined in (6.14). Then the error estimate holds for $\widetilde{\boldsymbol{E}}^k_R$ and $\widetilde{\boldsymbol{E}}^k_J$ with residuals of the expected orders and with additional higher order source terms. Since the perturbations of the error are of order $\mathcal{O}(h^3)$, the estimate in Theorem 6.6 remains true for $\boldsymbol{E}_R$ and $\boldsymbol{E}_J$, although the $\boldsymbol{R}$-component of the corrections is of order $\mathcal{O}(h^{-1})$. $\qquad\square$

The leading order of the error estimate is determined by three sums of the same order, each one considered by itself. But it might be that there are correlations between these sums and hence cancellations may appear. In the numerical experiments we shall see that there is a specific value for the relaxation parameter ω, for which we get improved convergence results. This fact can be described in the following way.

In the periodic case on FD grids we define

$$G := \frac{1}{4\omega} \begin{bmatrix} \theta A + 2\boldsymbol{Id}_N & \theta D \\ -D & -A + 2\boldsymbol{Id}_N \end{bmatrix} \in \mathbb{R}^{2N,2N},$$

with the matrices A and D given in (5.35) and $N = 2n$ or $N = 2n-1$. We consider the modified lattice Boltzmann matrix equation (5.32)

$$G\left(P^{k+1} - P^k\right) = G(K - \boldsymbol{Id}_{2N})P^k + \tau GG^k. \tag{6.15}$$

The matrix G has the same eigenvectors as K (see (5.89)). The corresponding eigenvalues of G are given by

$$\mu_0^+ := 1, \quad \mu_0^- := 0,$$

$$\mu_l^+ := \frac{1}{2\omega}\left(1 - \beta_l + V_l\right),$$
$$\mu_l^- := \frac{1}{2\omega}\left(1 - \beta_l - V_l\right), \qquad \text{for } l = 1,\ldots,n-1,$$

$$\mu_n^+ := \frac{1}{\omega}, \quad \mu_n^- := \frac{1-\omega}{\omega}, \quad \text{if } N = 2n,$$

with β_l and V_l given in (5.86). The interesting fact now is to find

$$G(K - \boldsymbol{Id}_{2N}) = \frac{1-\omega}{2\omega}\begin{bmatrix} A - 2\boldsymbol{Id}_N & \\ & A - 2\boldsymbol{Id}_N \end{bmatrix}.$$

This is the canonical matrix for the determination of the second order derivatives in combination with the numerical viscosity of the scheme. Hence, the equation (6.15) can be seen as a discretization of the heat equation with the mass matrix G. For $\omega = 1/2$, the R-component of the mass matrix G is the unit matrix. In order to fit the system (6.15) into a finite element setting, we have to find discrete basis function that render the mass matrix G and yield the central second order derivatives in the stiffness matrix. This problem is unsolved yet.

In the case of boundary conditions, analogous mass matrices are determined by the choice of the eigenvalues as above and with the same eigenvectors as the time evolution matrix K.

In the periodic case the residuals of (6.15) are defined by

$$\xi_l^k := \frac{1}{2}\left(r_{l+1}^{k+1} - r_{l+1}^k + r_{l-1}^{k+1} - r_{l-1}^k\right)$$
$$- \frac{1}{4\omega}\left(r_{l+1}^{k+1} - r_{l+1}^k - 2r_l^{k+1} + 2r_l^k + r_{l-1}^{k+1} - r_{l-1}^k\right)$$
$$+ \frac{2\omega - 1}{4\omega}\left(j_{l+1}^{k+1} - j_{l+1}^k - j_{l-1}^{k+1} + j_{l-1}^k\right) - \frac{1-\omega}{2\omega}\left(r_{l+1}^k - 2r_l^k + r_{l-1}^k\right)$$
$$- \frac{\tau}{4}\left(f_{l+2}^k + 2f_l^k + f_{l-2}^k\right) + \frac{\tau}{8\omega}\left(f_{l+2}^k - 2f_{l+1}^k + 2f_l^k - 2f_{l-1}^k + f_{l-2}^k\right)$$
$$+ \frac{2\omega - 1}{8\omega}\tau\left(f_{l+2}^k - 2f_l^k + f_{l-2}^k\right),$$

$$\eta_l^k := -\frac{1}{4\omega}\left(r_{l+1}^{k+1} - r_{l+1}^k - r_{l-1}^{k+1} + r_{l-1}^k\right)$$
$$- \frac{1}{4\omega}\left(j_{l+1}^{k+1} - j_{l+1}^k - 2j_l^{k+1} + 2j_l^k + j_{l-1}^{k+1} - j_{l-1}^k\right)$$
$$- \frac{1-\omega}{2\omega}\left(j_{l+1}^k - 2j_l^k + j_{l-1}^k\right) + \frac{\tau}{4\omega}\left(f_{l+1}^k - f_{l-1}^k\right).$$

If r is a solution to the heat equation, we find

$$\xi_l^k = h^4 \left((1 - \omega) \frac{2\omega^2 - 6\omega + 3}{24\omega^3} \partial_x^4 r_l^k + \frac{1 - 12\omega}{8\omega\gamma} \partial_x^2 f_l^k + \frac{1}{2} \partial_t f_l^k \right) + \mathcal{O}(h^6),$$

$$\eta_l^k = h^3 \left(\frac{1 - \omega}{8\omega^3} \partial_x^3 r_l^k + \frac{1}{4\omega^2\gamma} \partial_x f_l^k \right) + \mathcal{O}(h^5).$$

For the special choice $\omega = (3 - \sqrt{3})/2$, the coefficient of $\partial_x^4 r_l^k$ in ξ_l^k vanishes. For schemes without source terms, the residual becomes of order $\mathcal{O}(h^6)$. A pronounced improvement of the convergence can be seen in the numerical experiments in Section 7.2.

It is not clear how these results can be expressed in an error estimate. For low frequency solutions, the crucial eigenvalues μ_l^+ of $\boldsymbol{G}$ with $l \ll N/2$ are of order $\mathcal{O}(1)$, whereas we find $\mu_l^- = \mathcal{O}(h^2)$ for $l \ll N/2$. Hence, the norm of the pseudo-inverse of $\boldsymbol{G}$ is of order $\mathcal{O}(h^{-2})$.

CHAPTER 7

Numerical Results

In this chapter we want to confirm the analytical results of the previous chapters by performing numerical experiments. All computations are done by using $\mathrm{MATLAB}^{\circledR}$; see `http://www.mathworks.com`.

The investigations mainly face the question of the convergence of the lattice Boltzmann schemes. In Chapter 6, we proved second order convergence of the lattice Boltzmann solutions towards the density r, modeled by the heat equation, in the discrete $L^\infty(L^2)$-norm and second order convergence towards the flux $\partial_x r$ in the discrete $L^2(L^2)$-norms, both considered for a fixed end time T with respect to the grid size h. The numerical results imply that the second order convergence for the flux $\partial_x r$ also holds in the discrete $L^\infty(L^2)$-norm, whereas our theoretical results in Chapter 6 only ensure first order convergence in this norm.

7.1. Experimental Orders of Convergence

In a first step, we consider known solutions for the heat equation on bounded intervals. In our Example 1, this is the eigenfunction solution

$$r(t,x) := \exp(-4\nu\pi^2 t)\sin(2\pi x) \quad \text{in } (0,T] \times (0,1)$$

with the choice $\nu = 0.1$ for the viscosity. The solution up to the end time $T = 1.0$ is plotted in Figure 7.1. For the lattice Boltzmann solutions we examine the behavior of the errors on a sequence of grids for different kinds of boundary conditions. In each grid point, we gain information on the discrete solution and the exact solution.

Example 1

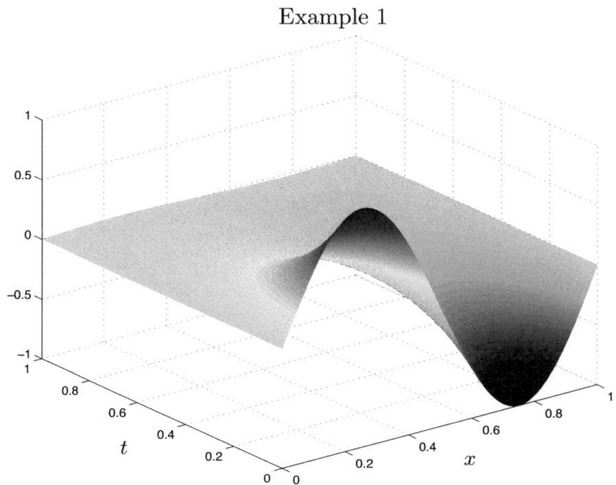

FIGURE 7.1. Solution $r(t,x) := \exp(-4\nu\pi^2 t)\sin(2\pi x)$ in the domain $(0,1] \times (0,1)$ with $\nu = 0.1$.

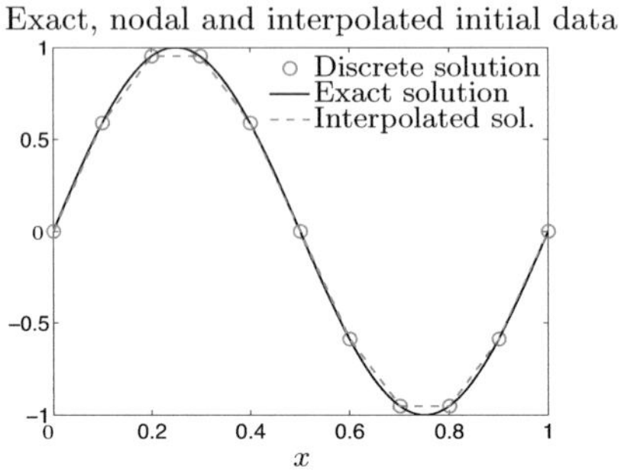

FIGURE 7.2. Nodal values, interpolated values and exact values of the initial data $r_0(x) := \sin(2\pi x)$ for Example 1 with $N = 10$.

Hence, we can determine the *nodal error*. By a linear interpolation we construct a globally continuous numerical solution. Hereby, we can determine a continuous error function, which is denoted by the *interpolated error* in the sequel. Linear interpolation turns out to be sufficient to preserve the nodal convergence properties. In Section 8.5, we see that the interpolated errors show a better performance on coupled grids than the nodal errors. Superconvergence properties are lost by the linear interpolation only in the case of a specific choice for the parameters, where the nodal errors show a convergence of fourth order. The corresponding results can be found in Section 7.2.

In Figure 7.2 the exact, the nodal and the interpolated initial data for the r-equation is plotted for the case $N = 10$. By performing several time steps, the errors increase. The nodal and the interpolated errors at time $t = 0.2$ are plotted in Figure 7.3 for a grid with $N = 50$. The nodal error is an envelope to the interpolated error. In each grid cell, there is a zero of the interpolated error.

From the nodal errors we can compute the discrete L^2-errors introduced in Chapter 5.9. The choice of the discrete L^2-norms depends on the grid and the employed boundary condition. For periodic boundary conditions on VC grids we use for example

$$E_N^k(r) := \left(\sum_{l=0}^{N-1} h|r(t_k, x_l) - R_l^k|^2 \right)^{1/2},$$

$$E_N^k(j) := \left(\sum_{l=0}^{N-1} h|j(t_k, x_l) - J_l^k|^2 \right)^{1/2}.$$

The linear interpolants of the vector valued discrete solutions $\boldsymbol{R}^k$ and $\boldsymbol{J}^k$ are denoted by $I(\boldsymbol{R}^k)$ and $I(\boldsymbol{J}^k)$. For the indices $l = 0, \ldots, N$, we have $I(\boldsymbol{R})_l^k := I(\boldsymbol{R}^k)(x_l) = R_l^k$ and $I(\boldsymbol{J})_l^k := I(\boldsymbol{J}^k)(x_l) = J_l^k$. By involving the linear interpolants $I(\boldsymbol{R}^k)$, $I(\boldsymbol{J}^k)$ and a finer grid $\{z_l\}_{l=0,\ldots,NK}$, where we introduce $K - 1$ additional equidistant points in each grid cell with $z_{lN} = x_l$, $l = 0, \ldots, N$, we can compute an approximation of the L^2-errors by the application of the trapezoidal rule for the integral expressions. For the indices $l = 0, \ldots, NK$, we define $I(\boldsymbol{R})_l^k := I(\boldsymbol{R}^k)(z_l)$

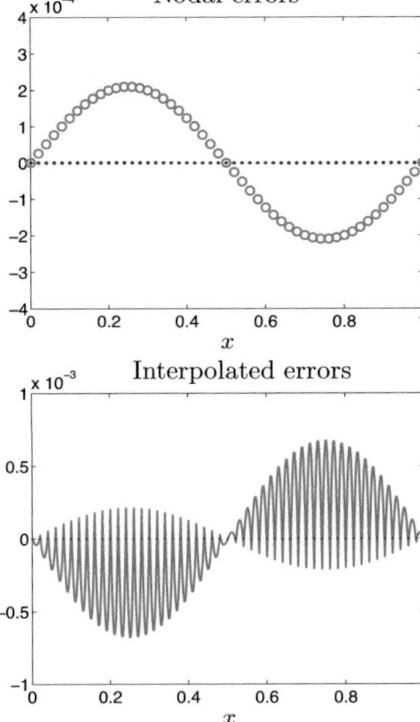

FIGURE 7.3. Nodal errors and interpolated errors of the lattice Boltzmann solution for the density at time $t = 0.2$ for Example 1 with $N = 50$.

and $I(\boldsymbol{J})_l^k := I(\boldsymbol{J}^k)(z_l)$. We end up with the errors

$$E^k(r) := \left(\sum_{l=0}^{NK-1} \frac{h}{K} |r(t_k, z_l) - I(\boldsymbol{R})_l^k|^2 \right)^{1/2},$$

$$E^k(j) := \left(\sum_{l=0}^{NK-1} \frac{h}{K} |j(t_k, z_l) - I(\boldsymbol{J})_l^k|^2 \right)^{1/2}.$$

We call these errors the *approximated L^2-errors*. In our applications, typical values are $K = 10, 20, 50, 100$, depending on the size of N.

In order to assess the performance of the lattice Boltzmann schemes, we examine the *experimental order of convergence* (EOC). The underlying assumption is the validity of error laws of the form

$$E_N^k(r) = K_r^k N^{-p}, \quad E_N^k(j) = K_j^k N^{-q}$$

for positive values p and q and N-independent constants K_r^k and K_j^k. The error constants (as well as the errors) depend on the time t_k. In logarithmic terms the error laws can be rewritten as linear polynomials with slopes $-p$ and $-q$ of the form

$$\log(E_N^k(r)) = \log(K_r^k) - p \log(N),$$

$$\log(E_N^k(j)) = \log(K_j^k) - q \log(N).$$

We determine the errors $E_N^M(r)$ and $E_N^M(j)$ for a fixed end time t_M on a sequence of grids. Then we compute the EOCs p and q and the error constants K_r^M and

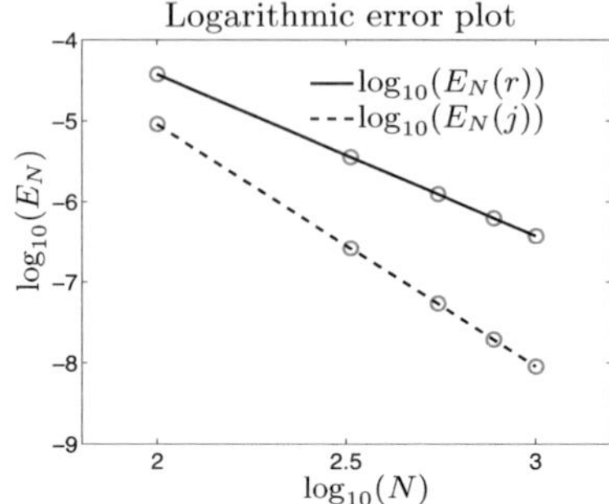

FIGURE 7.4. Logarithmic plot of the errors.

K_j^M by least square polynomial fitting. With the aid of the error laws we are able to estimate the fitted errors for all values of N in some limited range.

For Example 1 with end time $t_M = 0.2$, the logarithmic error plot for the lattice Boltzmann solutions on VC grids with density boundary conditions on a grid sequence with $N = 100, 325, 550, 775, 1000$ is shown in Figure 7.4. The fitted linear polynomials for the density r and the h-scaled flux j meet the fitting points quite closely. The slopes -2 for the density and -3 for the h-scaled flux are obvious. The upper time index k for the errors and the error constants are disregarded in the following. The missing index indicates the measurement at the given end time.

We investigate the experimental orders of convergence for the FD lattice Boltzmann schemes (5.1) on VC grids and (5.2) on CC grids depending on different boundary conditions. For all presented boundary conditions we proved L^2-stability for the solutions and second order convergence for the density. Initial conditions are chosen by $R_l^0 = r_0(x_l)$ and $J^0 = -h\partial_x r_0(x_l)/(2\omega)$, where the range of the indices depends on the type of the grid and the boundary conditions. The initial and boundary data are taken from the knowledge of the exact solution. For Example 1, there is no source term in the corresponding heat equation. The number of grid points is chosen by $N = 60, 145, 230, 315$ and 400. For the relaxation parameter $\omega = 0.7$ the detected values for the EOCs with respect to the discrete L^2-norms are displayed in Table 7.1. The EOCs are independent of the choice of the presented boundary conditions. We find second order convergence for the density r and third order convergence for the h-scaled flux j. This translates to second order convergence for the flux $\partial_x r$. The error constants K_r for the density errors are of moderate size. For all types of flux boundary conditions the errors and error constants are slightly increased. The best choice for flux boundary conditions is (5.13) on CC grids with $\delta = 0$. The errors presented in the table are the polynomial fitted errors on the finest grid with $N = 400$. The errors for the flux are smaller by one magnitude than the errors for the density because of the h-scaling of the flux j. The corresponding error constants are larger by factors of size 5 to 25 than those of the r-equation.

In Table 7.2 we consider the results of the FV lattice Boltzmann schemes (5.3) and (5.4) depending on the presented boundary conditions. All boundary conditions are proved to render L^2-stable solutions.

The results for the experimental orders of convergence are the same as for the FD schemes. The error constants for the density are larger by a factor of up to 10, whereas the error constants for the flux are smaller except for the last presented flux boundary conditions. Due to the changed space-time coupling in the FV context,

Boundary conditions	Grid	Density			Flux		
		EOC	K_r	$E_{400}(r)$	EOC	K_j	$E_{400}(j)$
Periodic (5.5)	VC	2.00	3.67e-1	2.34e-6	3.00	8.98e0	1.41e-7
Density (5.6)	VC	2.00	3.67e-1	2.34e-6	3.00	8.98e0	1.41e-7
Flux (5.8)	VC	2.00	1.63e0	1.01e-5	3.00	7.20e0	1.13e-7
Inflow (5.10)	VC	2.08	5.95e-1	2.37e-6	2.99	8.63e0	1.41e-7
Density (5.12), $\delta = 0$	CC	2.00	3.67e-1	2.34e-6	3.00	8.98e0	1.41e-7
Density (5.12), $\delta = 1$	CC	2.00	3.67e-1	2.34e-6	3.00	8.98e0	1.41e-7
Flux (5.13), $\delta = 0$	CC	2.00	7.33e-1	4.59e-6	3.00	1.15e1	1.80e-7
Flux (5.13), $\delta = 1$	CC	2.00	1.45e0	9.08e-6	3.00	7.14e0	1.12e-7

TABLE 7.1. Example 1: Experimental orders of convergence, error constants and fitted errors with respect to the discrete L^2-norms for the FD lattice Boltzmann schemes (5.1) and (5.2) on a sequence of grids with $N = 60, 145, 230, 315, 400$ with data $T = 0.2$, $\nu = 0.1$ and $\omega = 0.7$, depending on the boundary conditions.

Boundary conditions	Grid	Density			Flux		
		EOC	K_r	$E_{400}(r)$	EOC	K_j	$E_{400}(j)$
Periodic (5.5)	VC	2.00	3.23e0	2.00e-5	3.02	1.05e0	1.42e-8
Density (5.6)	VC	2.00	3.23e0	2.00e-5	3.02	1.05e0	1.42e-8
Flux (5.8)	VC	2.00	2.67e0	1.65e-5	3.01	4.17e0	6.04e-8
Inflow (5.10)	VC	1.99	3.09e0	2.00e-5	3.05	1.16e0	1.32e-8
Density (5.12), $\delta = 0$	CC	2.00	3.23e0	2.00e-5	3.02	1.05e0	1.42e-8
Density (5.12), $\delta = 1$	CC	2.00	3.23e0	2.00e-5	3.02	1.05e0	1.42e-8
Flux (5.14), $\delta = 0$	CC	2.00	3.85e0	2.39e-5	3.00	6.81e0	1.08e-7
Flux (5.14), $\delta = 1$	CC	2.00	4.89e0	3.05e-5	3.00	2.44e1	3.81e-7

TABLE 7.2. Example 1: Experimental orders of convergence, error constants and fitted errors with respect to the discrete L^2-norms for the FV lattice Boltzmann schemes (5.3) and (5.4) on a sequence of grids with $N = 60, 145, 230, 315, 400$ with data $T = 0.2$, $\nu = 0.1$ and $\omega = 0.7$, depending on the boundary conditions.

time steps are now larger by a factor $1/(1 - \omega) = 10/3$. This fact finds expression in larger errors for the density. However, the errors for the flux are smaller than those of the FD lattice Boltzmann solutions.

We remark that although all variables in these schemes and their derivations were interpreted in a finite volume context, all data and the errors are evaluated in a pointwise sense and not in the mean integral sense.

In Table 7.3 we examine the experimental orders of convergence with respect to the approximated L^2-errors for the FD lattice Boltzmann schemes (5.1) and (5.2). The results for the EOCs are the same as for the discrete L^2-errors. The errors and the error constants are more or less of the same magnitude. For the determination of the approximated L^2-errors we use $K = 20$ additional grid points per cell.

Qualitatively, the same results are obtained when considering the approximated L^2-errors for the FV lattice Boltzmann schemes (5.3) and (5.4); see Table 7.4.

In Example 1, we are in the situation of vanishing density boundary conditions. In order to examine nonhomogeneous density boundary values, we modify

Boundary conditions	Grid	Density			Flux		
		EOC	K_r	$E_{400}(r)$	EOC	K_j	$E_{400}(j)$
Periodic (5.5)	VC	2.00	8.34e-1	5.22e-6	3.00	4.86e0	7.65e-8
Density (5.6)	VC	2.00	8.34e-1	5.22e-6	3.00	4.86e0	7.65e-8
Flux (5.8)	VC	2.00	1.31e0	8.17e-6	3.00	5.60e0	8.78e-8
Inflow (5.10)	VC	1.97	7.13e-1	5.19e-6	2.99	4.56e0	7.64e-8
Density (5.12), $\delta = 0$	CC	2.00	8.36e-1	5.22e-6	3.09	8.37e0	7.82e-8
Density (5.12), $\delta = 1$	CC	2.00	8.36e-1	5.22e-6	3.09	8.37e0	7.82e-8
Flux (5.13), $\delta = 0$	CC	2.00	1.42e0	8.98e-6	3.05	1.04e1	1.17e-7
Flux (5.13), $\delta = 1$	CC	2.00	1.20e0	7.31e-6	3.03	6.28e0	8.04e-8

TABLE 7.3. Example 1: Experimental orders of convergence, error constants and fitted errors with respect to the approximated L^2-norms for the FD lattice Boltzmann schemes (5.1) and (5.2) on a sequence of grids with $N = 60, 145, 230, 315, 400$ with data $T = 0.2$, $\nu = 0.1$ and $\omega = 0.7$, depending on the boundary conditions.

Boundary conditions	Grid	Density			Flux		
		EOC	K_r	$E_{400}(r)$	EOC	K_j	$E_{400}(j)$
Periodic (5.5)	VC	2.00	4.29e0	2.67e-5	3.00	6.12e0	9.42e-8
Density (5.6)	VC	2.00	4.29e0	2.67e-5	3.00	6.12e0	9.42e-8
Flux (5.8)	VC	2.00	3.68e0	2.30e-5	3.00	7.85e0	1.21e-7
Inflow (5.10)	VC	2.00	4.15e0	2.67e-5	2.99	5.58e0	9.38e-8
Density (5.12), $\delta = 0$	CC	2.00	4.29e0	2.67e-5	3.02	6.67e0	9.44e-8
Density (5.12), $\delta = 1$	CC	2.00	4.29e0	2.67e-5	3.02	6.67e0	9.44e-8
Flux (5.14), $\delta = 0$	CC	2.00	4.87e0	3.03e-5	3.06	9.05e0	9.71e-8
Flux (5.14), $\delta = 1$	CC	2.00	5.07e0	3.14e-5	2.99	2.62e1	4.29e-7

TABLE 7.4. Example 1: Experimental orders of convergence, error constants and fitted errors with respect to the approximated L^2-norms for the FV lattice Boltzmann schemes (5.3) and (5.4) on a sequence of grids with $N = 60, 145, 230, 315, 400$ with data $T = 0.2$, $\nu = 0.1$ and $\omega = 0.7$, depending on the boundary conditions.

the solution under consideration. In Example 2, we investigate the numerical approximation of the eigenfunction solution

$$r(t, x) := \exp(-4\nu\pi^2 t)\cos(2\pi x) \quad \text{in } (0, T] \times (0, 1)$$

subject to the choice $\nu = 0.1$. For the end time $T = 1.0$ the solution is plotted in Figure 7.5.

As before, we use the initial conditions $R_l^0 = r_0(x_l)$ and $J^0 = -h\partial_x r_0(x_l)/(2\omega)$, where the initial and boundary data is taken from the knowledge of the exact solution. Like in Example 1, there is no source term in the corresponding heat equation.

The numerical results for the experimental orders of convergence are presented in Table 7.5 for the FD lattice Boltzmann schemes (5.1) and (5.2) and in Table 7.6 for the FV lattice Boltzmann schemes (5.3) and (5.4). Both tables contain the

Example 2

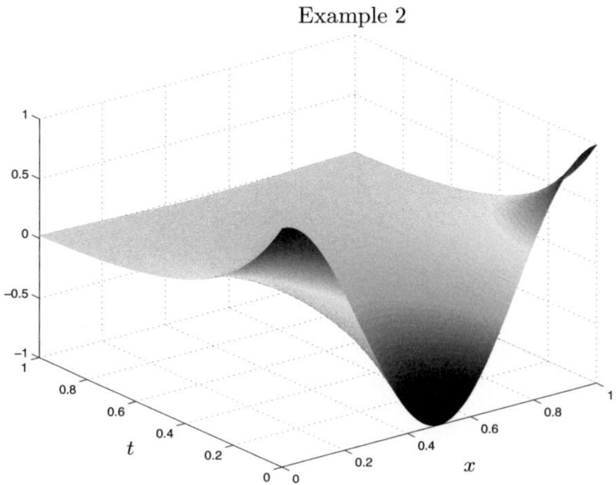

FIGURE 7.5. Solution $r(t,x) := \exp(-4\nu\pi^2)\cos(2\pi x)$ in the domain $(0,1] \times (0,1)$ with $\nu = 0.1$.

Boundary conditions	Grid	Density			Flux		
		EOC	K_r	$E_{400}(r)$	EOC	K_j	$E_{400}(j)$
Periodic (5.5)	VC	2.00	3.67e-1	2.34e-6	3.00	8.98e0	1.41e-7
Density (5.6)	VC	2.00	2.79e-1	1.78e-6	3.00	8.44e0	1.32e-7
Flux (5.8)	VC	2.00	3.67e-1	2.34e-6	3.00	8.98e0	1.41e-7
Inflow (5.10)	VC	2.00	2.87e-1	1.78e-6	3.00	8.56e0	1.33e-7
Density (5.12), $\delta = 0$	CC	2.00	3.15e-1	1.98e-6	3.00	7.99e0	1.26e-7
Density (5.12), $\delta = 1$	CC	2.00	2.27e0	1.39e-5	3.00	3.98e0	6.16e-8
Flux (5.13), $\delta = 0$	CC	2.00	3.67e-1	2.34e-6	3.00	8.98e0	1.41e-7
Flux (5.13), $\delta = 1$	CC	2.00	3.67e-1	2.34e-6	3.00	8.98e0	1.41e-7

TABLE 7.5. Example 2: Experimental orders of convergence, error constants and fitted errors with respect to the discrete L^2-norms for the FD lattice Boltzmann schemes (5.1) and (5.2) on a sequence of grids with $N = 60, 145, 230, 315, 400$ with data $T = 0.2$, $\nu = 0.1$ and $\omega = 0.7$, depending on the boundary conditions.

results with respect to the discrete L^2-norms. The results are pretty much the same as for Example 1.

In the same manner, high-frequency eigenfunction solutions of the heat equation are suited to be the objective of the convergence investigations. Due to the fast decay of the solutions and hence of the errors, smaller time scales or smaller viscosities have to be used. Otherwise, we get in conflict with the machine precision and the logarithmic error curves do no longer show linear behavior. For the initial data $r_0(x) := \sin(20\pi x)$ on a grid with $N = 100$ points and viscosity $\nu = 0.1$, we have to restrict to the end time $T = 0.05$ to achieve the desired results. Within these bounds the convergence orders are confirmed. For the high-frequency eigenfunction solutions, that is, the frequency is close to the number of the grid points, the lattice Boltzmann solutions do not show the expected decay. Hence, there is no

Boundary conditions	Grid	Density			Flux		
		EOC	K_r	$E_{400}(r)$	EOC	K_j	$E_{400}(j)$
Periodic (5.5)	VC	2.00	3.23e0	2.00e-5	3.02	1.05e0	1.42e-8
Density (5.6)	VC	2.00	2.45e0	1.52e-5	3.00	8.22e0	1.28e-7
Flux (5.8)	VC	2.00	3.23e0	2.00e-5	3.02	1.05e0	1.42e-8
Inflow (5.10)	VC	2.01	2.52e0	1.52e-5	2.98	7.02e0	1.27e-7
Density (5.12), $\delta = 0$	CC	2.00	3.13e0	1.95e-5	2.89	1.31e1	4.06e-7
Density (5.12), $\delta = 1$	CC	2.00	5.48e0	3.43e-5	2.82	1.62e1	7.34e-7
Flux (5.14), $\delta = 0$	CC	2.00	3.23e0	2.00e-5	3.02	1.05e0	1.42e-8
Flux (5.14), $\delta = 1$	CC	2.00	3.23e0	2.00e-5	3.02	1.05e0	1.42e-8

TABLE 7.6. Example 2: Experimental orders of convergence, error constants and fitted errors with respect to the discrete L^2-norms for the FV lattice Boltzmann schemes (5.3) and (5.4) on a sequence of grids with $N = 60, 145, 230, 315, 400$ with data $T = 0.2$, $\nu = 0.1$ and $\omega = 0.7$, depending on the boundary conditions.

convergence. This fact is subject of our investigations performed in the following sections.

A further possibility for the computation of the unknown inflow values at the boundary is the usage of data extrapolation from the interior of the domain. But as expected, the obtained results are unsatisfactory, because the boundary values are not involved into the computation. The extrapolation of either the missing distribution function or the density leads to first order convergence for the density and second order convergence for the h-scaled flux in the best case. Stable solutions are only obtained in the case of inflow boundary conditions with extrapolations. However, the performance is unsatisfactory. For Example 1, these results are gained for linear or cubic extrapolation. For Example 2, quadratic extrapolation seems to be more appropriate. This awareness has to be kept in mind when it comes to the grid coupling on nonuniform grids in Chapter 8.

The examples so far are furnished with vanishing source terms in the corresponding heat equation. In Example 3, we consider the time-periodic solution

$$r(t, x) := 256 \sin(4\pi t) x^4 (1 - x)^4 \quad \text{in } (0, T] \times (0, 1)$$

with nonvanishing source term. The solution up to the end time $T = 1.0$ is plotted in Figure 7.6. By the application of the heat equation we determine the necessary source term f. The initial and boundary data are chosen from the exact solution. Initial values are taken in the form $R_l^0 = r_0(x_l)$ and $J_l^0 = -h\partial_x r_0(x_l)/(2\omega)$, where the range of l depends on the grid and the boundary conditions.

The results of the convergence investigations for the FD lattice Boltzmann solutions are shown in Table 7.7. For all given boundary conditions, the errors and error constants are increased by a factor of up to 30 in comparison to the previous examples.

The long time behavior of the error is examined in Figure 7.7, where the nodal error of the VC lattice Boltzmann solution in $x = 0.5$ is displayed in the time interval $(0, 50]$. The number of 233100 time steps have to be computed to reach the end time $T = 50$ on a grid with $N = 100$, viscosity $\nu = 0.1$ and the relaxation parameter $\omega = 0.7$. Once the error is tuned in after $t \approx 4.0$, the amplitude of the error in the density remains on the same level. The error for the flux behaves similarly.

Example 3

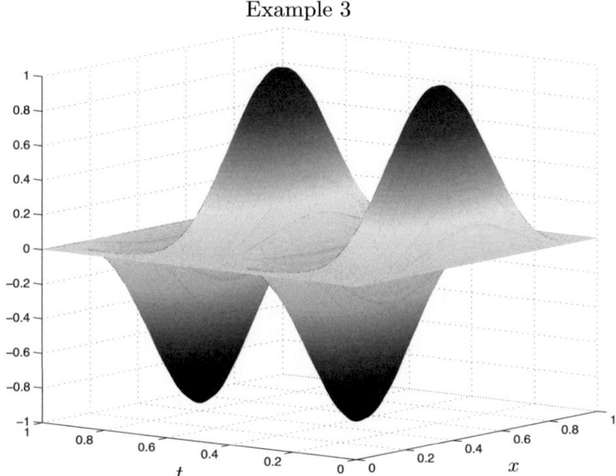

FIGURE 7.6. Solution $r(t,x) := 256\sin(4\pi t)x^4(1-x)^4$ in the domain $(0,1] \times (0,1)$ with $\nu = 0.1$.

Boundary conditions	Grid	Density			Flux		
		EOC	K_r	$E_{400}(r)$	EOC	K_j	$E_{400}(j)$
Density (5.6)	VC	2.00	1.04e1	6.48e-5	3.00	1.79e1	2.77e-7
Flux (5.8)	VC	2.00	1.07e1	6.64e-5	3.00	1.41e1	2.18e-7
Inflow (5.10)	VC	2.00	1.05e1	6.48e-5	2.99	1.70e1	2.77e-7
Density (5.12), $\delta = 0$	CC	2.00	1.04e1	6.47e-5	3.00	1.79e1	2.77e-7
Density (5.12), $\delta = 1$	CC	2.00	1.04e1	6.47e-5	3.00	1.79e1	2.77e-7
Flux (5.13), $\delta = 0$	CC	2.01	1.10e1	6.64e-5	2.97	1.19e1	2.17e-7
Flux (5.13), $\delta = 1$	CC	2.01	1.10e1	6.64e-5	2.97	1.19e1	2.17e-7

TABLE 7.7. Example 3: Experimental orders of convergence, error constants and fitted errors with respect to the discrete L^2-norms for the FD lattice Boltzmann schemes (5.1) and (5.2) on a sequence of grids with $N = 60, 145, 230, 315, 400$ with data $T = 0.2$, $\nu = 0.1$ and $\omega = 0.7$, depending on the boundary conditions.

7.2. Dependence on the Parameters

The performance of the lattice Boltzmann methods depends on the choice of the relaxation parameter ω. At first sight, it mainly decides on the space-time coupling. For a prescribed viscosity we found $\tau/h^2 = (1 - \omega)/(2\omega\nu)$ for the FD lattice Boltzmann schemes and $\tau/h^2 = 1/(2\omega\nu)$ for the FV lattice Boltzmann schemes. Hence, the time step size approximates 0 when ω reaches 1 in the case of the FD lattice Boltzmann schemes. This leads to a large number of necessary time steps for increasing ω. In Table 7.8 the time step size and the number of necessary time steps to reach the end time $T = 0.2$ on a grid with $N = 400$ points is shown. The FV lattice Boltzmann schemes appear to be preferable due to the smaller time steps.

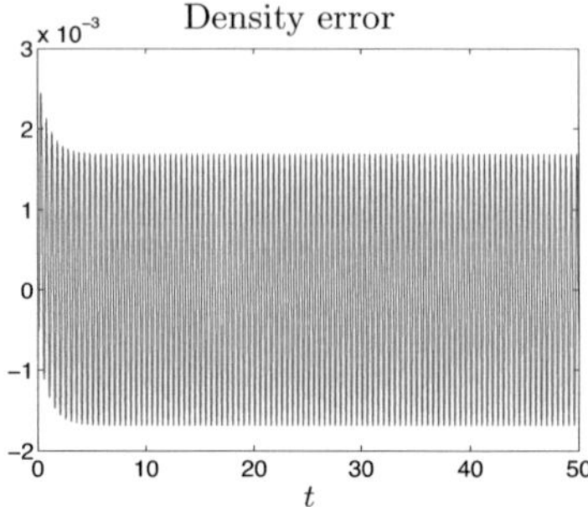

FIGURE 7.7. Example 3: Long time behavior of the density error in $x = 0.5$ on the time interval $(0, 50]$.

	FD lattice Boltzmann		FV lattice Boltzmann	
ω	τ	#Time steps	τ	#Time steps
0.2	1.25e-4	1601	1.56e-4	1281
0.3	7.29e-5	2743	1.04e-4	1921
0.4	4.69e-5	4267	7.81e-5	2560
0.5	3.13e-5	6400	6.25e-5	3201
0.6	2.08e-5	9601	5.21e-5	3840
0.7	1.34e-5	14934	4.46e-5	4480
0.8	7.81e-6	25600	3.91e-5	5121
0.9	3.47e-6	47601	3.47e-5	5760

TABLE 7.8. Dependence of the time steps on the relaxation parameter ω for the FD lattice Boltzmann schemes (5.1) and (5.2) and the FV lattice Boltzmann schemes (5.3) and (5.4) on a grid with $N = 400$, viscosity $\nu = 0.1$ and end time $T = 0.2$.

On the other hand, smaller time steps for the FD lattice Boltzmann schemes directly lead to smaller errors, as can be seen in Table 7.9, where the discrete L^2-errors, the error constants and the experimental orders of convergence are listed depending on ω. These results are achieved by applying Example 1 with periodic boundary conditions to grid sequences with $N = 60, 145, 230, 315$ and 400 on VCFD grids. In Table 7.10 the corresponding results for the VCFV lattice Boltzmann solutions are shown depending on ω.

Taking a closer look, we find that the steps in ω were chosen too large to resolve all subtleties of the convergence process. In Figure 7.8 the logarithms of the errors for the density and the flux are plotted with a step size $\Delta\omega = 0.01$ for Example 1. Here, the vanishing right hand side in the heat equation plays a crucial role. The left figure shows the decay of the discrete L^2-errors, in the right figure the approximated L^2-errors are visualized. On the basis of the discrete L^2-errors there is huge decay of the density error for a specific choice of ω. As pointed out at the end of Chapter 6, the minimal error is attained for the value $\omega^* := (3 - \sqrt{3})/2 \approx 0.63397$. For this specific value we find $E_{100}(r) = 9.51\text{e-}8$ and $E_{100}(j) = 1.05\text{e-}5$.

In Figure 7.9 the behavior of the discrete L^2-errors is plotted with ω-steps of size 1e-4 around the minimum.

	Density			Flux		
ω	EOC	K_r	$E_{400}(r)$	EOC	K_j	$E_{400}(j)$
0.2	2.01	4.08e1	2.45e-4	3.09	3.38e2	3.00e-6
0.3	2.00	1.30e1	7.99e-5	3.02	6.34e1	9.00e-7
0.4	2.00	4.86e0	3.00e-5	3.00	2.59e1	4.17e-7
0.5	2.00	1.69e0	1.04e-5	3.00	1.58e1	2.51e-7
0.6	2.01	2.87e-1	1.74e-6	3.00	1.13e1	1.79e-7
0.7	2.00	3.67e-1	2.34e-6	3.00	8.98e0	1.41e-7
0.8	2.00	6.70e-1	4.24e-6	3.00	7.56e0	1.19e-7
0.9	2.00	7.95e-1	5.02e-6	3.00	6.62e0	1.04e-7

TABLE 7.9. Example 1: Experimental orders of convergence, error constants and fitted errors with respect to the discrete L^2-norms for the VCFD lattice Boltzmann scheme (5.1) on a sequence of grids with $N = 60,\ 145,\ 230,\ 315,\ 400$ with data $T = 0.2$ and $\nu = 0.1$, depending on ω.

	Density			Flux		
ω	EOC	K_r	$E_{400}(r)$	EOC	K_j	$E_{400}(j)$
0.2	2.01	5.48e1	3.23e-4	3.13	4.11e2	3.01e-6
0.3	2.01	2.19e1	1.32e-4	3.00	4.87e1	7.49e-7
0.4	2.00	1.13e1	6.90e-5	3.00	1.50e1	2.32e-7
0.5	2.00	6.80e0	4.17e-5	2.94	3.22e0	7.23e-8
0.6	2.00	4.51e0	2.78e-5	2.86	3.06e-1	1.10e-8
0.7	2.00	3.23e0	2.00e-5	3.02	1.05e0	1.42e-8
0.8	2.00	2.48e0	1.53e-5	3.02	1.82e0	2.46e-8
0.9	2.00	1.99e0	1.24e-5	3.01	1.94e0	2.88e-8

TABLE 7.10. Example 1: Experimental orders of convergence, error constants and fitted errors with respect to the discrete L^2-norms for the VCFV lattice Boltzmann scheme (5.3) on a sequence of grids with $N = 60,\ 145,\ 230,\ 315,\ 400$ with data $T = 0.2$ and $\nu = 0.1$, depending on ω.

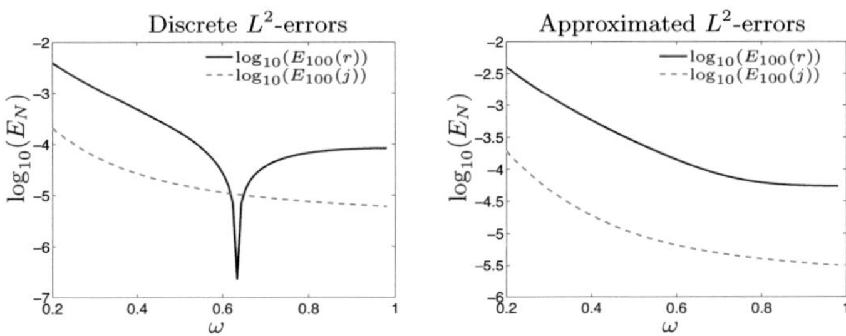

FIGURE 7.8. Example 1: Dependence of the errors on the relaxation parameter ω for the FD lattice Boltzmann solutions with vanishing source term and $N = 100$.

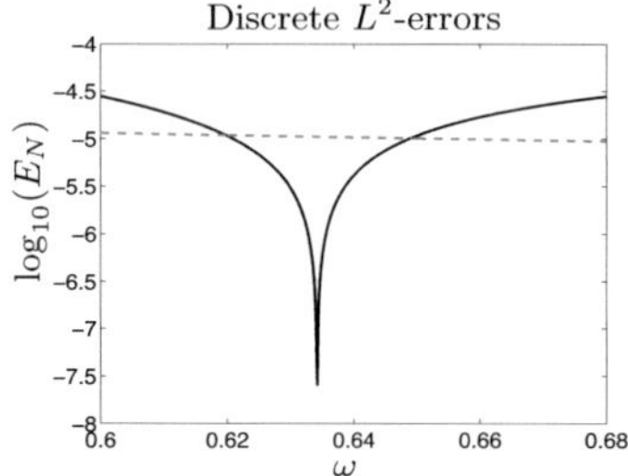

FIGURE 7.9. Example 1: Dependence of the errors on the relaxation parameter ω for the FD lattice Boltzmann solutions with vanishing source term and $N = 100$.

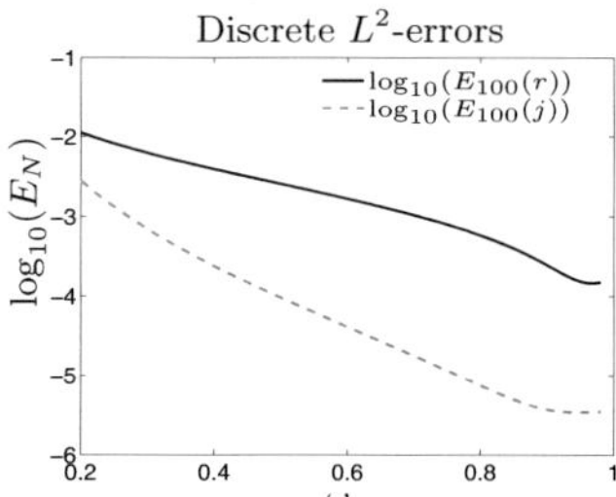

FIGURE 7.10. Example 3: Dependence of the errors on the relaxation parameter ω for the FD lattice Boltzmann solutions with nonvanishing source term and $N = 100$.

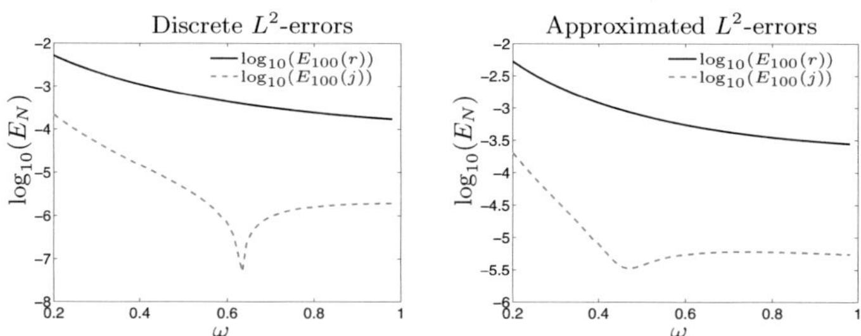

FIGURE 7.11. Example 1: Dependence of the errors on the relaxation parameter ω for the FV lattice Boltzmann solutions with vanishing source term and $N = 100$.

Since the decay of the errors cannot be seen in terms of the approximated L^2-errors, this is an effect owed to nodal superconvergence. The effect of improved convergence for the specific choice of ω is due to the vanishing right hand side in the heat equation. The appearance of source terms spoils this advantage. In Figure 7.10 the error curves for the FD lattice Boltzmann solutions of Example 3 with a nonzero source term are plotted.

Boundary conditions	Grid	Density			Flux		
		EOC	K_r	$E_{400}(r)$	EOC	K_j	$E_{400}(j)$
Periodic (5.5)	VC	4.00	9.47e0	3.72e-10	3.00	1.04e1	1.64e-7
Density (5.6)	VC	4.00	9.47e0	3.72e-10	3.00	1.04e1	1.64e-7
Flux (5.8)	VC	2.00	1.63e0	1.02e-5	3.00	8.54e0	1.34e-7
Inflow (5.10)	VC	2.97	6.69e0	1.24e-7	2.99	1.01e1	1.64e-7
Density (5.12), $\delta = 0$	CC	4.00	9.47e0	3.72e-10	3.00	1.04e1	1.64e-7
Density (5.12), $\delta = 1$	CC	4.00	9.47e0	3.72e-10	3.00	1.04e1	1.64e-7
Flux (5.13), $\delta = 0$	CC	2.00	1.01e0	6.26e-6	3.00	1.40e1	2.18e-7
Flux (5.13), $\delta = 1$	CC	2.00	1.81e0	1.13e-5	3.00	8.78e0	1.38e-7

TABLE 7.11. Example 1: Experimental orders of convergence, error constants and fitted errors with respect to the discrete L^2-norms for the FD lattice Boltzmann schemes (5.1) and (5.2) on a sequence of grids with $N = 60$, 145, 230, 315, 400 with data $T = 0.2$, $\nu = 0.1$ and $\omega = \omega^* := (3 - \sqrt{3})/2$, depending on the boundary conditions.

For the FV lattice Boltzmann solutions there is a similar effect, but here the decay of the error can be seen for the flux and not for the density; see Figure 7.11.

For the choice $\omega = \omega^*$, the experimental orders of convergence for the FD lattice Boltzmann solution of Example 1 are given in Table 7.11. We find fourth order convergence for the density in the case of periodic or density boundary conditions. For the inflow boundary conditions we gain third order convergence. No improvements can be achieved in the case of flux boundary conditions.

The situation changes, when considering Example 2 instead. From the results given in Table 7.12, we find that the convergence behavior on VC grids remains the same for periodic and density boundary conditions. For flux and inflow boundary conditions on VC grids, we find an improvement. On CC grids the convergence is of second order in the case of density boundary conditions, whereas in the case of flux boundary conditions we find fourth order convergence. This fact implies that the odd and even higher order derivatives of the density have to vanish at the boundaries to achieve improvements.

In Table 7.13 the experimental convergence orders for the FV lattice Boltzmann solutions are displayed for Example 1. For the specific choice $\omega = \omega^*$, we get improved convergence results for the flux. Further investigations show that the convergence can be improved up to fifth order, when choosing a slightly smaller value for ω close to 0.6322.

Numerical examples with different viscosities cannot be compared directly. The exponential decay of the solution changes as the viscosity is varied. For constant νT, the solution behavior remains unchanged. Hence, we have to consider different end times (but the same number of time steps), if we want to compare solutions with different viscosities. The errors and the error constants are invariant as long as νT is kept at a constant value.

| | | Density | | | Flux | | |
Boundary conditions	Grid	EOC	K_r	$E_{400}(r)$	EOC	K_j	$E_{400}(j)$
Periodic (5.5)	VC	4.00	9.47e0	3.72e-10	3.00	1.04e1	1.64e-7
Density (5.6)	VC	4.00	8.32e0	3.25e-10	3.00	1.04e1	1.64e-7
Flux (5.8)	VC	4.00	9.47e0	3.72e-10	3.00	1.04e1	1.64e-7
Inflow (5.10)	VC	3.99	7.65e0	3.25e-10	3.00	1.04e1	1.64e-7
Density (5.12), $\delta = 0$	CC	2.00	1.84e-1	1.17e-6	3.00	1.10e1	1.72e-7
Density (5.12), $\delta = 1$	CC	2.00	2.59e0	1.63e-5	3.00	4.22e0	6.46e-8
Flux (5.13), $\delta = 0$	CC	4.00	9.47e0	3.72e-10	3.00	1.04e1	1.64e-7
Flux (5.13), $\delta = 1$	CC	4.00	9.47e0	3.72e-10	3.00	1.04e1	1.64e-7

TABLE 7.12. Example 2: Experimental orders of convergence, error constants and fitted errors with respect to the discrete L^2-norms for the FD lattice Boltzmann schemes (5.1) and (5.2) on a sequence of grids with $N = 60, 145, 230, 315, 400$ with data $T = 0.2$, $\nu = 0.1$ and $\omega = \omega^* := (3 - \sqrt{3})/2$, depending on the boundary conditions.

| | | Density | | | Flux | | |
Boundary conditions	Grid	EOC	K_r	$E_{400}(r)$	EOC	K_j	$E_{400}(j)$
Periodic (5.5)	VC	2.00	4.00e0	2.47e-5	4.07	1.25e1	3.26e-10
Density (5.6)	VC	2.00	4.00e0	2.47e-5	4.07	1.25e1	3.26e-10
Flux (5.8)	VC	2.00	3.26e0	2.01e-5	3.01	5.91e0	8.67e-8
Inflow (5.10)	VC	1.99	3.78e0	2.46e-5	4.27	1.41e2	1.08e-9
Density (5.12), $\delta = 0$	CC	2.00	4.00e0	2.47e-5	4.07	1.25e1	3.26e-10
Density (5.12), $\delta = 1$	CC	2.00	4.00e0	2.47e-5	4.07	1.25e1	3.26e-10
Flux (5.14), $\delta = 0$	CC	2.00	4.53e0	2.78e-5	3.01	8.98e0	1.33e-7
Flux (5.14), $\delta = 1$	CC	2.00	5.57e0	3.48e-5	3.00	3.07e1	4.65e-7

TABLE 7.13. Example 1: Experimental orders of convergence, error constants and fitted errors with respect to the discrete L^2-norms for the FV lattice Boltzmann schemes (5.3) and (5.4) on a sequence of grids with $N = 60, 145, 230, 315, 400$ with data $T = 0.2$, $\nu = 0.1$ and $\omega = \omega^* := (3 - \sqrt{3})/2$, depending on the boundary conditions.

7.3. Dependence on the Initial Data

For the initial data there are two possible choices. The first and natural choice are the first order initial data for the lattice Boltzmann schemes given by

$$\boldsymbol{U}^0 := \frac{1}{2}[r_0(x_l)]_l - \frac{h}{4\omega}[\partial_x r_0(x_l)]_l,$$
$$\boldsymbol{V}^0 := \frac{1}{2}[r_0(x_l)]_l + \frac{h}{4\omega}[\partial_x r_0(x_l)]_l, \tag{7.1}$$

with $\boldsymbol{R}^0 = \boldsymbol{U}^0 + \boldsymbol{V}^0 = [r_0(x_l)]_l$ and $\boldsymbol{J}^0 = \boldsymbol{U}^0 - \boldsymbol{V}^0 = -h/(2\omega)[\partial_x r_0(x_l)]_l$. The indices l take their values depending on the underlying grid. The second choice is

Initial conditions	T	Density			Flux		
		EOC	K_r	$E_{400}(r)$	EOC	K_j	$E_{400}(j)$
0th order	0.1	2.00	3.54e0	2.23e-5	3.00	5.07e0	7.92e-8
1st order	0.1	1.99	2.67e-1	1.74e-6	3.00	1.21e1	1.90e-7
0th order	0.2	2.00	2.20e0	1.38e-5	3.00	2.58e0	4.03e-8
1st order	0.2	2.00	3.67e-1	2.34e-6	3.00	8.98e0	1.41e-7
0th order	0.5	2.00	5.01e-1	3.16e-6	3.02	1.89e-2	2.65e-10
1st order	0.5	2.00	2.85e-1	1.79e-6	3.00	3.52e0	5.52e-8
0th order	1.0	2.00	3.00e-2	1.90e-7	3.00	1.76e-1	2.75e-9
1st order	1.0	2.00	7.95e-2	4.97e-7	3.00	6.69e-1	1.05e-8

TABLE 7.14. Example 2: Experimental orders of convergence, error constants and fitted errors with respect to the discrete L^2-norms of the VCFD lattice Boltzmann scheme (5.1) on a sequence of grids with $N = 60, 145, 230, 315, 400$ for $\omega = 0.7$, depending on the initial conditions and different end times T.

given by the zeroth order initial values

$$\boldsymbol{U}^0 := \frac{1}{2}[r_0(x_l)]_l,$$
$$\boldsymbol{V}^0 := \frac{1}{2}[r_0(x_l)]_l, \tag{7.2}$$

with $\boldsymbol{R}^0 = \boldsymbol{U}^0 + \boldsymbol{V}^0 = [r_0(x_l)]_l$ and $\boldsymbol{J}^0 = \boldsymbol{U}^0 - \boldsymbol{V}^0 = \boldsymbol{0}$.

The examination of the experimental orders of convergence shows that the initial data have no effect on the experimental orders of convergence. In Table 7.14 the results of the application of the VCFD lattice Boltzmann scheme (5.1) to Example 2 with relaxation parameter $\omega = 0.7$ are displayed depending on the initial conditions and different end times T. For both types of initial conditions we find second order convergence for the density and third order convergence for the h-scaled flux in the discrete L^2-norms.

In Figure 7.12 the nodal errors in the density are plotted up to the end time $T = 1.0$ on a grid with $N = 100$ points for Example 2. Here, the first order initial conditions (7.1) are used. Up to the time $t \approx 0.2$ the error increases, before the exponential decay becomes effective.

In Figure 7.13 the nodal errors in the density are displayed for the zeroth order initial conditions (7.2). Here, the exponential decay starts with the first time step, but there is an initial layer. The sign of the error is opposite compared to the previous choice. The nodal errors are larger by two orders of magnitudes.

The nodal errors for the flux are displayed in Figures 7.14 and 7.15. In Figure 7.14 the first order initial conditions (7.1) find application. We detect the expected exponential decay. For zeroth order initial conditions (7.2) there is an initial layer at $t = 0$ that is visualized in Figure 7.15. For a finer resolution on a short time scale the same errors are plotted in Figures 7.16 and 7.17.

The evaluation of the Fourier representations of the lattice Boltzmann solution leads to the insight that there are initial oscillations of the error in the flux. These oscillations can be seen in the numerical solutions in Figures 7.16 and 7.17.

For further investigations, we define the step dependent numerical viscosities

$$\nu_r^k := \frac{\gamma|\Omega|^2}{L^2\pi^2 h^2}\left(1 - \frac{\|\boldsymbol{R}^k\|_\infty}{\|\boldsymbol{R}^{k-1}\|_\infty}\right), \quad \nu_j^k := \frac{\gamma|\Omega|^2}{L^2\pi^2 h^2}\left(1 - \frac{\|\boldsymbol{J}^k\|_\infty}{\|\boldsymbol{J}^{k-1}\|_\infty}\right)$$

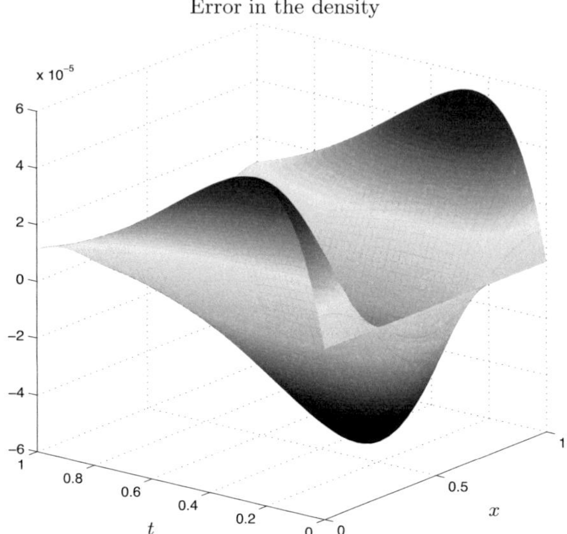

FIGURE 7.12. Example 2: Error in the density with 1st order initial conditions.

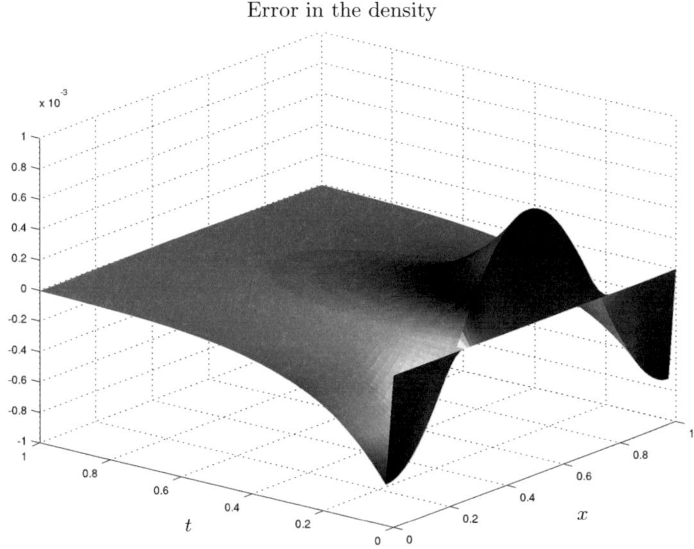

FIGURE 7.13. Example 2: Error in the density with 0th order initial conditions.

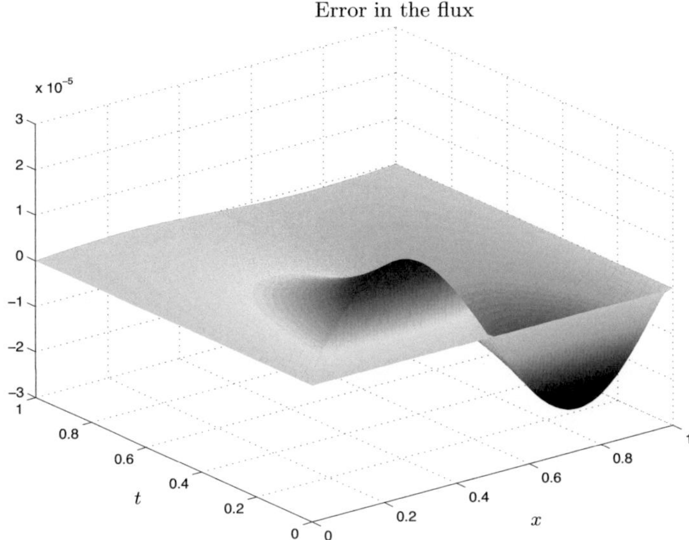

FIGURE 7.14. Example 2: Error in the flux with 1st order initial conditions.

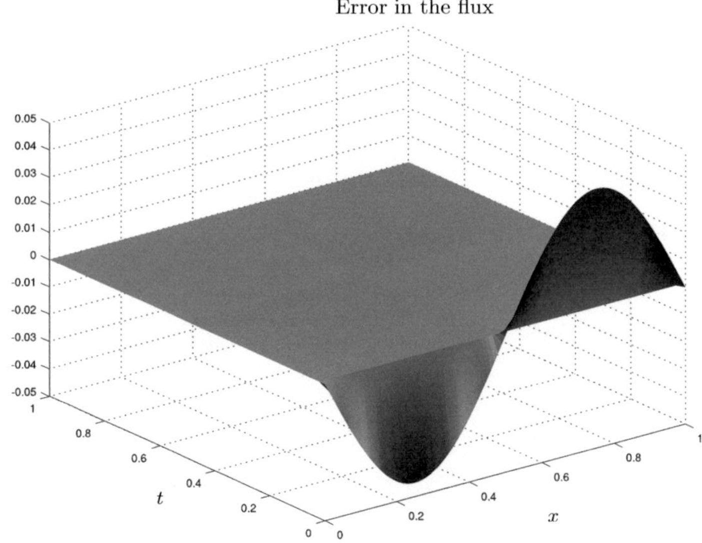

FIGURE 7.15. Example 2: Error in the flux with 0th order initial conditions.

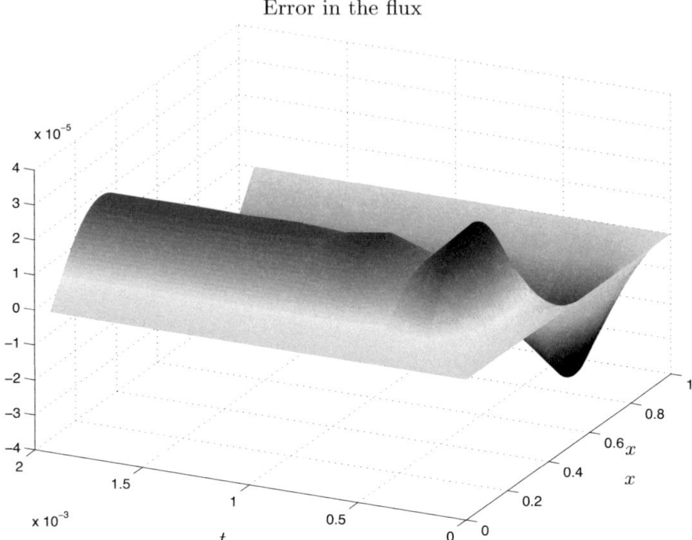

FIGURE 7.16. Example 2: Error in the flux with 1st order initial
conditions on a short time scale.

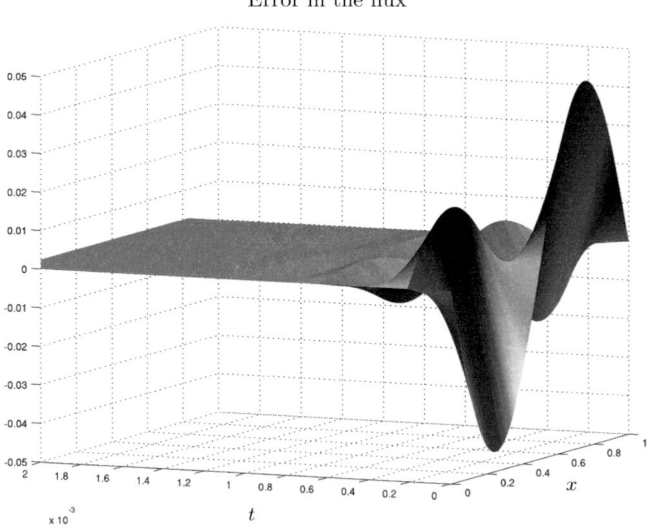

FIGURE 7.17. Example 2: Error in the flux with 0th order initial
conditions on a short time scale.

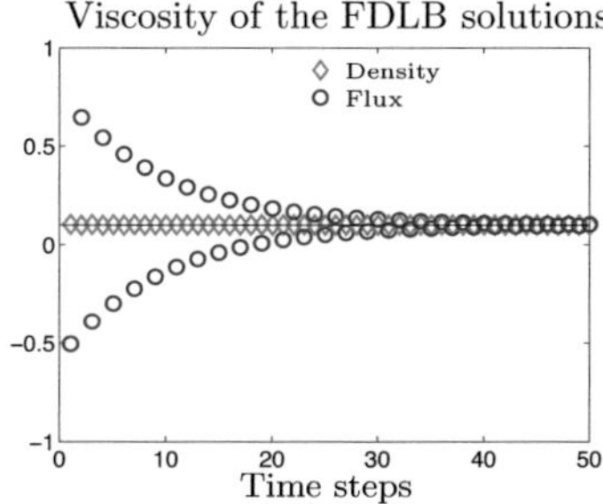

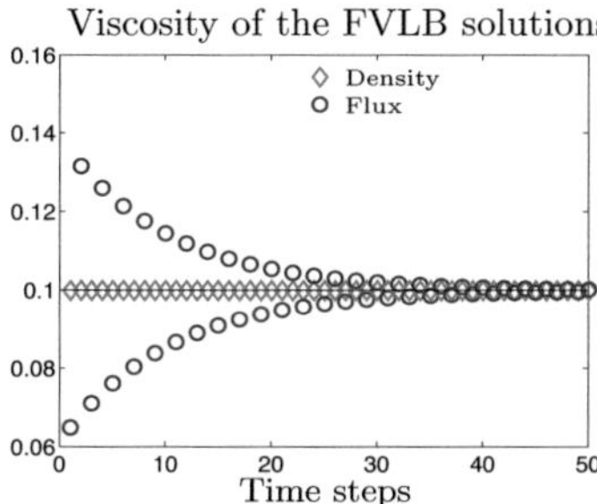

FIGURE 7.18. Example 1: Numerical viscosity of the density and the flux depending on the time steps for $\omega = 0.95$ and $N = 50$.

for $k = 1, \ldots, M$. Here, L is the frequency of the eigenfunction solution. These formulas are motivated by the knowledge that in each time step the eigenfunction solution with frequency L shall decay with the rate $1 - \nu L^2 \pi^2 / (\gamma N^2)$. The desired values are $(1 - \omega)\gamma / (2\omega)$ for the FD lattice Boltzmann solutions and $\gamma / (2\omega)$ for the FV lattice Boltzmann solutions.

In Figure 7.18 the achieved values of ν_r^k and ν_j^k are plotted depending on the time steps k. The underlying test problem is Example 1 on a VC grid with $N = 50$ and $L = 2$. The relaxation parameter is chosen as $\omega = 0.95$. We find an oscillating behavior for the fluxes for both types of schemes. The correct numerical viscosities are reached after some time steps, when the oscillating parts belonging to the eigenvalue λ_2^- have vanished (see Sections 5.10 and 5.11). The density is equipped with the right numerical viscosity from the first time step on.

Smaller values for the relaxation parameter lead to a faster decay of the oscillations. By performing a Taylor series analysis of the Fourier representations of the FD lattice Boltzmann solutions, we gain an iteration formula for the numerical viscosities of the form

$$c^0 := 0,$$

$$\left. \begin{aligned} v^k &:= (3 - 4\omega + 12\omega c^k)/6, \\ c^{k+1} &:= c^k - v^k + (1 - \omega)/(2\omega), \end{aligned} \right\} \quad \text{for } k = 0, \ldots, M,$$

where v^k is an approximation of ν_j^k / γ. The results of these iterations are plotted in Figure 7.19 depending on the relaxation parameter $\omega \in [1/2, 1)$. For each time step, there is a solid line approaching the right viscosity $(1 - \omega)/(2\omega)$ for increasing k. In the first approximation we get the lowest straight line. Most oscillations occur close to $\omega = 1$. For $\omega = 1/2$, the right value is reached in the second time step.

All computations put forward on the previous pages are done with the choice of the first order initial data for the flux, that is, $j_0 := -h \partial_x r_0 / (2\omega)$. If we start with the initial data $j_0 := 0$, then the correct numerical viscosity is also reached after some time steps. This can be seen in Figure 7.20.

7.4. Numerical Reference Solutions

If we want to apply the lattice Boltzmann schemes to further test problems, we need numerical reference solutions in order to rate the quality of the solutions. Numerical methods for the computation of the reference solutions are introduced in this section.

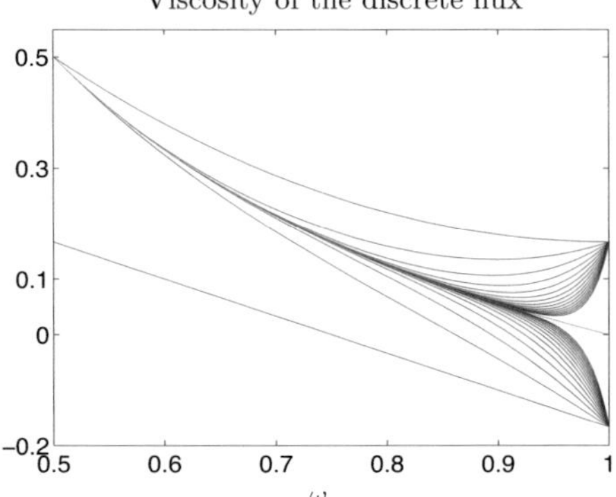

FIGURE 7.19. Numerical viscosity of the flux (FDLB) depending
on the relaxation parameter ω for the first 30 time steps for $N = 50$.

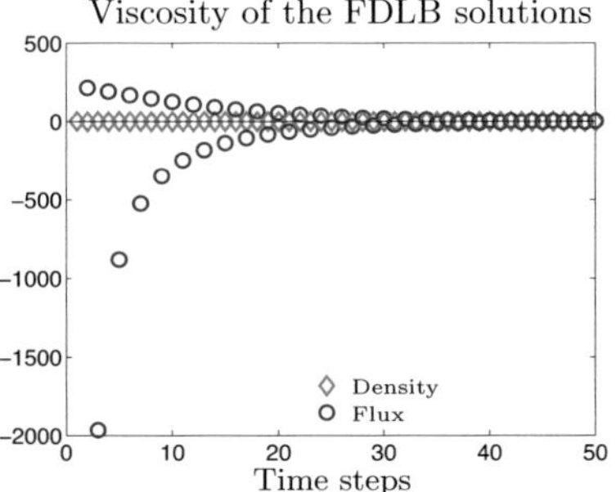

FIGURE 7.20. Example 1: Numerical viscosity of the density and
the flux depending on the time steps for $\omega = 0.95$ and $N = 50$ for
the initial condition $\boldsymbol{J}^0 = \boldsymbol{0}$.

In the case of periodic, Dirichlet or Neumann boundary conditions for the heat
equation, numerical reference solutions for the density and the flux can be obtained
by solving the discrete equations

$$(\boldsymbol{Id} - \sigma\kappa\boldsymbol{A})\boldsymbol{R}^{k+1} = (\boldsymbol{Id} + (1 - \sigma)\kappa\boldsymbol{A})\boldsymbol{R}^k + \tau\boldsymbol{F}^k + \sigma\kappa\boldsymbol{R}_B^{k+1} + (1 - \sigma)\kappa\boldsymbol{R}_B^k,$$

$$(\boldsymbol{Id} - \sigma\kappa\boldsymbol{B})\boldsymbol{J}^{k+1} = (\boldsymbol{Id} + (1 - \sigma)\kappa\boldsymbol{B})\boldsymbol{J}^k - \frac{\tau h}{2\omega}\boldsymbol{DF}^k + \sigma\kappa\boldsymbol{J}_B^{k+1} + (1 - \sigma)\kappa\boldsymbol{J}_B^k,$$

$$(7.3)$$

for $k = 0, \ldots, M - 1$ with

$$\kappa := \nu\frac{\tau}{h^2}$$

and $\sigma \in \{0, 1/2, 1\}$. The schemes in (7.3) are called the σ-*schemes* in the following.
Here, the heat equation for the density and the additional heat equation for the flux
are discretized with one sided forward difference quotients for the time derivatives
and central difference quotients for the second order spatial derivatives. For $\sigma = 0$,
this is the *explicit Euler scheme*, for $\sigma = 1$ the scheme is recognized as the *implicit
Euler scheme* and for $\sigma = 1/2$ we gain the well-known *Crank-Nicolson scheme*. In

the latter two cases, the computational effort is increased, since we have to solve an additional linear system of equations in each time step.

For periodic boundary conditions (2.6) on VC grids, we choose

$$
\boldsymbol{A} := \boldsymbol{B} := \begin{bmatrix} -2 & 1 & & & 1 \\ 1 & -2 & 1 & & \\ & & \ddots & & \\ & & 1 & -2 & 1 \\ 1 & & & 1 & -2 \end{bmatrix} \in \mathbb{R}^{N,N}, \quad \boldsymbol{F}^k := \begin{bmatrix} f_0^k \\ \vdots \\ f_{N-1}^k \end{bmatrix}, \quad \boldsymbol{DF}^k := \begin{bmatrix} \partial_x f_0^k \\ \vdots \\ \partial_x f_{N-1}^k \end{bmatrix},
$$

and $\boldsymbol{R}_B^k := \boldsymbol{J}_B^k := \boldsymbol{0}$.

For Dirichlet boundary conditions (2.3) for the heat equation on VC grids, we choose

$$
\boldsymbol{A} := \begin{bmatrix} -2 & 1 & & \\ 1 & -2 & 1 & \\ & & \ddots & \\ & & 1 & -2 & 1 \\ & & & 1 & -2 \end{bmatrix} \in \mathbb{R}^{N-1,N-1}, \quad \boldsymbol{F}^k := \begin{bmatrix} f_1^k \\ \vdots \\ f_{N-1}^k \end{bmatrix}, \quad \boldsymbol{R}_B^k := \begin{bmatrix} r_L^k \\ 0 \\ \vdots \\ 0 \\ r_R^k \end{bmatrix},
$$

$$
\boldsymbol{B} := \begin{bmatrix} -2 & 2 & & \\ 1 & -2 & 1 & \\ & & \ddots & \\ & & 1 & -2 & 1 \\ & & & 2 & -2 \end{bmatrix} \in \mathbb{R}^{N+1,N+1}, \quad \boldsymbol{DF}^k := \begin{bmatrix} \partial_x f_0^k \\ \vdots \\ \partial_x f_N^k \end{bmatrix}, \quad \boldsymbol{J}_B^k := \begin{bmatrix} -2hdj_L^k \\ 0 \\ \vdots \\ 0 \\ 2hdj_R^k \end{bmatrix}.
$$

Dirichlet values for the density are given by $r_L^k := r^{\mathrm{Di}}(t_k, x_L)$, $r_R^k := r^{\mathrm{Di}}(t_k, x_R)$ and Neumann values for the flux are chosen by $dj_L^k := h(f(t_k, x_L) - \partial_t r^{\mathrm{Di}}(t_k, x_L))/(2\omega\nu)$, $dj_R^k := h(f(t_k, x_R) - \partial_t r^{\mathrm{Di}}(t_k, x_R))/(2\omega\nu)$ by invoking the heat equation.

For Dirichlet boundary conditions on CC grids, we use

$$
\boldsymbol{A} := \begin{bmatrix} -3 & 1 & & \\ 1 & -2 & 1 & \\ & & \ddots & \\ & & 1 & -2 & 1 \\ & & & 1 & -3 \end{bmatrix} \in \mathbb{R}^{N,N}, \quad \boldsymbol{F}^k := \begin{bmatrix} f_{1/2}^k \\ \vdots \\ f_{N-1/2}^k \end{bmatrix}, \quad \boldsymbol{R}_B^k := \begin{bmatrix} 2r_L^k \\ 0 \\ \vdots \\ 0 \\ 2r_R^k \end{bmatrix},
$$

$$
\boldsymbol{B} := \begin{bmatrix} -1 & 1 & & \\ 1 & -2 & 1 & \\ & & \ddots & \\ & & 1 & -2 & 1 \\ & & & 1 & -1 \end{bmatrix} \in \mathbb{R}^{N,N}, \quad \boldsymbol{DF}^k := \begin{bmatrix} \partial_x f_{1/2}^k \\ \vdots \\ \partial_x f_{N-1/2}^k \end{bmatrix}, \quad \boldsymbol{J}_B^k := \begin{bmatrix} -hdj_L^k \\ 0 \\ \vdots \\ 0 \\ hdj_R^k \end{bmatrix},
$$

with r_L^k, r_R^k, dj_L^k and dj_R^k chosen as in the previous case.

For Neumann boundary conditions (2.4) on VC grids, we choose

$$
\boldsymbol{A} := \begin{bmatrix} -2 & 2 & & \\ 1 & -2 & 1 & \\ & & \ddots & \\ & & 1 & -2 & 1 \\ & & & 2 & -2 \end{bmatrix} \in \mathbb{R}^{N+1,N+1}, \quad \boldsymbol{F}^k := \begin{bmatrix} f_0^k \\ \vdots \\ f_N^k \end{bmatrix}, \quad \boldsymbol{R}_B^k := \begin{bmatrix} -2hdr_L^k \\ 0 \\ \vdots \\ 0 \\ 2hdr_R^k \end{bmatrix},
$$

$$\boldsymbol{B} := \begin{bmatrix} -2 & 1 & & & \\ 1 & -2 & 1 & & \\ & & \ddots & & \\ & & 1 & -2 & 1 \\ & & & 1 & -2 \end{bmatrix} \in \mathbb{R}^{N-1,N-1}, \quad \boldsymbol{DF}^k := \begin{bmatrix} \partial_x f_1^k \\ \vdots \\ \partial_x f_{N-1}^k \end{bmatrix}, \quad \boldsymbol{J}_B^k := \begin{bmatrix} j_L^k \\ 0 \\ \vdots \\ 0 \\ j_R^k \end{bmatrix}.$$

Neumann values for the density are given by $dr_L^k := r^{\mathrm{Neu}}(t_k, x_L)$, $dr_R^k := r^{\mathrm{Neu}}(t_k, x_R)$ and Dirichlet values for the flux are chosen by $j_L^k := -hr^{\mathrm{Neu}}(t_k, x_L)/(2\omega)$, $j_R^k := -hr^{\mathrm{Neu}}(t_k, x_R)/(2\omega)$.

For Neumann boundary conditions on CC grids, we use

$$\boldsymbol{A} := \begin{bmatrix} -1 & 1 & & & \\ 1 & -2 & 1 & & \\ & & \ddots & & \\ & & 1 & -2 & 1 \\ & & & 1 & -1 \end{bmatrix} \in \mathbb{R}^{N,N}, \quad \boldsymbol{F}^k := \begin{bmatrix} f_{1/2}^k \\ \vdots \\ f_{N-1/2}^k \end{bmatrix}, \quad \boldsymbol{R}_B^k := \begin{bmatrix} -hdr_L^k \\ 0 \\ \vdots \\ 0 \\ hdr_R^k \end{bmatrix},$$

$$\boldsymbol{B} := \begin{bmatrix} -3 & 1 & & & \\ 1 & -2 & 1 & & \\ & & \ddots & & \\ & & 1 & -2 & 1 \\ & & & 1 & -3 \end{bmatrix} \in \mathbb{R}^{N,N}, \quad \boldsymbol{DF}^k := \begin{bmatrix} \partial_x f_{1/2}^k \\ \vdots \\ \partial_x f_{N-1/2}^k \end{bmatrix}, \quad \boldsymbol{J}_B^k := \begin{bmatrix} 2j_L^k \\ 0 \\ \vdots \\ 0 \\ 2j_R^k \end{bmatrix},$$

with dr_L^k, dr_R^k, j_L^k and j_R^k chosen as in the previous case.

The explicit and implicit Euler schemes are known to have consistency of order $\mathcal{O}(h^2 + \tau)$. For the choice $\tau \sim h^2$, one can prove convergence behavior of the same order. Reduced consistency at the boundaries does not lead to a decreased order of convergence for the presented discretizations; see Ref. [**26**, Chapter 5].

For the FD lattice Boltzmann schemes (5.1) and (5.2), in each timestep $4N$ multiplications are required. The number of total time steps is

$$M = \left\lfloor \frac{T}{\tau} \right\rfloor + 1 \approx \frac{\nu T}{|\Omega|^2} \frac{2\omega}{1-\omega} N^2.$$

Consequently, the total computational effort is of order $\mathcal{O}(N^3)$. The FV lattice Boltzmann schemes require $8N$ multiplications in each time step. However, the number of time steps is only

$$M = \left\lfloor \frac{T}{\tau} \right\rfloor + 1 \approx \frac{\nu T}{|\Omega|^2} 2\omega N^2.$$

Hence, the ratio of the computational costs of both approaches is $2(1 - \omega)$. The explicit Euler method requires $6N$ multiplications in each time step. If we use the LU-factorizations of the tridiagonal system matrices (see Ref. [**25**]), the Crank-Nicolson scheme and the implicit Euler scheme require $14N$ operations. For these methods the choice of the time step is only restricted by stability considerations.

For all introduced numerical methods the number of degrees of freedom DOF in space and time is given by $\mathrm{DOF} = CN^3$ for a given constant C. The error tolerance tol is chosen as $\mathrm{tol} = 1/N^2 \sim h^2$. We find the relation $\mathrm{DOF} = \mathrm{tol}^{-3/2}$. Hence, the decrease of the error by a factor of two requires the raise of the computational effort by a factor of $2\sqrt{2} \approx 2.8$.

In Figure 7.21 the experimental orders of convergence of the explicit Euler scheme is displayed in comparison to those of the lattice Boltzmann solutions. The underlying test problem is Example 1. We find a slightly smaller error constant

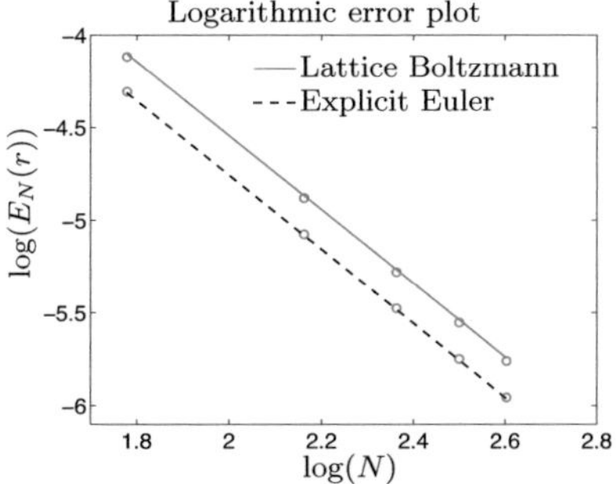

FIGURE 7.21. Example 1: Logarithmic error plot for the explicit Euler scheme and the FD lattice Boltzmann scheme.

for the explicit Euler schemes. The corresponding results for the Crank-Nicolson scheme and the implicit Euler scheme are very similar.

The consistency of the Crank-Nicolson scheme is of order $\mathcal{O}(h^2 + \tau^2)$, if the right hand side is evaluated at time $t_{k+1/2}$ instead of t_k. Hence, the choice $\tau \sim h$ seems to be appropriate to reduce the computational effort. Less time steps have to be performed. But due to the lack of maximum principles and L^∞-stability, the choice $\tau \sim h$ leads to very poor results. We shall elucidate this point in a numerical example.

Maximum principles and L^∞-stability for the discrete equations (7.3) can be verified by following the concept of inverse monotone M-matrices. A matrix $\boldsymbol{M} = [m_{ij}]_{i,j} \in \mathbb{R}^{N,N}$ is called L-matrix, if we have $m_{ij} \leq 0$ for all $i \neq j$ $(i,j = 1,\ldots,N)$ and $m_{ii} > 0$ for $i = 1,\ldots,N$. An L-matrix $\boldsymbol{M}$ is called M-matrix, if it is nonsingular with $\boldsymbol{M}^{-1} = [m_{ij}^{-1}]_{i,j}$ and $m_{ij}^{-1} \geq 0$ for $i,j = 1,\ldots,N$. Then for $\boldsymbol{x} \leq \boldsymbol{y}$ with $\boldsymbol{x}, \boldsymbol{y} \in \mathbb{R}^N$ we find $\boldsymbol{M}^{-1}\boldsymbol{x} \leq \boldsymbol{M}^{-1}\boldsymbol{y}$, where the less or equal sign has to be understood in the componentwise sense.

The matrix $\boldsymbol{M}$ is called *diagonally dominant*, if we have

$$\sum_{j=1,j\neq i}^{N} |m_{ij}| < |m_{ii}| \quad \text{for } i = 1,\ldots,N.$$

Diagonally dominant L-matrices are M-matrices (see Ref. [22]). Hence, we find that the matrices on the left hand side of (7.3) are M-matrices (or the unit matrix) for all given choices of $\boldsymbol{A}$ and $\boldsymbol{B}$. Maximum principles can now be assured, if the matrix on the right hand side is non-negative in the componentwise sense. We find the condition $1 - (1-\sigma)\kappa|m| \geq 0$, where $|m|$ is the modulus of the largest diagonal entry of the corresponding matrix $\boldsymbol{A}$ or $\boldsymbol{B}$. We find $|m| = 2$ or $|m| = 3$ in our applications. For $\sigma \in \{0, 1/2\}$ we arrive at the step size restriction

$$\nu\frac{\tau}{h^2} \leq \frac{1}{(1-\sigma)|m|}.$$

For the implicit Euler schemes with $\sigma = 1$ there is no restriction on the step size, but an equivalent choice is necessary to attain second order for the consistency and the convergence.

The step size restrictions have to be compared with the lattice Boltzmann space-time coupling, where we have

$$\kappa := \nu\frac{\tau}{h^2} = \begin{cases} \dfrac{1-\omega}{2\omega}, & \text{for the FD schemes,} \\[2mm] \dfrac{1}{2\omega}, & \text{for the FV schemes.} \end{cases}$$

The proof of a maximum principle can be found in Ref. [**23**, Section 100]. Stability in the L^∞-norm can be proved in an analogous way. See also Ref. [**22**].

The eigenvectors of the matrices $\boldsymbol{A}$ and $\boldsymbol{B}$ are given by (5.88) in the periodic case, (5.100) and (5.101) in the boundary case on VC grids and (5.107) and (5.108) in the boundary case on CC grids. The corresponding eigenvalues are $2\cos(2y_j)-2$ in the periodic case and $2\cos(y_j)-2$ in the boundary case, where we use $y_j := j\pi/N$. Hence, we find the eigenvalues of the discrete equations (7.3) in the form

$$\lambda_j^0 = 1 + 2\kappa(c_j - 1) \quad \text{for } \sigma = 0,$$

$$\lambda_j^{1/2} = \frac{1 + \kappa(c_j - 1)}{1 - \kappa(c_j - 1)} \quad \text{for } \sigma = 1/2,$$

$$\lambda_j^1 = \frac{1}{1 - 2\kappa(c_j - 1)} \quad \text{for } \sigma = 1,$$

where we use $c_j := \cos(y_j)$. For the low-frequency eigenvalues we find

$$\lambda_j^\sigma \approx 1 - \kappa y_j^2 + \mathcal{O}(y_j^4) \quad \text{for } j \ll N \text{ and } \sigma \in \{0, 1/2, 1\}.$$

The same behavior for $j \ll N$ is attained for the eigenvalues $\lambda_j^+ = (1 - \omega)c_j + (2\omega - 1 + (1 - \omega)^2 c_j^2)^{1/2}$ of the FD lattice Boltzmann discretizations and $\lambda_j^+ = (1 - \omega/2)c_j - \omega/2 + (((1 - \omega/2)c_j - \omega/2)^2 + \omega(1 + c_j) - 1)^{1/2}$ of the FV lattice Boltzmann discretizations with regard to the corresponding space-time couplings. Hence, we find in all cases the correct approximation of the time kernel of the heat equation, that is,

$$(\lambda_j)^k \approx \exp(-\nu j^2\pi^2 t_k/|\Omega|^2) \quad \text{for } j \ll N.$$

The more decisive point is the decay of the high-frequency parts. Here, we find

$$\lambda_j^0 \approx 1 - 4\kappa \quad \text{for } \sigma = 0 \text{ and } j \approx N,$$

$$\lambda_j^{1/2} \approx \frac{1 - 2\kappa}{1 + 2\kappa} \quad \text{for } \sigma = 1/2 \text{ and } j \approx N,$$

$$\lambda_j^1 \approx \frac{1}{1 + 4\kappa} \quad \text{for } \sigma = 1 \text{ and } j \approx N.$$

In order to achieve stability in the L^2-sense, these expressions have to be larger or equal than -1. For $\sigma = 0$, we have to require $\kappa \leq 1/2$. For $\sigma = 1/2$ and $\sigma = 1$, we find unconditional stability with respect to the L^2-norm. For the explicit Euler schemes we observe problems in the performance for nonsmooth data and $\kappa = 1/2$. The high-frequency parts in the solution are not negligible in this case. Due to the missing decay, these parts lead to spurious oscillations. Numerical evidence for this phenomenon is presented in the next section.

The same problem appears for the FD lattice Boltzmann discretizations, where the eigenvalues λ_j^- tend to -1 when j approaches N. In the case of the FV lattice Boltzmann discretizations, the eigenvalues λ_j^+ and λ_j^- take imaginary values and tend to -1; see Figures 5.3 and 5.7. For the choice $\omega = 1/2$, the discretization of the density stemming from the FD lattice Boltzmann equations (5.1) and (5.2) coincides with the explicit Euler scheme with $\kappa = 1/2$. These schemes show oscillations for nonsmooth data. Whereas the explicit Euler scheme can be cured by decreasing

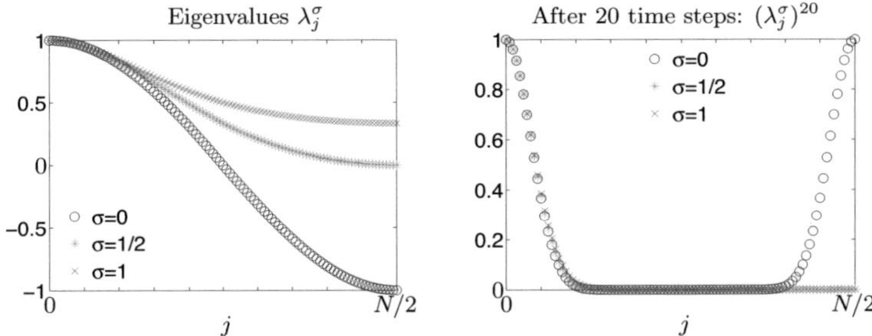

FIGURE 7.22. Behavior of the eigenvalues λ_j^σ of the σ-schemes for $N = 100$ and $\kappa = 1/2$ and the behavior after $k = 20$ time steps.

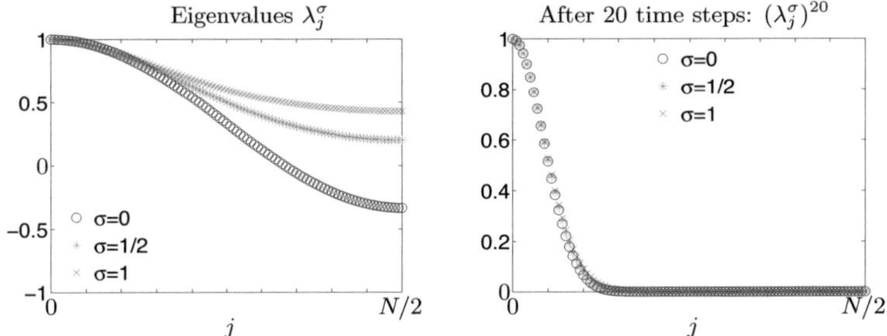

FIGURE 7.23. Behavior of the eigenvalues λ_j^σ of the σ-schemes for $N = 100$ and $\kappa = 1/3$ and the behavior after $k = 20$ time steps.

the time step in order to keep the eigenvalues away from -1, we are trapped in the fixed space-time coupling for the lattice Boltzmann schemes.

In Figure 7.22 and 7.23 we examine the behavior of the eigenvalues of the σ-schemes (7.3) for $\sigma \in \{0, 1/2, 1\}$. In Figure 7.22 we choose $\kappa = 1/2$, which corresponds to the case $\omega = 1/2$ for the FD lattice Boltzmann schemes with the same time step. In the plot on the right hand side we see the decay after 20 time steps. This figure has to be compared with Figure 5.6 for the FD lattice Boltzmann schemes. In Figure 7.23 we consider smaller time steps with $\kappa = 1/3$. For FD lattice Boltzmann schemes the same time step requires $\omega = 3/4$. While the explicit Euler scheme now shows a fast decay for the high-frequency parts, the FD lattice Boltzmann schemes are not equipped with this feature (see Figure 5.6).

Numerical reference solutions can also be computed with the aid of the two step *Du Fort-Frankel scheme*

$$R_l^{k+1} = R_l^{k-1} + 2\kappa \left(R_{l+1}^k - R_l^{k+1} - R_l^{k-1} + R_{l-1}^k \right) + 2\tau f_l^k .$$

The Du Fort-Frankel scheme is an unconditionally stable explicit scheme that is second order consistent to the heat equation and to the telegraph equation presented in Chapter 4. The space-time coupling $\tau \sim h^2$ is required; see Ref. [26] and Ref. [33].

In Example 4 we investigate the solution of the heat equation with the concentrated initial data

$$r_0(x) := \sin^{100}(\pi x) \quad \text{in } \Omega.$$

Example 4, Density

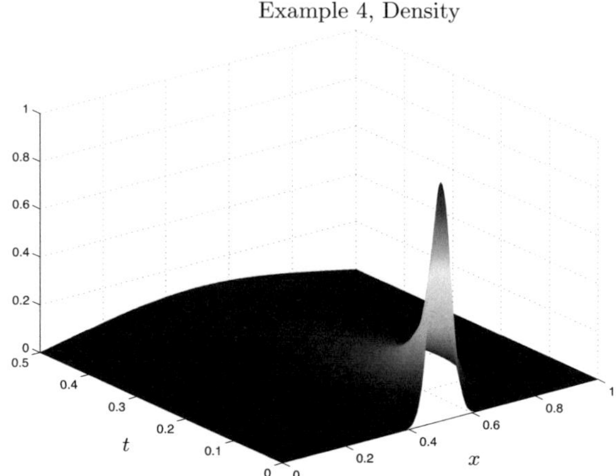

FIGURE 7.24. Lattice Boltzmann solution for the density with the concentrated initial data $r_0(x) := \sin^{100}(\pi x)$ in the domain $(0, 1/2] \times (0, 1)$ with $\nu = 0.1$ and $N = 200$.

Example 4, Flux

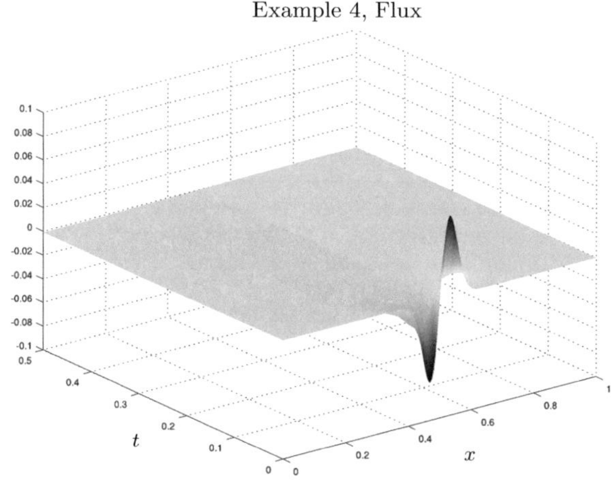

FIGURE 7.25. Lattice Boltzmann solution for the flux with the initial data $r_0(x) := \sin^{100}(\pi x)$ in the domain $(0, 1/2] \times (0, 1)$ with $\nu = 0.1$ and $N = 200$.

The lattice Boltzmann solutions are plotted in Figures 7.24 and 7.25.

The application of the Crank-Nicolson scheme with the space-time coupling $\tau \sim h$ leads to wrong results. In Figure 7.26 the Crank-Nicolson solution for Example 4 with time step $\tau = 2h$ is plotted. At the peak there is an unphysical oscillation. In Ref. [**23**, Section 100], the author attributes this behavior to the lack of maximum principles. As for the lattice Boltzmann schemes, the coupling $\tau \sim h^2$ is required.

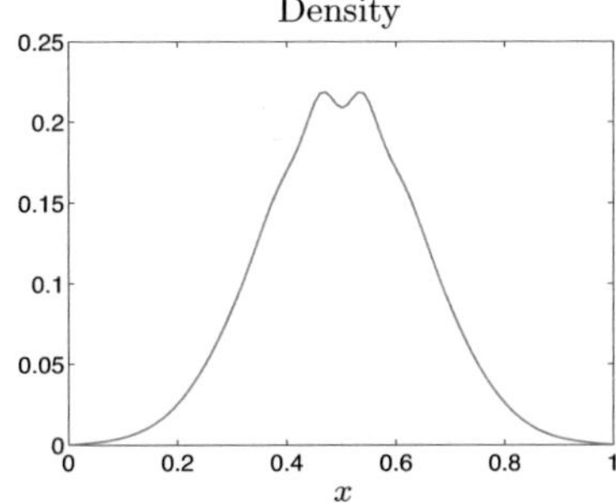

FIGURE 7.26. Example 4: Crank-Nicolson solution at time $t = 0.1$ for the time step size $\tau = 2h$ and $N = 100$.

σ-scheme	Density			Flux		
	EOC	K_r	$E_{400}(r)$	EOC	K_j	$E_{400}(j)$
Explicit Euler	2.00	3.83e-1	2.46e-6	3.00	8.34e0	1.32e-7
Crank-Nicolson	2.00	2.94e-1	1.84e-6	3.00	2.38e0	3.82e-8
Implicit Euler	2.00	9.79e-1	6.15e-6	3.00	6.68e0	1.04e-7

TABLE 7.15. Example 4: Experimental orders of convergence, error constants and fitted errors with respect to the discrete L^2-norms. The VCFD lattice Boltzmann solution are compared with the solutions of the σ-schemes on the same grid, where we use a sequence of grids with $N = 60,\ 145,\ 230,\ 315,\ 400$, the end time $T = 0.1$ and $\omega = 0.7$.

In Table 7.15 we examine the error behavior of the FD lattice Boltzmann solution for Example 4. All errors are determined on the basis of the reference solutions of the σ-schemes on the same grids. We find the expected second order convergence for the density and third order convergence for the h-scaled flux.

7.5. Nonsmooth Data

In many (multi-dimensional) applications the given data do not have the required regularity. In our test problems we consider initial data representing a shock or a concentrated peak.

In Example 5, the initial data for the heat equation is given by the shock

$$r_0(x) := \begin{cases} 0, & \text{for } x \in (0, x_S), \\ 1, & \text{for } x \in [x_S, 1) \end{cases}$$

for fixed $x_S \in (0, 1)$. The corresponding VC lattice Boltzmann solution for the density with vanishing density boundary conditions, $x_S = 0.7$, $\nu = 0.1$, $\omega = 0.7$, $N = 100$ and the end time $T = 0.02$ is shown in Figure 7.27. In Figure 7.28 the corresponding flux is displayed. In Figure 7.27 no oscillations are observed in the solution for the density. The flux shows some minor initial oscillations (as for smooth initial data).

As reference solutions we choose the Crank-Nicolson solutions on the same grid. The errors for the density and the flux with respect to the Crank-Nicolson solutions

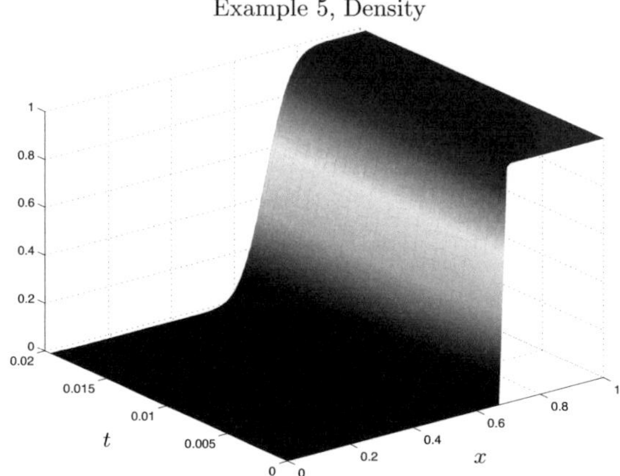

FIGURE 7.27. Lattice Boltzmann solution for the density with the initial shock in $(0, 0.02] \times (0, 1)$ with $\nu = 0.1$, $\omega = 0.7$ and $N = 200$.

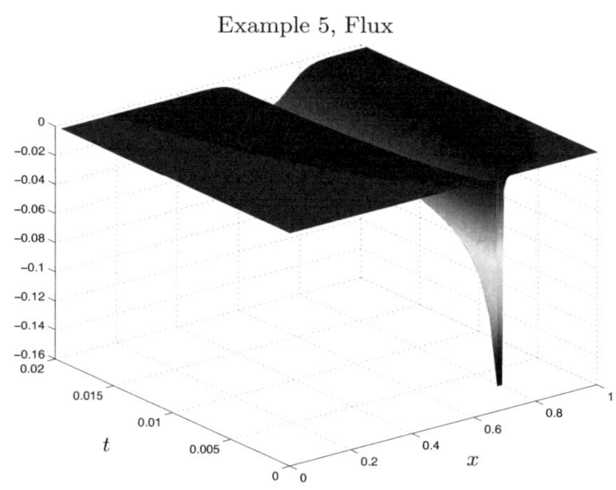

FIGURE 7.28. Lattice Boltzmann solution for the flux with the initial shock in $(0, 0.02] \times (0, 1)$ with $\nu = 0.1$, $\omega = 0.7$ and $N = 200$.

are displayed in Figures 7.29 and 7.30. The error in the density changes sign in every grid point around the shock.

If we consider the errors with respect to the Crank-Nicolson solutions on the same grid on a sequence of grids, then the experimental orders of convergence are not the same as for smooth data. There is only first order convergence for the density and the h-scaled flux. Due to the missing decay of the high-frequency parts, the lattice Boltzmann solutions are responsible for this failure.

Example 5, Error in density

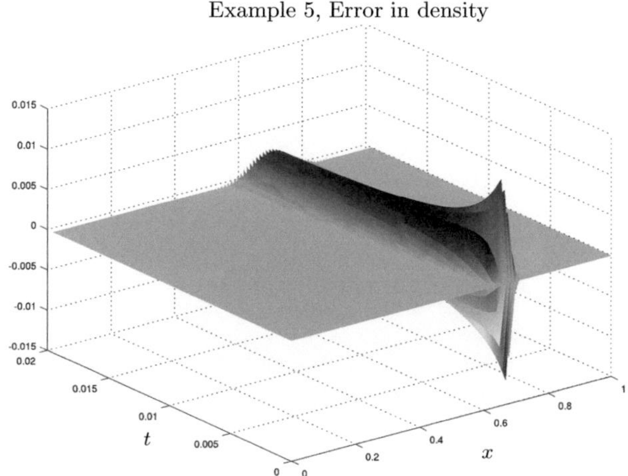

FIGURE 7.29. Error of the lattice Boltzmann solution for the density with respect to the Crank-Nicolson solution.

Example 5, Error in flux

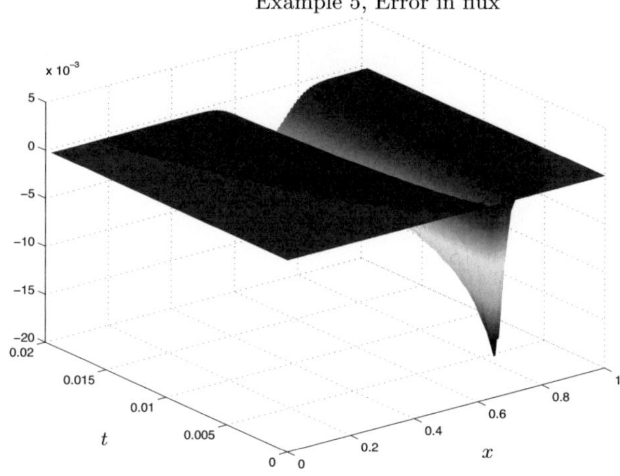

FIGURE 7.30. Error of the lattice Boltzmann solution for the flux with respect to the Crank-Nicolson solution.

In Example 6, we examine the lattice Boltzmann solutions, where the initial data

$$
r_0(x) := \begin{cases} \dfrac{x - x_{S-K}}{x_S - x_{S-K}} & \text{for } x \in (x_{S-K}, x_S), \\ \dfrac{x_{S+K} - x}{x_{S+K} - x_S} & \text{for } x \in (x_S, x_{S+K}), \\ 0 & \text{else}, \end{cases}
$$

Example 6

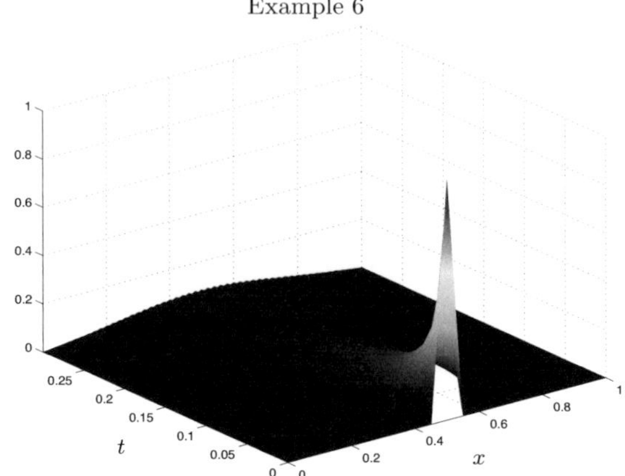

FIGURE 7.31. Crank-Nicolson solution for the density.

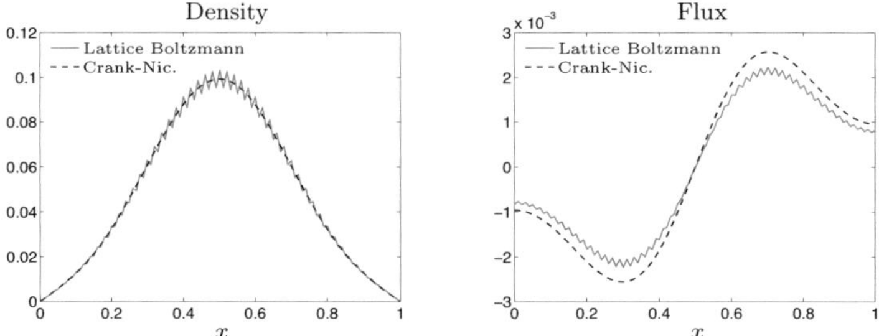

FIGURE 7.32. Lattice Boltzmann solutions and Crank-Nicolson solutions with the same time step size for the density and the flux at time $t = 0.2$ with $N = 100$ and $\omega = 0.7$.

for $K \in \mathbb{N}$ and $S \in \{K, \ldots, N - K\}$ find application. This is a hat function with peak in x_S and a basis of length $2Kh$.

The Crank-Nicolson solution for the density with $\nu = 0.1$ and vanishing density boundary conditions on a VC grid with $N = 100$ is plotted in Figure 7.31.

Taking a look at the lattice Boltzmann solutions, we find disturbing oscillations, whereas the Crank-Nicolson solutions or the explicit Euler solutions are smooth. In Figure 7.32 the lattice Boltzmann solutions for $\omega = 0.7$ and the Crank-Nicolson solutions for the density and the flux are plotted at time $t = 0.2$. In this example we find first order convergence for the density and second order convergence for the h-scaled flux, where the errors are taken with respect to the Crank-Nicolson solutions on the same grids.

7.6. Lattice Boltzmann or σ-Schemes?

In this section we shall briefly summarize the advantages and disadvantages of the lattice Boltzmann schemes and compare them with the presented σ-schemes.

The first point is the fact that all lattice Boltzmann methods are explicit methods, which only work with the constraint of step size restrictions. But if we take a look at the implicit Euler scheme, we find that due to the consistency the same step size restriction has to be applied here. For the Crank-Nicolson schemes, larger time step can be chosen from a theoretical point of view. However, the oscillations in the solution show that this is not a good choice in most of the applications. Because of the larger numerical effort of the implicit schemes, the lattice Boltzmann methods can be seen in competition with the explicit Euler scheme only. The number of required multiplications in each time step is smaller for the lattice Boltzmann schemes. However, the total number of timesteps may differ. For all methods the total computational amount is of order $\mathcal{O}(N^3)$.

For vanishing source terms, the lattice Boltzmann schemes on VC grids are superior due to the fourth order convergence for the density for the specific choice of the relaxation parameter.

For the application of the explicit Euler scheme, the data of the right hand side have to be differentiated for the flux equation. For the lattice Boltzmann methods there is no differentiation of the source data required. For Dirichlet boundary conditions, the heat equation as the limiting equation has to be employed for the data of the σ-schemes.

The decisive disadvantage of the lattice Boltzmann schemes is the distribution of the eigenvalues. For all parameter choices there are eigenvalues close to -1. This fact leads to oscillations for nonsmooth data. In the case of the explicit Euler scheme, this effect only appears for the choice $\nu\tau/h^2 = 1/2$. For smaller time steps this lack can be repaired. For the lattice Boltzmann schemes there is no comparable mending. Even in the presence of transport terms this behavior does not change, as we shall see in the next section, where we examine lattice Boltzmann methods for the advection-diffusion equation.

7.7. The Advection-Diffusion Equation

The linear Ruijgrok-Wu model is a velocity discrete system of the form

$$\begin{aligned}
\partial_t u^\epsilon + \frac{1}{\epsilon}\partial_x u^\epsilon + \frac{1}{2\nu\epsilon^2}(u^\epsilon - v^\epsilon) - \frac{A}{2\nu\epsilon}(u^\epsilon + v^\epsilon) &= \frac{1}{2}f \quad \text{in } \Omega_T, \\
\partial_t v^\epsilon - \frac{1}{\epsilon}\partial_x v^\epsilon - \frac{1}{2\nu\epsilon^2}(u^\epsilon - v^\epsilon) + \frac{A}{2\nu\epsilon}(u^\epsilon + v^\epsilon) &= \frac{1}{2}f \quad \text{in } \Omega_T.
\end{aligned} \tag{7.4}$$

As before we define the density $r^\epsilon := u^\epsilon + v^\epsilon$ and the scaled flux $\kappa^\epsilon := (u^\epsilon - v^\epsilon)/\epsilon$. The fluid-dynamic limit equation of this system is the advection-diffusion equation

$$\partial_t r + A\partial_x r - \nu\partial_x^2 r = f, \tag{7.5}$$

where the flux is defined by $Ar - \nu\partial_x r$. Numerical problems for the solution of the advection-diffusion equation appear, when the ratio A/ν is large. This is known as the convection dominated case. In typical situations boundary layers have to be treated. In fact, in the lattice Boltzmann approach we have to treat a double singularly perturbed problem then. In the following, we assume that the ratio A/ν is of moderate size.

The lattice-Boltzmann equations on VC grids for this advection-diffusion equation are obtained in the form

$$\begin{aligned}
U_{l+1}^{k+1} = U_l^k - \omega(U_l^k - V_l^k) + \omega A\frac{\tau}{h}(U_l^k + V_l^k) + \frac{\tau}{2}F_l^k, \quad l = 0, \ldots, N-1, \\
V_{l-1}^{k+1} = V_l^k + \omega(U_l^k - V_l^k) - \omega A\frac{\tau}{h}(U_l^k + V_l^k) + \frac{\tau}{2}F_l^k, \quad l = 1, \ldots, N.
\end{aligned} \tag{7.6}$$

Boundary conditions have to be supplied for U_0^{k+1} and V_N^{k+1}. The space-time coupling has to be chosen of the form $\gamma = h^2/\tau = 2\omega\nu/(1 - \omega)$. Hence, we find

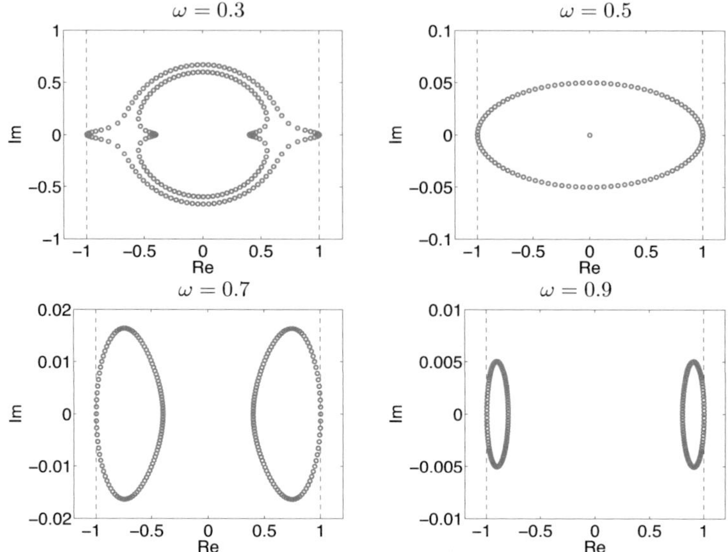

FIGURE 7.33. Eigenvalues of the time evolution matrix of the VC lattice Boltzmann schemes for the advection-diffusion equation with $\nu = 0.1$, $A = 1.0$ and $N = 100$ in dependence on ω.

$\omega A\tau/h = (1 - \omega)Ah/(2\nu)$. The complex eigenvalues of the corresponding time-evolution matrix are plotted in Figure 7.33 for $\nu = 0.1$, $A = 1.0$ and $N = 100$. For this choice of the parameters we gain L^2-stable solutions. The stability gets lost, when the *local Péclet number* Ah/ν becomes larger (Ref. [**26**]).

For all values of ω, there are several eigenvalues close to -1, which is also attained as a eigenvalue. This leads to the conjecture that there is an oscillatory behavior for the lattice Boltzmann solutions in the case of nonsmooth data.

The lattice Boltzmann solution for the advection-diffusion equation (7.5) with $A = 10$ and $\nu = 0.1$ for the nonsmooth initial data from Example 6 is plotted in Figure 7.34. Taking a closer look at the solution at time $t = 0.02$, we find the expected oscillations in the density; see Figure 7.35.

7.8. The Viscous Burgers Equation

The nonlinear Ruijgrok-Wu model

$$\partial_t u^\epsilon + \frac{1}{\epsilon}\partial_x u^\epsilon + \frac{1}{2\nu\epsilon^2}(u^\epsilon - v^\epsilon) - \frac{B}{\nu\epsilon}u^\epsilon v^\epsilon = \frac{1}{2}f \quad \text{in } \Omega_T,$$
$$\partial_t v^\epsilon - \frac{1}{\epsilon}\partial_x v^\epsilon - \frac{1}{2\nu\epsilon^2}(u^\epsilon - v^\epsilon) + \frac{B}{\nu\epsilon}u^\epsilon v^\epsilon = \frac{1}{2}f \quad \text{in } \Omega_T \tag{7.7}$$

is a velocity discrete model with the viscous Burgers equations

$$\partial_t r + \frac{B}{2}\partial_x r^2 - \nu\partial_x^2 r = f \tag{7.8}$$

as the fluid-dynamic limit equation. The flux is defined by $B/(2\gamma)r^2 - 1/(2\omega)\partial_x r$.

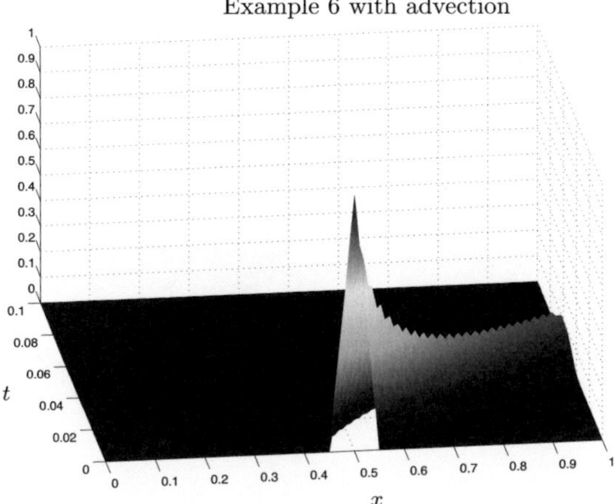

FIGURE 7.34. Example 6: Density solution of the advection-diffusion equation with nonsmooth initial data.

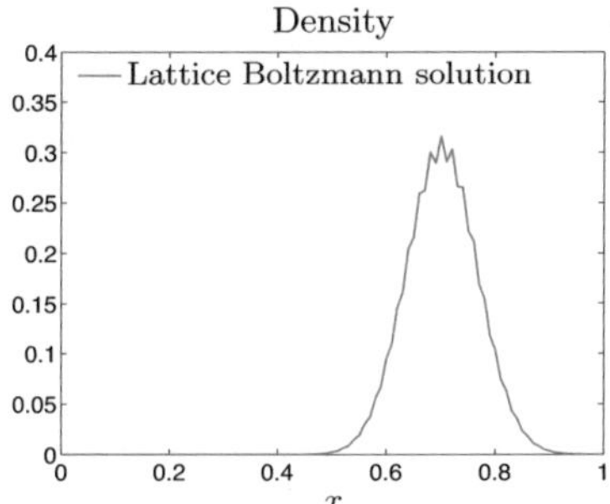

FIGURE 7.35. Example 6: Oscillations in the density solution of the advection-diffusion equation at time $t = 0.02$.

The lattice Boltzmann equations on VC grids for the viscous Burgers equation read

$$U_{l+1}^{k+1} = U_l^k - \omega(U_l^k - V_l^k) + 2\omega B \frac{\tau}{h} U_l^k V_l^k + \frac{\tau}{2} F_l^k, \quad l = 0, \ldots, N-1,$$
$$V_{l-1}^{k+1} = V_l^k + \omega(U_l^k - V_l^k) - 2\omega B \frac{\tau}{h} U_l^k V_l^k + \frac{\tau}{2} F_l^k, \quad l = 1, \ldots, N,$$

$$(7.9)$$

with additional boundary conditions for U_0^{k+1} and V_N^{k+1}. The space-time coupling has to be chosen by $\gamma = h^2/\tau = 2\omega\nu/(1-\omega)$.

As test example we consider the moving shock solution

$$r(t, x) := \frac{1}{2}\left(1 - \tanh\left(\frac{2x - t - 1}{8\nu}\right)\right) \quad \text{in } (0, T] \times (0, 1)$$

of the viscous Burgers equation (7.8) with $B = 1$. We choose a small viscosity with $\nu = 0.01$. The solution for the density with end time $T = 0.5$ is plotted in Figure 7.36.

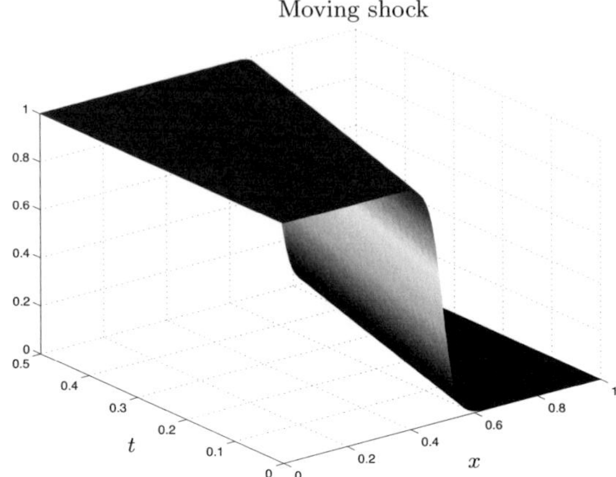

FIGURE 7.36. Moving shock solution of the viscous Burgers equation.

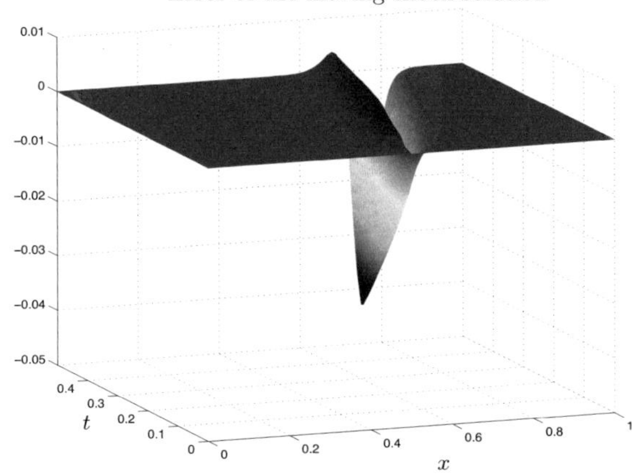

FIGURE 7.37. Error of the moving shock solution of the viscous
Burgers equation.

The nodal errors of the lattice Boltzmann solution are plotted in Figure 7.37.
The error increases as the time advances. For the density boundary conditions we
find second order convergence for the density and third order convergence for the
h-scaled flux.

CHAPTER 8

Lattice Boltzmann Methods on Coupled Grids

Lattice Boltzmann methods are designed for uniform grids. The symmetry of the grids is a basic ingredient for the performance. But due to the space-time coupling, local refinements of the grid are desirable. This is the basis for the setup of adaptive methods. In order to preserve the influence of the symmetry, only refinements with locally constant grid sizes can be chosen. The grid sizes of two neighboring grids should differ by a factor of two and the refinement zones have to possess a minimal number of cells in their width.

At the intersections of the refinement zones with grid sizes $h = h_C$ and $h = h_F = h_C/2$ intersection conditions for the distribution functions have to be prescribed. Since we are interested in the approximation of the macroscopic mass density r and its derivative $\partial_x r$, we postulate the discrete continuity of $U + V$ and $(U - V)/h$ at the intersections, that is, the corresponding quantities have to coincide on the coarse and on the fine grid in selected points of the intersection zones. Due to the approximation properties

$$U \approx \frac{1}{2}[r(x_l)]_l - \frac{h}{4\omega}[\partial_x r(x_l)]_l,$$

$$V \approx \frac{1}{2}[r(x_l)]_l + \frac{h}{4\omega}[\partial_x r(x_l)]_l,$$

the distribution functions u and v cannot be assumed to be continuous at the zone intersections because of their dependence on the grid size h. The approximation properties lead to a dependence on the grid size for the initial data and for the inflow boundary conditions. The data for the distribution functions have to be computed with respect to the grid size. Since the parameters ν and γ do not (or should not) depend on the grid size, the relaxation parameter ω does not depend on the grid size. By refining and coarsening the grid, the grid size h changes. Since $U + V$ and $(U - V)/h$ should be left unchanged, we have to rescale the distribution functions U and V. For the refinement and the coarsening, we impose invariance of $U + V$ and $(U - V)/h$. Hence, we get the transformation conditions

$$R^C = U^C + V^C = U^F + V^F = R^F,$$

$$J^C = U^C - V^C = 2U^F - 2V^F = 2J^F.$$

The upper indices indicate the affiliation of the variables to the coarse grid (C) or to the fine grid (F). For the distribution functions we find the converse relations

$$U^C = \frac{1}{2}\left(R^C + J^C\right) = \frac{1}{2}\left(R^F + 2J^F\right) = \frac{3}{2}U^F - \frac{1}{2}V^F,$$

$$V^C = \frac{1}{2}\left(R^C - J^C\right) = \frac{1}{2}\left(R^F - 2J^F\right) = -\frac{1}{2}U^F + \frac{3}{2}V^F,$$

$$U^F = \frac{1}{2}\left(R^F + J^F\right) = \frac{1}{2}\left(R^C + \frac{1}{2}J^C\right) = \frac{3}{4}U^C + \frac{1}{4}V^C,$$

$$V^F = \frac{1}{2}\left(R^F - J^F\right) = \frac{1}{2}\left(R^C - \frac{1}{2}J^C\right) = \frac{1}{4}U^C + \frac{3}{4}V^C.$$

(a) Single overlapping VC grids

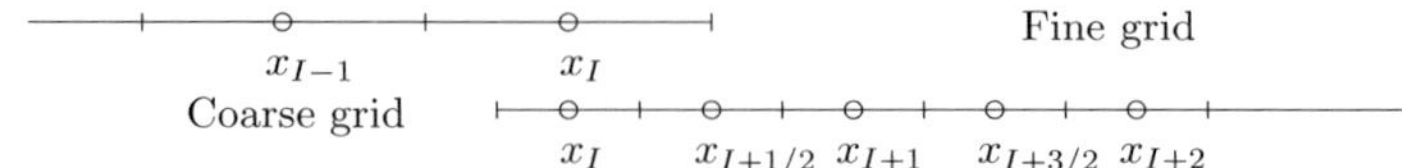

(b) Double overlapping VC grids

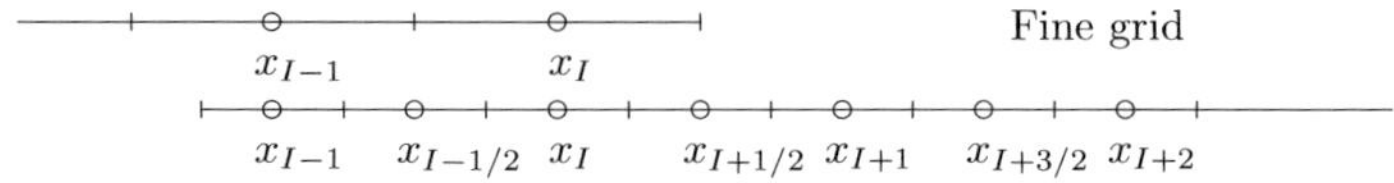

FIGURE 8.1. Grid intersections for VC grids.

(a) Single overlapping CC grids

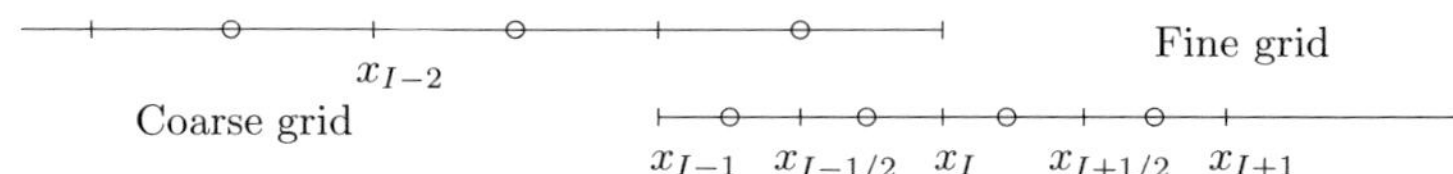

(b) Double overlapping CC grids

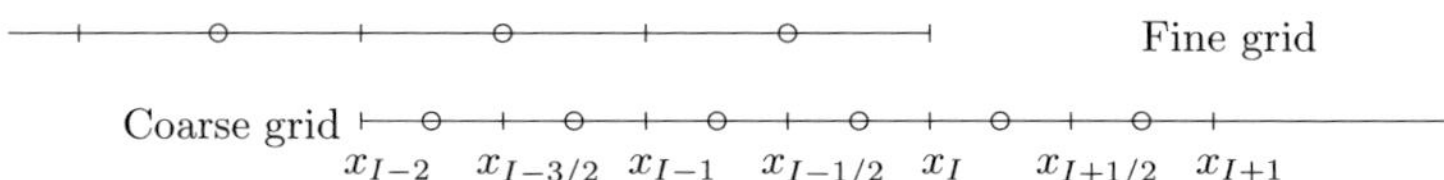

FIGURE 8.2. Grid intersections for CC grids.

For the refinement, the data for $U + V$ and $U - V$ have to be interpolated and transformed. For the coarsening of the grid, the data have to be restricted to the nodes of the coarse grid and transformed.

Due to the space time coupling $\tau = h^2/\gamma$, four time steps have to be performed on grids with halved grid size in order to arrive at the same physical time that is reached after one time step on the nonrefined grid. At the grid intersections time interpolations are required for the intersection conditions.

In computations with a fixed end time T, a global refinement of the grid decreases the error by a factor of four. This result is implied by the quadratic convergence with respect to the grid size that is approved from the theoretical and experimental results in this work. However, the computational effort is raised by a factor of eight due to the necessary space-time coupling.

8.1. Overlapping Grids

For the exchange of the data at the intersection zones, the coarse and the fine grids have to overlap to some extent. We describe some of the possible choices. For VC grids and for CC grids we consider single and double overlapping grids as depicted in Figure 8.1 and in Figure 8.2.

In the situation of double overlapping grids one can restrict to the interior grid points, where all information on the macroscopic quantities is available. On single overlapping grids the data of both grids have to be used at the same time.

The numerical results on single overlapping grids do not yield the desired performance in all situations. For Example 8 in Section 8.5, the errors are amplified at the intersection and the density error takes its maximum there. Similar observations are found for some of the coupling conditions on double overlapping grids. The density intersection conditions on double overlapping grids turn out to be favorable, since they render the expected improvement for the computations.

8.2. Intersection Conditions

For the treatment at the intersection zones, there are three possibilities to determine the missing inflow values.

(1) For single overlapping VC grids, the outermost points of the coarse and the fine grid coincide. Hence, the two outflow values of the coarse and the fine grid can be used to determine the two missing inflow values. If the coarse grid is on the left hand side, then the outflow values U_{Out}^C and V_{Out}^F are known. The inflow values U_{In}^F and V_{In}^C can be computed by

$$U_{\text{In}}^F = \frac{1}{3} V_{\text{Out}}^F + \frac{2}{3} U_{\text{Out}}^C,$$
$$V_{\text{In}}^C = \frac{4}{3} V_{\text{Out}}^F - \frac{1}{3} U_{\text{Out}}^C.$$

If the coarse grid is on the right hand side, then the inflow values V_{In}^F and U_{In}^C can be determined by employing the known outflow values V_{Out}^C and U_{Out}^F in the form

$$V_{\text{In}}^F = \frac{1}{3} U_{\text{Out}}^F + \frac{2}{3} V_{\text{Out}}^C,$$
$$U_{\text{In}}^C = \frac{4}{3} U_{\text{Out}}^F - \frac{1}{3} V_{\text{Out}}^C.$$

See Figure 8.3.

(2) For double overlapping VC grids, the missing inflow values on the fine grid can be determined by using the information that is given on the coarse grid and the other way round by restricting to interior points. For the determination of the inflow values on the fine grid, there are three possibilities given:

  i) Determine R^C on the coarse grid and use the density intersection condition $R^F = R^C$.

  ii) Determine J^C on the coarse grid and use the flux intersection condition $J^F = J^C/2$.

  iii) Determine R^C and J^C on the coarse grid and use the inflow intersection conditions $U^F = R^C/2 + J^C/4$ or $V^F = R^C/2 - J^C/4$, respectively.

For the determination of the inflow value on the coarse grid, there are four possibilities:

  i) Determine R^F on the fine grid and use the density intersection condition $R^C = R^F$.

  i) Determine J^F on the fine grid and use the flux intersection condition $J^C = 2J^F$.

  iii) Determine R^F and J^F on the fine grid and use the inflow intersection conditions $U^C = R^F/2 + J^F$ or $V^C = R^F/2 - J^F$, respectively.

  iv) Determine R^F and J^F on the fine grid and update the inflow and the outflow value on the coarse grid by using $U^C = R^F/2 + J^F$ and $V^C = R^F/2 - J^F$.

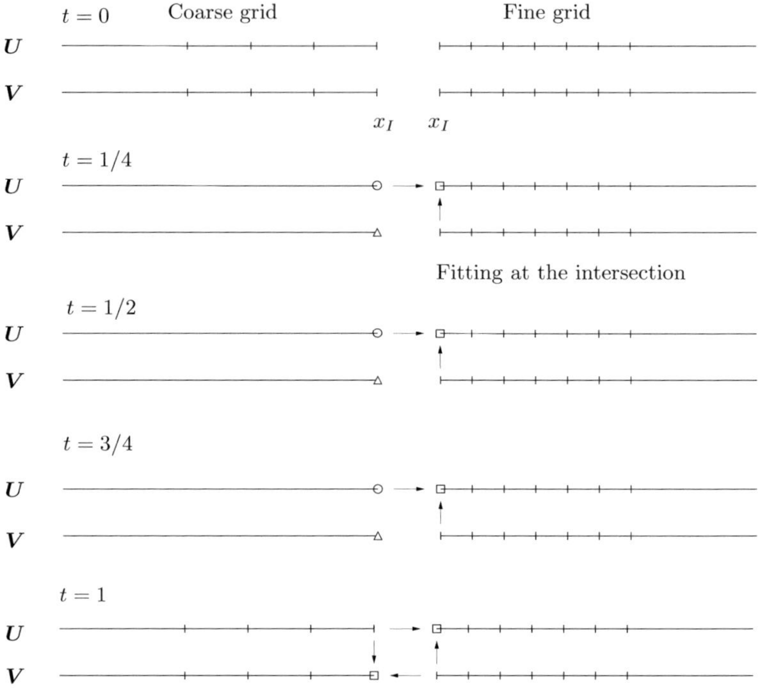

One time step on the coarse grid.   Four time steps on the fine grid.

Time interpolation of the outflow value on the coarse grid at the intersection.

| Known value | Determine value by the intersection cond. |
| Time interpolated value | Value of no interest |

FIGURE 8.3. Fitting at the intersections for single overlapping VC grids.

(3) For single overlapping grids, the missing inflow values can be extrapolated from the interior of the domain.

The extrapolation of the inflow values from the interior of the domain leads to reduced orders of convergence in the boundary situations. The same poor performance is attained at the grid intersections. Hence, we do not pursue the approach in (3).

On CC grids the loci of the coarse and the fine grid points do not coincide at the intersection zone. Hence, a spatial interpolation is required. These additional interpolations render unsatisfactory convergence results.

The general algorithm is the same for all cases. Assume that all data is given at the lattice Boltzmann time $t = 0$. At first, one lattice Boltzmann step is performed on the coarse grid to arrive at lattice Boltzmann time $t = 1$. The data from the coarse grid at the intersection has to be interpolated in time to the intermediate times $t = 1/4$, $t = 1/2$ and $t = 3/4$. From this interpolated data the inflow values on the coarse grid are determined. Now there are four lattice Boltzmann steps to

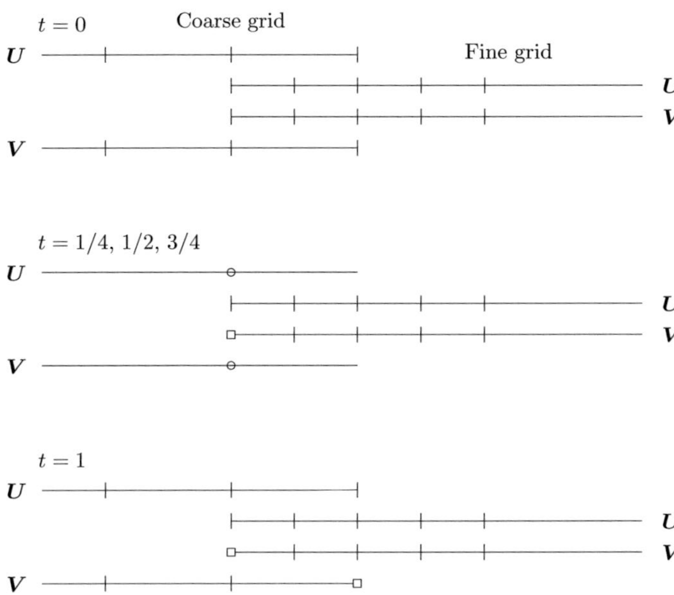

| Known value
□ Determine value by the intersection conditions
○ Time interpolated value

FIGURE 8.4. Fitting at the intersections for double overlapping VC grids.

be done on the fine grid to arrive at the time $t = 1$. Then the missing inflow value on the coarse grid at the time $t = 1$ is completed by the use of the data of the fine grid.

8.3. Algorithms for VC Grid Intersections

In the following, capital letters U, V, R and J denote the grid functions on the coarse grid, and small letters u, v, r and j denote the grid functions on the fine grid. In the descriptions of the algorithms we distinguish if the coarse grid is located at the left hand side of the fine grid (coarse - fine) or if it is located on the right hand side of the fine grid (fine - coarse).

LB-algorithm for single overlapping VC grid intersections :
See Figure 8.1(a). The intersection point is x_I.

(1) Let the solutions U^0, V^0 on the coarse grid, and u^0, v^0 on the fine grid be given for the time $t = 0$.
(2) Perform one LB-step on the coarse grid to get U^1, V^1. The inflow value on the coarse grid at the intersection is missing.
(3) Interpolate the outflow values on the coarse grid to the intermediate times $t = 1/4$, $t = 1/2$ and $t = 3/4$:

coarse - fine	fine - coarse
$U_I^0,\ U_I^1 \to U_I^{k/4}, \quad k = 1, 2, 3,$	$V_I^0,\ V_I^1 \to V_I^{k/4}, \quad k = 1, 2, 3,$

(4) Perform four LB-steps on the fine grid. Determine the inflow value on the fine grid in each step:

coarse - fine	fine - coarse
$u_I^{k/4} = v_I^{k/4}/3 + 2U_I^{k/4}/3,$ $k = 1, 2, 3, 4,$	$v_I^{k/4} = u_I^{k/4}/3 + 2V_I^{k/4}/3,$ $k = 1, 2, 3, 4.$

(5) Update the unknown inflow value on the coarse grid:

coarse - fine	fine - coarse
$V_I^1 = 4v_I^1/3 - U_I^1/3$	$U_I^1 = 4u_I^1/3 - V_I^1/3.$

LB-algorithm for double overlapping VC grid intersections:
See Figure 8.1(b). The intersection point is x_I. Extend the fine grid into the coarse grid.

(1) Let the solutions $\boldsymbol{U}^0$, $\boldsymbol{V}^0$ on the coarse grid, and $\boldsymbol{u}^0$, $\boldsymbol{v}^0$ on the fine grid be given for the time $t = 0$.

(2) Perform one LB-step on the coarse grid to get $\boldsymbol{U}^1$, $\boldsymbol{V}^1$. The inflow value on the coarse grid at the intersection is missing.

(3) Restrict to the inner point x_{I-1} (coarse - fine) or x_{I+1} (fine - coarse) of the coarse grid.

(4) Interpolate the macroscopic values R^0, J^0 and R^1, J^1 to the intermediate times $t = 1/4$, $t = 1/2$ and $t = 3/4$:

coarse - fine	fine - coarse
$R_{I-1}^0,\ R_{I-1}^1 \to R_{I-1}^{k/4}, \quad k = 1, 2, 3,$	$R_{I+1}^0,\ R_{I+1}^1 \to R_{I+1}^{k/4}, \quad k = 1, 2, 3,$
$J_{I-1}^0,\ J_{I-1}^1 \to J_{I-1}^{k/4}, \quad k = 1, 2, 3,$	$J_{I+1}^0,\ J_{I+1}^1 \to J_{I+1}^{k/4}, \quad k = 1, 2, 3.$

(5) Perform four LB-steps on the fine grid. Determine the inflow value on the fine grid in each step by using the

    i) density intersection conditions

coarse - fine
$u_{I-1}^{k/4} + v_{I-1}^{k/4} = R_{I-1}^{k/4}, \quad k = 1, 2, 3, 4,$

fine - coarse
$u_{I+1}^{k/4} + v_{I+1}^{k/4} = R_{I+1}^{k/4}, \quad k = 1, 2, 3, 4,$

  ii) flux intersection conditions

coarse - fine
$u_{I-1}^{k/4} - v_{I-1}^{k/4} = J_{I-1}^{k/4}/2, \quad k = 1, 2, 3, 4,$

fine - coarse
$u_{I+1}^{k/4} - v_{I+1}^{k/4} = J_{I+1}^{k/4}/2, \quad k = 1, 2, 3, 4,$

 iii) inflow intersection conditions

coarse - fine
$u_{I-1}^{k/4} = R_{I-1}^{k/4}/2 + J_{I-1}^{k/4}/4, \quad k = 1, 2, 3, 4,$

$$\text{fine - coarse}$$

$$v_{I+1}^{k/4} = R_{I+1}^{k/4}/2 - J_{I+1}^{k/4}/4, \quad k = 1, 2, 3, 4.$$

(6) Determine the inflow value on the coarse grid by the macroscopic variables r_I^1 and j_I^1 in the inner point of the fine grid by using the
    i) density intersection condition $U_I^1 + V_I^1 = r_I^1$,
    ii) flux intersection condition $U_I^1 - V_I^1 = 2j_I^1$,
    iii) inflow intersection condition $U_I^1 = r_I^1/2 + j_I^1$ (coarse - fine) or $V_I^1 = r_I^1/2 - j_I^1$ (fine - coarse),
    iv) update the inflow and outflow value on the coarse grid by taking $U_I^1 = r_I^1/2 + j_I^1$ and $V_I^1 = r_I^1/2 - j_I^1$.

In both proposed algorithms we use linear time interpolations at the intersection.

8.4. Algorithms for CC Grid Intersections

LB-algorithm for single overlapping CC grid intersections:
See Figure 8.2(a). The intersection point is $x_{I-1/2}$.

(1) Let the solutions $\boldsymbol{U}^0$, $\boldsymbol{V}^0$ on the coarse grid, and $\boldsymbol{u}^0$, $\boldsymbol{v}^0$ on the fine grid be given for the time $t = 0$.
(2) Perform one LB-step on the coarse grid to get $\boldsymbol{U}^1$ and $\boldsymbol{V}^1$. The inflow value on the coarse grid at the intersection is missing.
(3) Interpolate the outflow values on the coarse grid to the intermediate times $t = 1/4$, $t = 1/2$ and $t = 3/4$:

$$\text{coarse - fine}$$

$$U_{I-1/2}^0, U_{I-1/2}^1 \rightarrow U_{I-1/2}^{k/4}, \quad k = 1, 2, 3,$$

$$\text{fine - coarse}$$

$$V_{I+1/2}^0, V_{I+1/2}^1 \rightarrow V_{I+1/2}^{k/4}, \quad k = 1, 2, 3.$$

(4) Perform four LB-steps on the fine grid. Determine the inflow values on fine grid in each step:

$$\text{coarse - fine}$$

$$u_{I-3/4}^{k/4} = (v_{I-3/4}^{k/4} + v_{I-1/4}^{k/4})/3 + 4U_{I-1/2}^{k/4}/3 - u_{I-1/4}^{k/4}, \quad k = 1, 2, 3, 4,$$

$$\text{fine - coarse}$$

$$v_{I+3/4}^{k/4} = (u_{I+1/4}^{k/4} + u_{I+3/4}^{k/4})/3 + 4V_{I+1/2}^{k/4}/3 - v_{I+1/4}^{k/4}, \quad k = 1, 2, 3, 4.$$

(5) Update the unknown inflow value on the coarse grid:

$$\text{coarse - fine}$$

$$V_{I-1/2}^1 = 2(v_{I-3/4}^1 + v_{I-1/4}^1)/3 - U_{I-1/2}^1/3,$$

$$\text{fine - coarse}$$

$$U_{I+1/2}^1 = 2(u_{I+1/4}^1 + u_{I+3/4}^1)/3 - V_{I+1/2}^1/3.$$

LB-algorithm for double overlapping CC grid intersections:
See Figure 8.2(b). The intersection point is $x_{I-1/2}$. Extend the fine grid into the coarse grid.

(1) Let the solutions $\boldsymbol{U}^0$, $\boldsymbol{V}^0$ on the coarse grid, and $\boldsymbol{u}^0$, $\boldsymbol{v}^0$ on the fine grid be given for the time $t = 0$.

(2) Perform one LB-step on the coarse grid to get U^1, V^1. The inflow value on the coarse grid at the intersection is missing.

(3) Restrict to the inner value $x_{I-3/2}$ (coarse - fine) or $x_{I+3/2}$ (fine - coarse) of the coarse grid.

(4) Interpolate the macroscopic quantities R^0, J^0 and and R^1, J^1 to the intermediate times $t = 1/4$, $t = 1/2$ and $t = 3/4$:

coarse - fine

$$R^0_{I-3/2},\ R^1_{I-3/2} \to R^{k/4}_{I-3/2}, \quad k = 1, 2, 3,$$

$$J^0_{I-3/2},\ J^1_{I-3/2} \to J^{k/4}_{I-3/2}, \quad k = 1, 2, 3,$$

fine - coarse

$$R^0_{I+3/2},\ R^1_{I+3/2} \to R^{k/4}_{I+3/2}, \quad k = 1, 2, 3,$$

$$J^0_{I+3/2},\ J^1_{I+3/2} \to J^{k/4}_{I+3/2}, \quad k = 1, 2, 3.$$

(5) Perform four LB-steps on the fine grid. Determine the inflow value on the fine grid in each step by using the

    i) density intersection conditions

coarse - fine

$$(u^{k/4}_{I-7/4} + v^{k/4}_{I-7/4} + u^{k/4}_{I-5/4} + v^{k/4}_{I-5/4})/2 = R^{k/4}_{I-3/2}, \quad k = 1, 2, 3, 4,$$

fine - coarse

$$(u^{k/4}_{I+5/4} + v^{k/4}_{I+5/4} + u^{k/4}_{I+7/4} + v^{k/4}_{I+7/4})/2 = R^{k/4}_{I+3/2}, \quad k = 1, 2, 3, 4,$$

    ii) flux intersection condition

coarse - fine

$$(u^{k/4}_{I-7/4} - v^{k/4}_{I-7/4} + u^{k/4}_{I-5/4} - v^{k/4}_{I-7/2})/2 = J^{k/4}_{I-3/2}/2, \quad k = 1, 2, 3, 4,$$

fine - coarse

$$(u^{k/4}_{I+5/4} - v^{k/4}_{I+5/4} + u^{k/4}_{I+7/4} - v^{k/4}_{I+7/4})/2 = J^{k/4}_{I+3/2}/2, \quad k = 1, 2, 3, 4,$$

    iii) inflow intersection conditions

coarse - fine

$$(u^{k/4}_{I-7/4} + u^{k/4}_{I-5/4})/2 = R^{k/4}_{I-3/2}/2 + J^{k/4}_{I-3/2}/4, \quad k = 1, 2, 3, 4,$$

fine - coarse

$$(v^{k/4}_{I+5/4} + v^{k/4}_{I+7/2})/2 = R^{k/4}_{I+3/2}/2 - J^{k/4}_{I+3/2}/4, \quad k = 1, 2, 3, 4.$$

(6) Determine the inflow value on the coarse grid by the inner information of fine grid by using the

    i) density intersection conditions

coarse - fine

$$U^1_{I-1/2} + V^1_{I-1/2} = (u^1_{I-3/4} + v^1_{I-3/4} + u^1_{I-1/4} + v^1_{I-1/4})/2,$$

fine - coarse

$$U^1_{I+1/2} + V^1_{I+1/2} = (u^1_{I+1/4} + v^1_{I+1/4} + u^1_{I+3/4} + v^1_{I+3/4})/2,$$

ii) flux intersection conditions

$$\text{coarse - fine}$$

$$U^1_{I-1/2} - V^1_{I-1/2} = u^1_{I-3/4} - v^1_{I-3/4} + u^1_{I-1/4} - v^1_{I-1/4},$$

$$\text{fine - coarse}$$

$$U^1_{I+1/2} - V^1_{I+1/2} = u^1_{I+1/4} - v^1_{I+1/4} + u^1_{I+3/4} - v^1_{I+3/4},$$

iii) inflow intersection conditions

$$\text{coarse - fine}$$

$$U^1_{I-1/2} = (r^1_{I-3/4} + r^1_{I-1/4})/4 + (j^1_{I-3/4} + j^1_{I-1/4})/2,$$

$$\text{fine - coarse}$$

$$V^1_{I+1/2} = (r^1_{I+1/4} + r^1_{I+3/4})/4 - (j^1_{I+1/4} + j^1_{I+3/4})/2,$$

iv) update the inflow and outflow values on the coarse grid by taking $U^1_{I-1/2} = (r^1_{I-3/4} + r^1_{I-1/4})/4 + (j^1_{I-3/4} + j^1_{I-1/4})/2$ and $V^1_{I+1/2} = (r^1_{I+1/4} + r^1_{I+3/4})/4 - (j^1_{I+1/4} + j^1_{I+3/4})/2$.

In both proposed algorithms we use linear time interpolations at the intersection.

8.5. Numerical Results

For the investigation of the performance of the proposed grid coupling algorithms, we consider Example 7 given by the eigenfunction solution

$$r(t, x) := e^{-\nu \pi^2 t} \sin(\pi x) \quad \text{in } (0, T] \times (0, 1),$$

with the data $\nu = 0.1$ and $T = 0.2$. The interval $(0, 1)$ is divided at the intersection point $x_I = 0.5$. At the left hand side $(0, x_I)$ of the interval $(0, 1)$ we define a coarse grid with grid size h, and on the right hand side $(x_I, 1)$ we define a fine grid with grid size $h/2$. The grids are overlapping in the single or double sense as displayed in Figure 8.1 and Figure 8.2. On both grids we examine the experimental orders of convergence (EOC) for the nodal L^2-errors and the approximated L^2-errors. We use a grid sequence with $N = 50, 134, 217, 300$ and the relaxation parameter $\omega = 0.7$ in combination with density boundary conditions for the left boundary of the coarse grid and the right boundary of the fine grid. At the intersection point $x_I = 0.5$ we employ the proposed intersection conditions. For single overlapping grids we use the continuity condition for the macroscopic quantities. For double overlapping grids we combine the density, flux or inflow intersection conditions for the fine grid with the density, flux, inflow or update intersection conditions for the coarse grid. The errors on the coarse and the fine grid are computed independently.

In nearly all of the examined situations the EOCs show the expected behavior, that is, we find second order convergence for the density and third order convergence for the h-scaled flux with respect to the nodal L^2-errors and the approximated L^2-errors. Only for the CCFV lattice Boltzmann schemes (5.4) with density intersection conditions we obtain weak performance. The results for the EOCs are displayed in Tables 8.1 - 8.4. In Table 8.1 the results are listed for the VCFD lattice Boltzmann schemes (5.1) depending on several intersection conditions. Table 8.2 contains the results for the CCFD schemes (5.2). The corresponding results for the VCFV lattice Boltzmann schemes (5.3) and the CCFV lattice Boltzmann schemes (5.4) are presented in Table 8.3 and in Table 8.4.

Further investigations show that the maintenance of the EOCs is not the decisive criterion for the performance of the grid coupling algorithms. Computational

Nodal L^2-errors				Approximated L^2-errors			
Coarse Grid		Fine Grid		Coarse Grid		Fine Grid	
Density	Flux	Density	Flux	Density	Flux	Density	Flux
Single overlap with continuity intersection conditions							
1.98	3.00	1.98	3.00	2.00	3.00	2.01	3.00
Double overlap with density/density intersection conditions							
1.98	3.02	1.99	3.00	2.00	3.00	2.01	2.99
Double overlap with flux/flux intersection conditions							
1.99	3.00	2.04	3.01	2.01	3.00	2.01	3.02
Double overlap with inflow/inflow intersection conditions							
1.97	3.00	1.98	2.99	2.00	3.00	2.01	2.99
Double overlap with density/update intersection conditions							
1.98	3.00	1.99	3.00	2.00	3.00	2.01	2.99
Double overlap with flux/update intersection conditions							
1.94	3.00	1.89	2.99	2.00	3.00	2.02	2.98
Double overlap with inflow/update intersection conditions							
1.97	3.00	1.96	2.99	2.01	3.00	2.02	2.99

TABLE 8.1. Example 7: Experimental orders of convergence on coupled VCFD grids. Intersection point is $x_I = 0.5$ with data $\nu = 0.1$, $T = 0.2$, $\omega = 0.7$ and grid sequences with $N = 50, 134, 217, 300$ with respect to the nonrefined grid.

Nodal L^2-errors				Approximated L^2-errors			
Coarse Grid		Fine Grid		Coarse Grid		Fine Grid	
Density	Flux	Density	Flux	Density	Flux	Density	Flux
Single overlap with continuity intersection conditions							
2.06	3.00	1.99	2.99	2.01	3.01	2.03	2.99
Double overlap with density/density intersection conditions							
2.06	3.00	2.00	2.99	2.01	3.01	2.03	2.99
Double overlap with flux/flux intersection conditions							
1.98	3.00	2.07	3.02	2.01	3.00	2.02	3.04
Double overlap with inflow/inflow intersection conditions							
2.08	3.00	2.00	2.98	2.01	3.01	2.03	2.98
Double overlap with density/update intersection conditions							
2.06	3.00	2.00	2.99	2.01	3.01	2.03	2.99
Double overlap with flux/update intersection conditions							
2.14	3.00	1.96	2.97	2.02	3.00	2.05	2.96
Double overlap with inflow/update intersection conditions							
2.09	3.00	1.99	2.98	2.01	3.00	2.04	2.98

TABLE 8.2. Example 7: Experimental orders of convergence on coupled CCFD grids. Intersection point is $x_I = 0.5$ with data $\nu = 0.1$, $T = 0.2$, $\omega = 0.7$ and grid sequences with $N = 50, 134, 217, 300$ with respect to the nonrefined grid.

Nodal L^2-errors				Approximated L^2-errors			
Coarse Grid		Fine Grid		Coarse Grid		Fine Grid	
Density	Flux	Density	Flux	Density	Flux	Density	Flux
Single overlap with continuity intersection conditions							
2.00	3.00	2.00	3.00	2.00	3.00	2.00	3.00
Double overlap with density/density intersection conditions							
1.99	3.00	2.01	3.00	2.00	3.00	2.01	3.01
Double overlap with flux/flux intersection conditions							
2.00	3.00	2.02	3.01	2.00	3.00	2.01	3.02
Double overlap with inflow/inflow intersection conditions							
2.00	3.00	2.02	3.01	2.00	3.00	2.01	3.02
Double overlap with density/update intersection conditions							
2.00	3.00	2.01	3.01	2.00	3.00	2.01	3.01
Double overlap with flux/update intersection conditions							
2.00	3.00	2.02	3.01	2.01	3.00	2.01	3.02
Double overlap with inflow/update intersection conditions							
2.00	3.00	2.01	3.01	2.00	3.00	2.01	3.01

TABLE 8.3. Example 7: Experimental orders of convergence on coupled VCFV grids. Intersection point is $x_I = 0.5$ with data $\nu = 0.1$, $T = 0.2$, $\omega = 0.7$ and grid sequences with $N = 50$, 134, 217, 300 with respect to the nonrefined grid.

Nodal L^2-errors				Approximated L^2-errors			
Coarse Grid		Fine Grid		Coarse Grid		Fine Grid	
Density	Flux	Density	Flux	Density	Flux	Density	Flux
Single overlap with continuity intersection conditions							
2.00	3.00	1.99	3.12	2.01	3.00	2.01	3.11
Double overlap with density/density intersection conditions							
2.00	3.02	0.99	1.87	2.00	3.03	1.02	1.87
Double overlap with flux/flux intersection conditions							
2.00	3.00	2.03	3.02	2.01	3.00	2.03	3.04
Double overlap with inflow/inflow intersection conditions							
2.00	3.00	2.00	2.96	2.00	3.01	2.02	2.97
Double overlap with density/update intersection conditions							
2.00	3.00	0.99	1.87	2.00	3.01	1.02	1.87
Double overlap with flux/update intersection conditions							
2.01	3.00	2.05	3.00	2.01	3.00	2.03	3.01
Double overlap with inflow/update intersection conditions							
2.00	3.00	2.00	2.96	2.00	3.01	2.01	2.96

TABLE 8.4. Example 7: Experimental orders of convergence on coupled CCFV grids. Intersection point is $x_I = 0.5$ with data $\nu = 0.1$, $T = 0.2$, $\omega = 0.7$ and grid sequences with $N = 50$, 134, 217, 300 with respect to the nonrefined grid.

Nodal L^2-errors					
Coarse Grid		Fine Grid		Ratios	
$\|e_R\|_2^C$	$\|e_J\|_2^C$	$\|e_R\|_2^F$	$\|e_J\|_2^F$	$\|e_R\|_2^C/\|e_R\|_2^F$	$\|e_J\|_2^C/\|e_J\|_2^F$
Single overlap with continuity intersection conditions					
3.49e-6	1.75e-6	1.71e-6	2.29e-7	2.04	7.66
Double overlap with density/density intersection conditions					
3.47e-6	1.75e-6	1.74e-6	2.26e-7	1.99	7.75
Double overlap with flux/flux intersection conditions					
6.75e-6	1.83e-6	2.39e-6	2.14e-7	2.83	8.52
Double overlap with inflow/inflow intersection conditions					
3.44e-6	1.75e-6	1.69e-6	2.26e-7	2.03	7.76
Double overlap with density/update intersection conditions					
3.46e-6	1.75e-6	1.73e-6	2.26e-7	2.00	7.75
Double overlap with flux/update intersection conditions					
3.23e-6	1.75e-6	1.45e-6	2.23e-7	2.23	7.84
Double overlap with inflow/update intersection conditions					
3.38e-6	1.75e-6	1.63e-6	2.25e-7	2.07	7.78

TABLE 8.5. Example 7: Ratio of the nodal L^2-errors on the coupled coarse and fine VCFD grids.

advantages can only be achieved, if the errors on the fine grid decrease as implied by the convergence orders. Since our test solution is symmetric with respect to the grid intersection, we expect in the optimal case that the errors in the density are smaller by a factor of four on the fine grid and the errors for the h-scaled flux are smaller by a factor of eight. Otherwise, the additional expense for the computation of the solutions on coupled grids is not justified.

Expressed in formulas, we wish for

$$\frac{\|e_R\|_2^C}{\|e_R\|_2^F} = 4, \quad \frac{\|e_J\|_2^C}{\|e_J\|_2^F} = 8.$$

Here, $\|\cdot\|_2^C$ denotes the norm on the coarse grid and $\|\cdot\|_2^F$ denotes the norm on the fine grid. We consider the nodal L^2-norms, as well as the approximated L^2-norms.

The nodal L^2-errors for the solutions of the VCFD lattice Boltzmann schemes are displayed in Table 8.5 depending on the intersection conditions. While the errors of the h-scaled flux nearly show the expected ratio, the errors for the density only render a ratio of approximately two.

The reason for this unsatisfactory behavior is founded in the nonlocal nature of the errors. In Figure 8.5 the errors on the uniform grid are displayed. The largest error for the density is attained in the intersection point x_I. Hence, for the refined grid the error on the fine grid is coupled to the error on the coarse grid; see Figure 8.6. The decay of the error on the fine grid is restricted. A better result is attained for the flux, because the flux error has a zero at the intersection point and hence the errors are decoupled to some extent. In the error plots for the flux we use the grid independent scaling $(U - V)/h$, such that these are the errors for the macroscopic quantity that approximates $\partial_x r$ up to a constant.

In Table 8.6 we examine the behavior of the approximated L^2-errors for the VCFD lattice Boltzmann solutions. It is quite astonishing that the approximated L^2-errors for the density show the right decay on the finer grid. In order to clarify

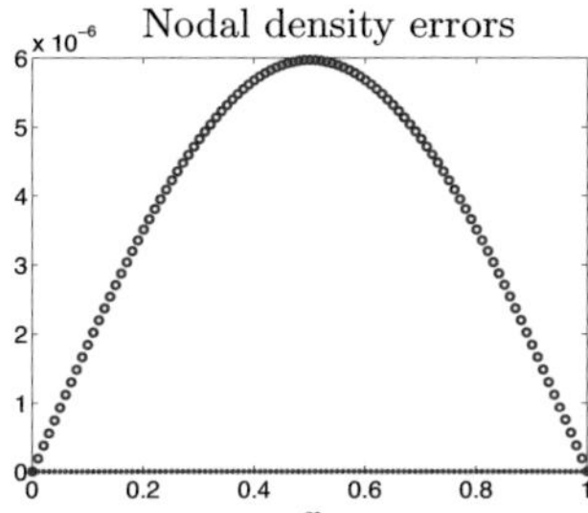

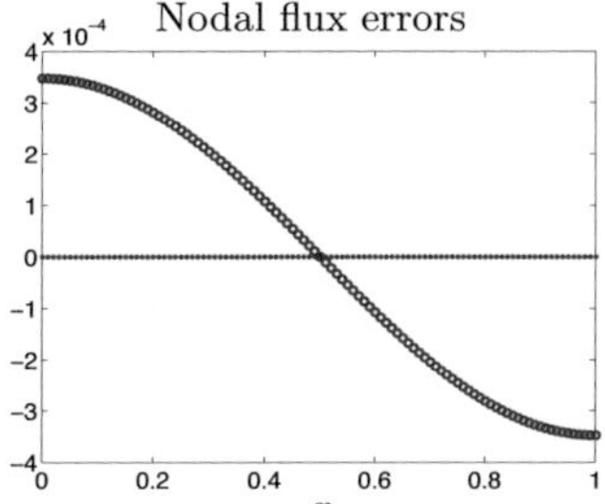

FIGURE 8.5. Example 7: Nodal errors on a uniform grid for $N = 100$.

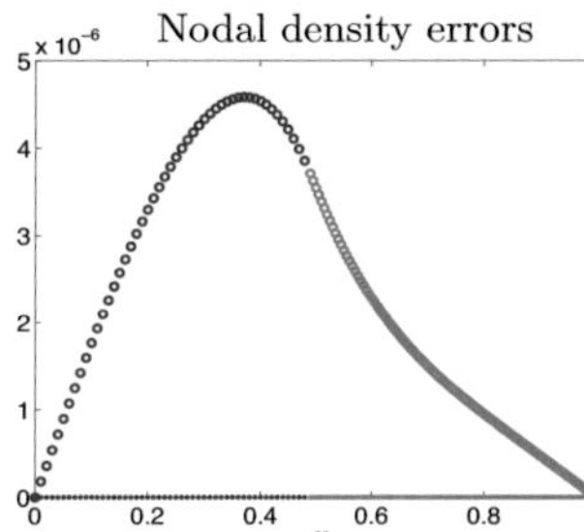

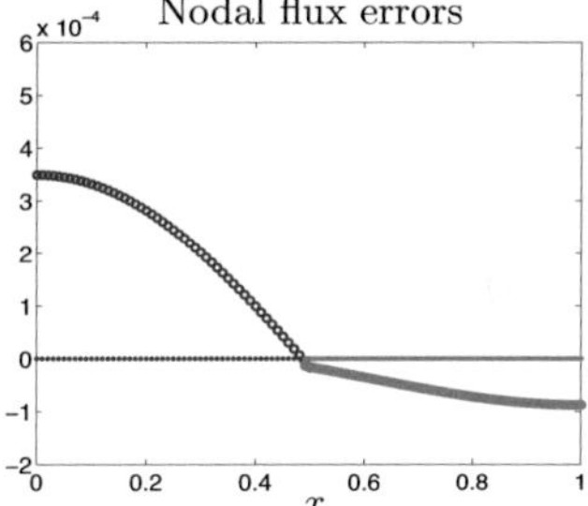

FIGURE 8.6. Example 7: Nodal errors on coupled grids.

Approximated L^2-errors					
Coarse Grid		Fine Grid		Ratios	
$\|e_R\|_2^C$	$\|e_J\|_2^C$	$\|e_R\|_2^F$	$\|e_J\|_2^F$	$\|e_R\|_2^C/\|e_R\|_2^F$	$\|e_J\|_2^C/\|e_J\|_2^F$
Single overlap with continuity intersection conditions					
4.95e-5	8.41e-7	1.17e-5	1.14e-7	4.23	7.37
Double overlap with density/density intersection conditions					
4.95e-5	8.41e-7	1.18e-5	1.13e-7	4.20	7.45
Double overlap with flux/flux intersection conditions					
4.69e-5	9.11e-7	1.48e-5	1.26e-7	3.17	7.24
Double overlap with inflow/inflow intersection conditions					
4.95e-5	8.40e-7	1.18e-5	1.12e-7	4.19	7.48
Double overlap with density/update intersection conditions					
4.95e-5	8.40e-7	1.18e-5	1.13e-7	4.19	7.46
Double overlap with flux/update intersection conditions					
4.97e-5	8.37e-7	1.20e-5	1.09e-7	4.14	7.64
Double overlap with inflow/update intersection conditions					
4.96e-5	8.39e-7	1.19e-5	1.12e-7	4.17	7.52

TABLE 8.6. Example 7: Ratio of the approximated L^2-errors on the coupled coarse and fine VCFD grids.

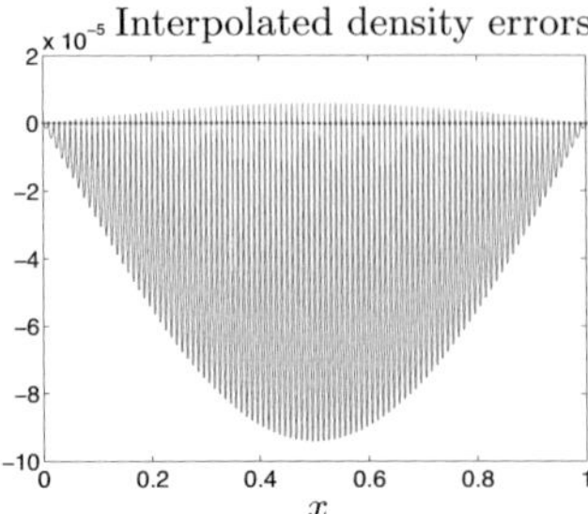

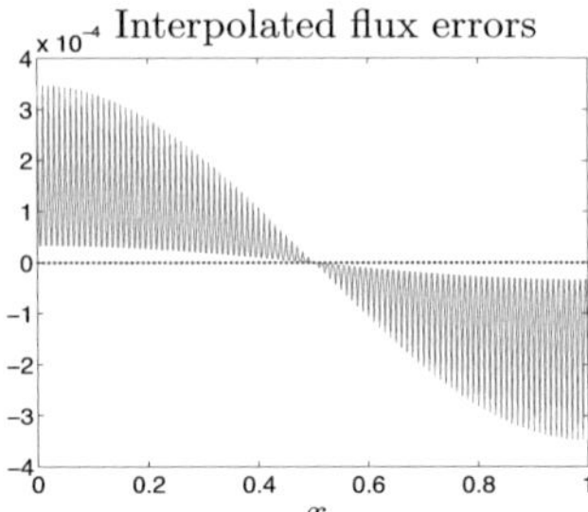

FIGURE 8.7. Example 7: Interpolated errors on a uniform grid for $N = 100$.

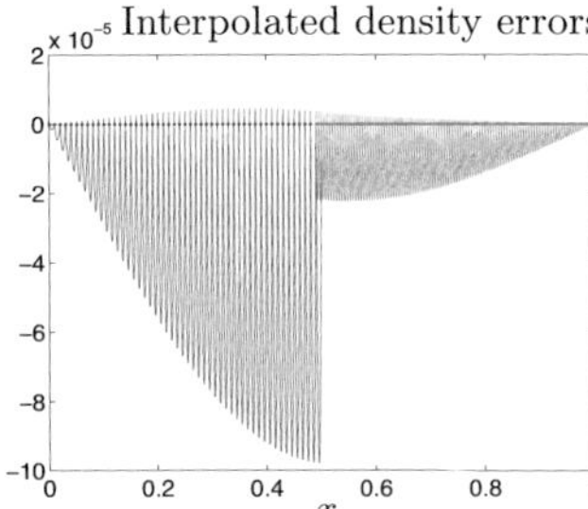

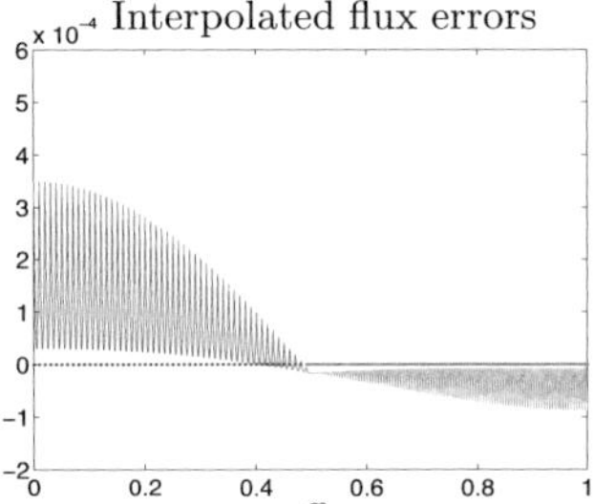

FIGURE 8.8. Example 7: Interpolated errors on coupled grids.

this result, we regard the interpolated errors of this problem. In Figure 8.7 the interpolated errors on a uniform grid are displayed. Figure 8.8 shows the interpolated errors on the coupled grids. The decay of the errors on the finer grid is pronounced.

In Example 8, we consider the eigenfunction solution

$$r(t, x) := e^{-\nu \pi^2 t} \cos(\pi x) \quad \text{in } (0, T] \times (0, 1),$$

with the data $\nu = 0.1$ and $T = 0.2$. This solution has a zero at the intersection point and hence, a zero is expected for the density error at the intersection point due to the symmetry of the problem. This fact should be expressed in a better error decay for the density on the coupled grids. A worse performance is expected for the flux error. The nodal errors on the uniform grid are depicted in Figure 8.9, the nodal errors on the coupled grids can be found in Figure 8.10. The solutions are computed with the density intersection conditions that do not ensure continuity of the flux. Hence, the flux error shows a jump at the intersection.

In Figure 8.11 we examine the interpolated errors for the VCFD lattice Boltzmann solutions on a uniform grid. In Figure 8.12 the interpolated errors on the coupled grids are displayed.

The behavior of the nodal L^2-errors on the coupled grids is listed in Table 8.7. Only for the density/density intersection conditions the optimal results can be achieved. Better results are obtained for the approximated L^2-errors; see Table 8.8. The nodal errors with respect to different intersection conditions are displayed in the Figures 8.13 - 8.19. For the continuity intersection conditions on single overlapping grids, the maximal density error is attained in the intersection point; see Figure 8.13. For the flux error there is only a slight improvement on the finer grid.

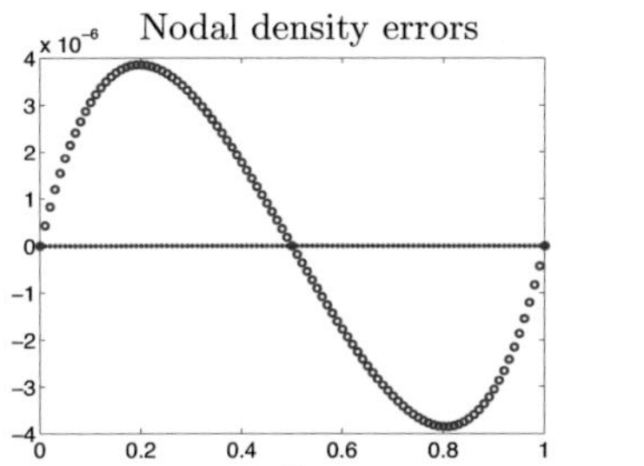
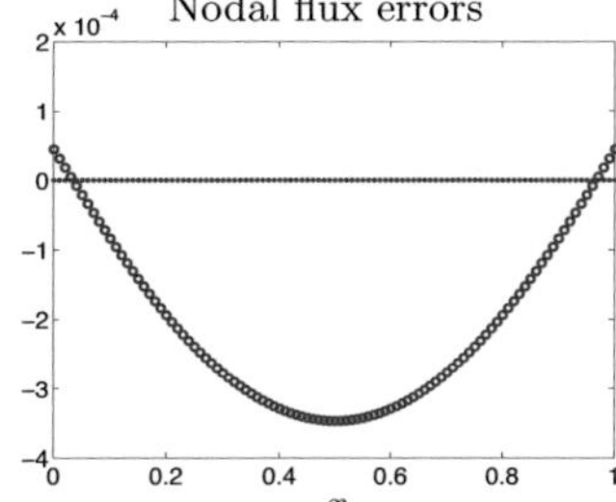

FIGURE 8.9. Example 8: Nodal errors on a uniform grid for $N = 100$.

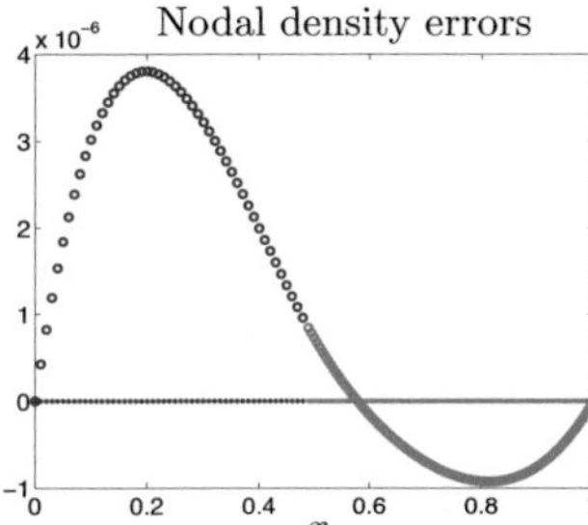
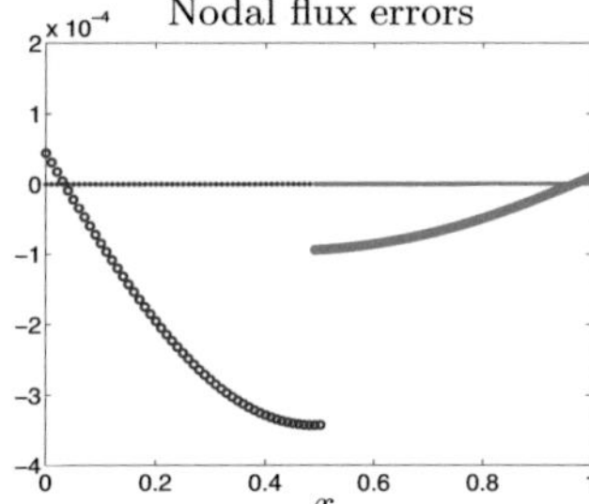

FIGURE 8.10. Example 8: Nodal errors on coupled grids.

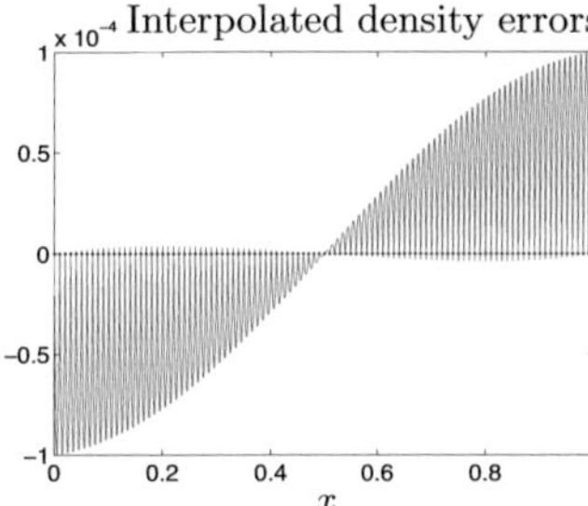
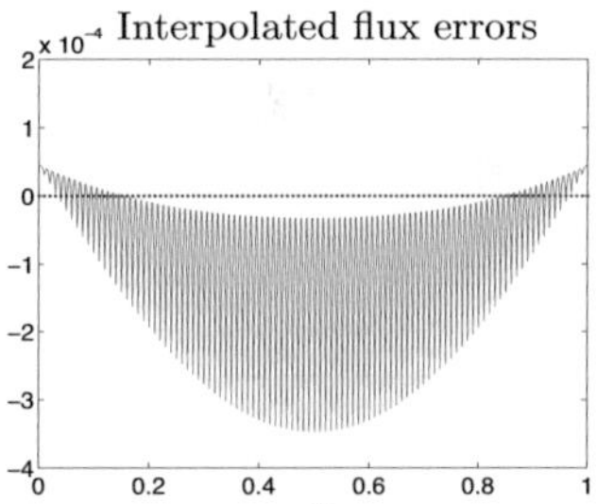

FIGURE 8.11. Example 8: Interpolated errors on the uniform grid.

These results are quite unsatisfactory. The same has to be said for the flux/flux
intersection conditions on double overlapping grids; see Figure 8.15. A marginal
improvement for the density errors can be found in the case of the inflow/inflow
intersection conditions in Figure 8.16. By far the best results are obtained for the
density/density intersection conditions on double-overlapping grids in Figure 8.14.
The density/update intersection conditions in Figure 8.17, the flux/update intersection conditions in Figure 8.18 and the density/update intersection conditions in
Figure 8.19 do also not render convincing results. Hence, the favored choice are
the density/density intersection conditions on double overlapping VC grids. All of

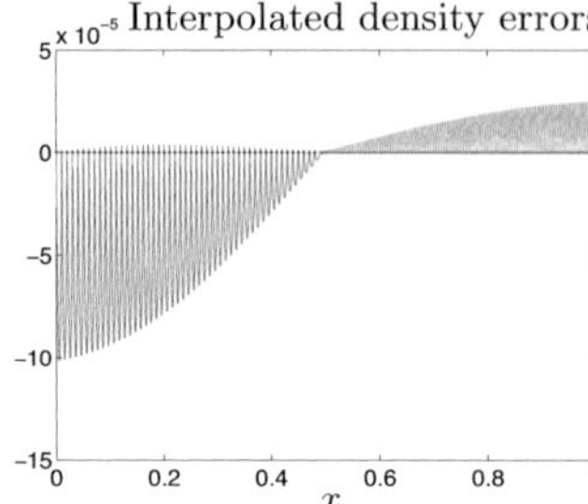

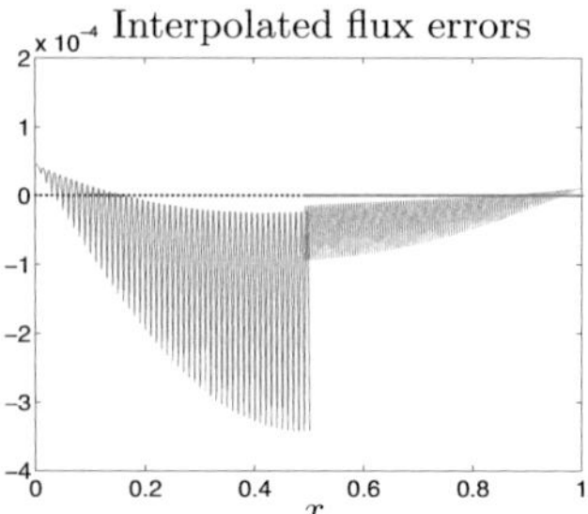

FIGURE 8.12. Example 8: Interpolated errors on coupled grids.

Nodal L^2-errors					
Coarse Grid		Fine Grid		Ratios	
$\|e_R\|_2^C$	$\|e_J\|_2^C$	$\|e_R\|_2^F$	$\|e_J\|_2^F$	$\|e_R\|_2^C/\|e_R\|_2^F$	$\|e_J\|_2^C/\|e_J\|_2^F$
Single overlap with continuity intersection conditions					
9.60e-6	1.35e-6	7.92e-6	4.18e-7	1.21	3.23
Double overlap with density/density intersection conditions					
2.75e-6	1.72e-6	6.47e-7	2.26e-7	4.24	7.59
Double overlap with flux/flux intersection conditions					
9.94e-6	1.34e-6	8.10e-6	4.21e-7	1.23	3.18
Double overlap with inflow/inflow intersection conditions					
6.12e-6	1.52e-6	4.22e-6	3.26e-7	1.45	4.66
Double overlap with density/update intersection conditions					
3.19e-6	1.84e-6	3.18e-6	1.55e-7	1.00	11.87
Double overlap with flux/update intersection conditions					
9.58e-6	1.35e-6	8.43e-6	4.28e-7	1.14	3.16
Double overlap with inflow/update intersection conditions					
2.93e-6	1.67e-6	9.57e-7	2.42e-7	3.07	6.89

TABLE 8.7. Example 8: Ratio of the nodal L^2-errors on the coupled coarse and fine VCFD grids.

the results remain unchanged, if we consider quadratic time interpolation instead of linear time interpolation. Hence there is no need for this additional effort.

The examination of the grid coupling algorithms for the VCFV lattice Boltzmann scheme (5.3) leads to similar results as for the VCFD lattice Boltzmann scheme (5.1) that has been considered so far. Example 7 serves as the test problem. The results for the decay of the nodal L^2-errors can be found in Table 8.9. The density errors decay by a factor of two, whereas the flux errors decay by a factor of eight. For the approximated L^2-errors we only find a decay of the density errors by a factor of three; see Table 8.10. In comparison to the results in Table 8.6 for VCFD grids, this is a deterioration.

The performance of the grid coupling algorithms on CC grids turns out to be insufficient. This has to be attributed to the additional spatial interpolations. The results for the nodal and the approximated L^2-errors on CCFD grids are presented in Table 8.11 and in Table 8.12. The nodal L^2-errors for the density on the fine

Approximated L^2-errors					
Coarse Grid		Fine Grid		Ratios	
$\|e_R\|_2^C$	$\|e_J\|_2^C$	$\|e_R\|_2^F$	$\|e_J\|_2^F$	$\|e_R\|_2^C/\|e_R\|_2^F$	$\|e_J\|_2^C/\|e_J\|_2^F$
Single overlap with continuity intersection conditions					
4.90e-5	5.86e-7	1.66e-5	2.93e-7	2.93	2.00
Double overlap with density/density intersection conditions					
5.04e-5	8.18e-7	1.25e-5	1.11e-7	4.02	7.38
Double overlap with flux/flux intersection conditions					
4.90e-5	5.60e-7	1.65e-5	2.96e-7	2.98	1.96
Double overlap with inflow/inflow intersection conditions					
4.95e-5	6.74e-7	1.41e-5	2.03e-7	3.50	3.35
Double overlap with density/update intersection conditions					
5.13e-5	9.35e-7	1.21e-5	6.92e-8	4.23	13.51
Double overlap with flux/update intersection conditions					
4.90e-5	5.86e-7	1.67e-5	3.02e-7	2.94	1.94
Double overlap with inflow/update intersection conditions					
5.03e-5	7.94e-7	1.27e-5	1.24e-7	3.95	6.39

TABLE 8.8. Example 8: Ratio of the approximated L^2-errors on the coupled coarse and fine VCFD grids.

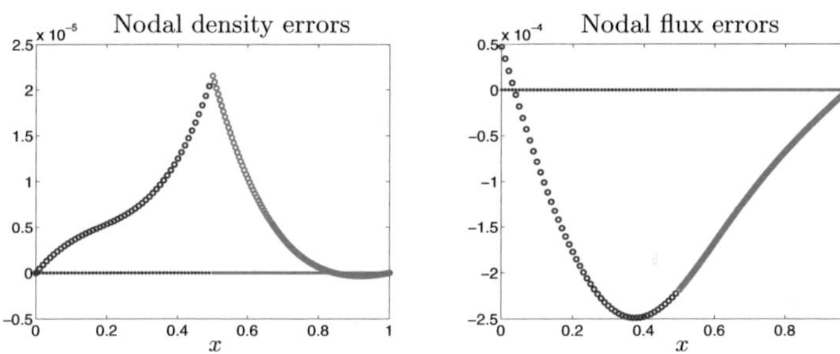

FIGURE 8.13. Example 8: Nodal errors on single overlapping VCFD grids with continuity intersection conditions.

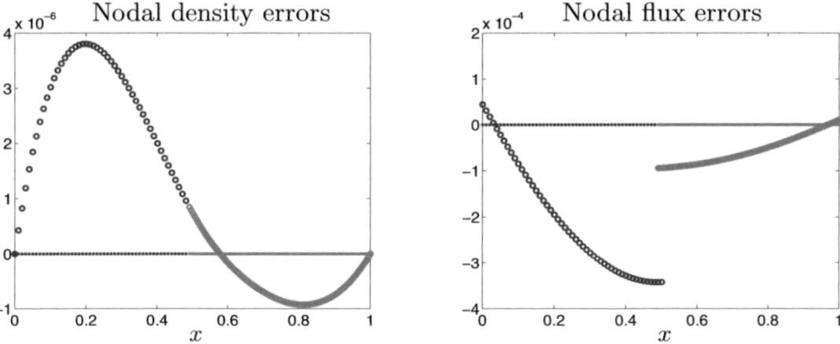

FIGURE 8.14. Example 8: Nodal errors on double overlapping VCFD grids with density/density intersection conditions.

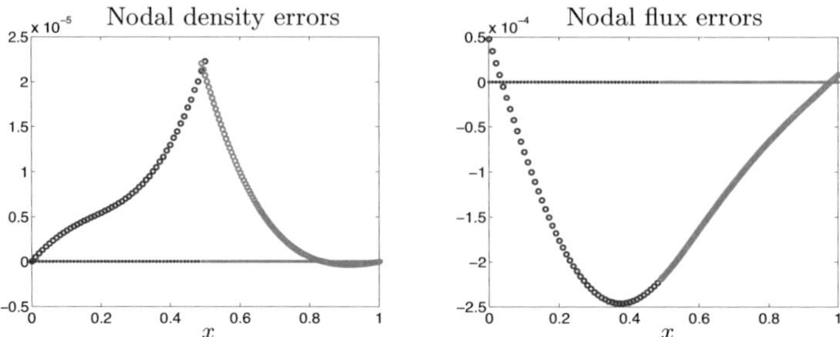

FIGURE 8.15. Example 8: Nodal errors on double overlapping grids with flux/flux intersection conditions.

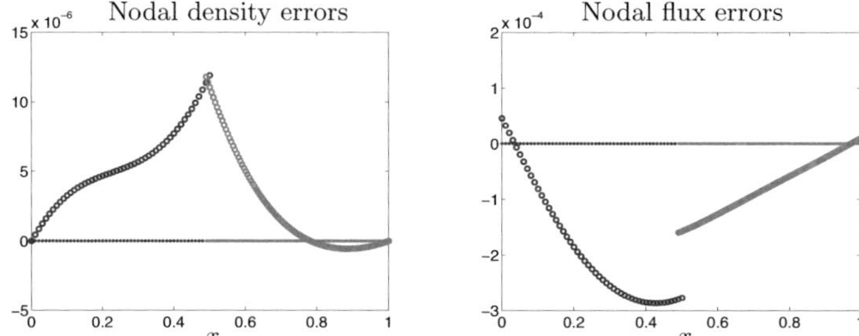

FIGURE 8.16. Example 8: Nodal errors on double overlapping grids with inflow/inflow intersection conditions.

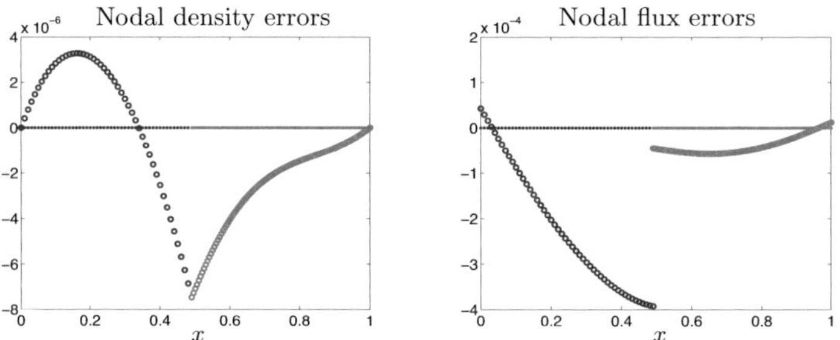

FIGURE 8.17. Example 8: Nodal errors on double overlapping grids with density/update intersection conditions.

grid are larger than those on the coarse grid. The approximated L^2-errors for the flux do not render the expected results.

The situation is a little bit improved for the coupled CCFV grids. The results for the nodal and the approximated L^2-errors are listed in the Table 8.13 and in Table 8.14.

The application of quadratic or cubic data interpolation in space at the intersection does not render improved results.

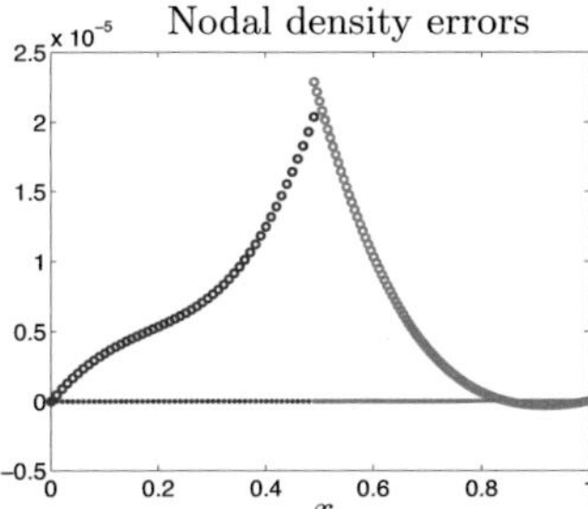
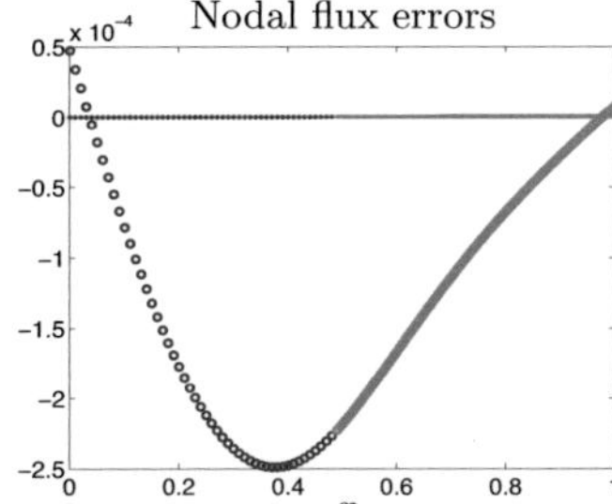

FIGURE 8.18. Example 8: Nodal errors on double overlapping grids with flux/update intersection conditions.

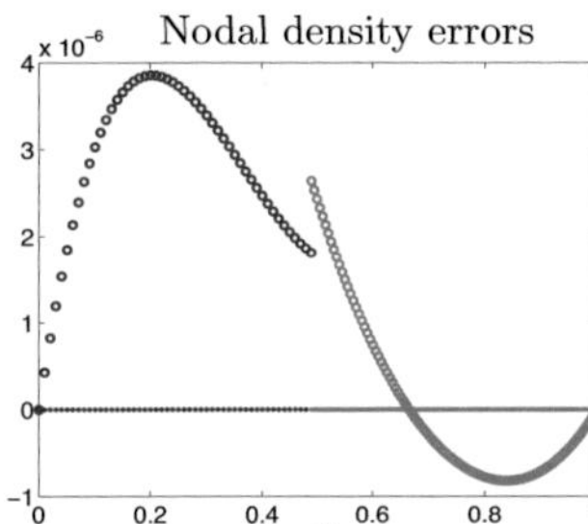
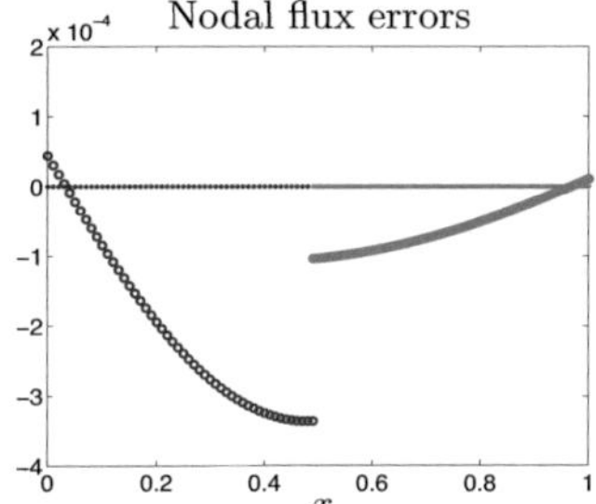

FIGURE 8.19. Example 8: Nodal errors on double overlapping grids with inflow/update intersection conditions.

Nodal L^2-errors					
Coarse Grid		Fine Grid		Ratios	
$\|e_R\|_2^C$	$\|e_J\|_2^C$	$\|e_R\|_2^F$	$\|e_J\|_2^F$	$\|e_R\|_2^C/\|e_R\|_2^F$	$\|e_J\|_2^C/\|e_J\|_2^F$
Single overlap with continuity intersection conditions					
2.82e-6	2.50e-6	1.35e-5	3.02e-7	2.09	8.27
Double overlap with density/density intersection conditions					
2.79e-6	2.50e-6	1.37e-5	3.14e-7	2.04	7.99
Double overlap with flux/flux intersection conditions					
2.95e-6	2.45e-6	1.24e-5	2.90e-7	2.39	8.47
Double overlap with inflow/inflow intersection conditions					
2.81e-6	2.50e-6	1.38e-5	3.07e-7	2.03	8.16
Double overlap with density/update intersection conditions					
2.80e-6	2.51e-6	1.37e-5	3.14e-7	2.04	8.00
Double overlap with flux/update intersection conditions					
2.81e-6	2.50e-6	1.39e-5	3.07e-7	2.02	8.14
Double overlap with inflow/update intersection conditions					
2.80e-6	2.50e-6	1.38e-5	3.06e-7	2.04	8.18

TABLE 8.9. Example 7: Ratio of the nodal L^2-errors on the coupled coarse and fine VCFV grids.

Approximated L^2-errors					
Coarse Grid		Fine Grid		Ratios	
$\|e_R\|_2^C$	$\|e_J\|_2^C$	$\|e_R\|_2^F$	$\|e_J\|_2^F$	$\|e_R\|_2^C/\|e_R\|_2^F$	$\|e_J\|_2^C/\|e_J\|_2^F$
Single overlap with continuity intersection conditions					
7.98e-5	1.52e-6	2.59e-5	2.16e-7	3.08	7.04
Double overlap with density/density intersection conditions					
7.96e-5	1.53e-6	2.62e-5	2.24e-7	3.04	6.81
Double overlap with flux/flux intersection conditions					
8.12e-5	1.47e-6	2.51e-5	1.91e-7	3.24	7.66
Double overlap with inflow/inflow intersection conditions					
7.97e-5	1.52e-6	2.63e-5	2.23e-7	3.02	6.82
Double overlap with density/update intersection conditions					
7.96e-5	1.53e-6	2.62e-5	2.24e-7	3.03	6.81
Double overlap with flux/update intersection conditions					
7.98e-5	1.52e-6	2.64e-5	2.24e-7	3.02	6.78
Double overlap with inflow/update intersection conditions					
7.97e-5	1.52e-6	2.63e-5	2.23e-7	3.03	6.86

TABLE 8.10. Ratio of the approximated L^2-errors on the coupled coarse and fine VCFV grids.

Nodal L^2-errors					
Coarse Grid		Fine Grid		Ratios	
$\|e_R\|_2^C$	$\|e_J\|_2^C$	$\|e_R\|_2^F$	$\|e_J\|_2^F$	$\|e_R\|_2^C/\|e_R\|_2^F$	$\|e_J\|_2^C/\|e_J\|_2^F$
Single overlap with continuity intersection conditions					
3.43e-6	1.64e-6	7.37e-6	3.00e-7	0.46	5.46
Double overlap with density/density intersection conditions					
3.49e-6	1.64e-6	7.56e-6	2.98e-7	0.46	5.50
Double overlap with flux/flux intersection conditions					
7.50e-6	1.85e-6	3.59e-6	2.24e-7	2.09	8.25
Double overlap with inflow/inflow intersection conditions					
3.57e-6	1.64e-6	7.47e-6	2.96e-7	0.48	5.53
Double overlap with density/update intersection conditions					
3.50e-6	1.64e-6	7.53e-6	2.98e-7	0.46	5.50
Double overlap with flux/update intersection conditions					
4.05e-6	1.63e-6	6.92e-6	2.88e-7	0.59	5.65
Double overlap with inflow/update intersection conditions					
3.69e-6	1.64e-6	7.33e-6	2.95e-7	0.50	5.54

TABLE 8.11. Ratio of the nodal L^2-errors on the coupled coarse and fine CCFD grids.

8.6. Concentrated Errors

In the problems considered so far, the error functions had a global spread. Hence, the solutions on coupled grids suffered from global pollution errors. A considerable decay of the errors can only be expected, if the refined grids are adapted to the error distributions. In the following, we consider a test problem, where the

Approximated L^2-errors					
Coarse Grid		Fine Grid		Ratios	
$\|e_R\|_2^C$	$\|e_J\|_2^C$	$\|e_R\|_2^F$	$\|e_J\|_2^F$	$\|e_R\|_2^C/\|e_R\|_2^F$	$\|e_J\|_2^C/\|e_J\|_2^F$
Single overlap with continuity intersection conditions					
5.51e-5	7.81e-7	8.11e-6	1.93e-7	6.79	4.04
Double overlap with density/density intersection conditions					
5.52e-5	7.79e-7	8.17e-6	1.93e-7	6.75	4.04
Double overlap with flux/flux intersection conditions					
4.67e-5	9.45e-7	1.60e-5	1.51e-7	2.93	6.26
Double overlap with inflow/inflow intersection conditions					
5.52e-5	7.79e-7	8.20e-6	1.91e-7	6.73	4.08
Double overlap with density/update intersection conditions					
5.52e-5	7.78e-7	8.19e-6	1.93e-7	6.74	4.03
Double overlap with flux/update intersection conditions					
5.57e-5	7.76e-7	8.47e-6	1.82e-7	6.58	4.25
Double overlap with inflow/update intersection conditions					
5.53e-5	7.77e-7	8.28e-6	1.89e-7	6.68	4.10

TABLE 8.12. Ratio of the approximated L^2-errors on the coupled coarse and fine CCFD grids.

Nodal L^2-errors					
Coarse Grid		Fine Grid		Ratios	
$\|e_R\|_2^C$	$\|e_J\|_2^C$	$\|e_R\|_2^F$	$\|e_J\|_2^F$	$\|e_R\|_2^C/\|e_R\|_2^F$	$\|e_J\|_2^C/\|e_J\|_2^F$
Single overlap with continuity intersection conditions					
3.33e-5	2.37e-6	1.07e-5	1.94e-6	3.12	1.22
Double overlap with density/density intersection conditions					
3.31e-5	2.40e-6	1.12e-4	3.20e-5	0.30	0.07
Double overlap with flux/flux intersection conditions					
2.90e-5	2.48e-6	1.36e-5	3.02e-7	2.14	8.20
Double overlap with inflow/inflow intersection conditions					
3.33e-5	2.37e-6	1.55e-5	1.54e-6	2.14	1.54
Double overlap with density/update intersection conditions					
3.30e-5	2.38e-6	1.12e-4	3.20e-5	0.30	0.07
Double overlap with flux/update intersection conditions					
3.37e-5	2.36e-6	8.83e-6	3.14e-7	3.82	7.52
Double overlap with inflow/update intersection conditions					
3.33e-5	2.37e-6	1.55e-5	1.54e-6	2.15	1.54

TABLE 8.13. Ratio of the nodal L^2-errors on the coupled coarse and fine CCFV grids.

error is mainly located on the refined grid. The initial profile with a peak is depicted in Figure 8.20.

The lattice Boltzmann solutions for the viscosity $\nu = 0.1$ and the end time $T = 0.2$ are plotted in Figure 8.21 on two uniform grids with $N = 50$ and $N = 100$. We use the VCFD lattice Boltzmann schemes (5.1) with vanishing density boundary

Approximated L^2-errors					
Coarse Grid		Fine Grid		Ratios	
$\|e_R\|_2^C$	$\|e_J\|_2^C$	$\|e_R\|_2^F$	$\|e_J\|_2^F$	$\|e_R\|_2^C/\|e_R\|_2^F$	$\|e_J\|_2^C/\|e_J\|_2^F$
Single overlap with continuity intersection conditions					
8.57e-5	1.39e-6	2.15e-5	1.13e-6	3.98	1.23
Double overlap with density/density intersection conditions					
8.54e-5	1.41e-6	6.82e-5	1.86e-5	1.25	7.58
Double overlap with flux/flux intersection conditions					
8.14e-5	1.50e-6	2.64e-5	2.15e-7	3.08	6.99
Double overlap with inflow/inflow intersection conditions					
8.56e-5	1.39e-6	2.27e-5	8.99e-7	3.77	1.55
Double overlap with density/update intersection conditions					
8.53e-5	1.40e-6	6.82e-5	1.86e-5	1.25	0.08
Double overlap with flux/update intersection conditions					
8.61e-5	1.38e-6	2.19e-5	2.04e-7	3.93	6.78
Double overlap with inflow/update intersection conditions					
8.56e-5	1.39e-6	2.27e-5	8.98e-7	3.77	1.55

TABLE 8.14. Ratio of the approximated L^2-errors on the coupled coarse and fine CCFV grids.

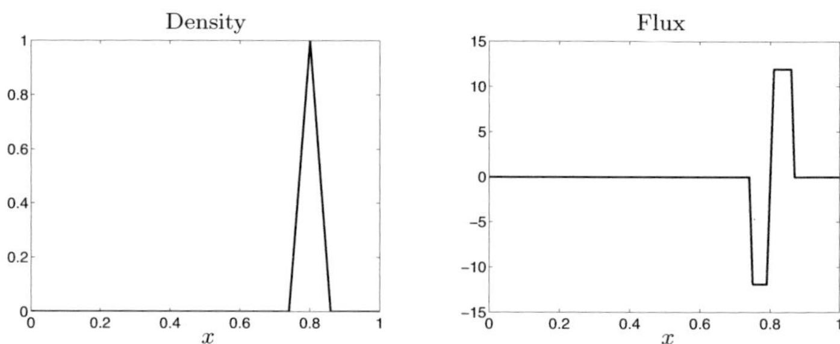

FIGURE 8.20. Initial profile with a concentrated peak.

conditions and density intersection conditions on double overlapping grids. As expected, in the case of nonsmooth data, the lattice Boltzmann solutions on the coarse grid with $N = 50$ show obvious oscillations. The oscillations are considerably smaller on the finer grid with $N = 100$.

We define the intersection point $x_I = 0.4$ and introduce a coarse grid on the left hand side with grid size $h = 1/50$. On the right hand side of the interval we define a finer grid with $h = 1/100$. The lattice Boltzmann solutions on the coupled grids are displayed in Figure 8.22. Even on the coarse grid no oscillations are observed in the solutions.

In the domain of the coarse grid the error is expected to be small. The main part of the error is located in the domain of the fine grid. As time advances, a portion of the error is transported onto the coarse grid. The nodal errors of the density and the flux on the coarse and on the fine grid are shown in Figure 8.23. In addition, the errors on the uniform grid with $N = 100$ are displayed. The errors are measured with respect to the lattice Boltzmann solution on a uniform grid with

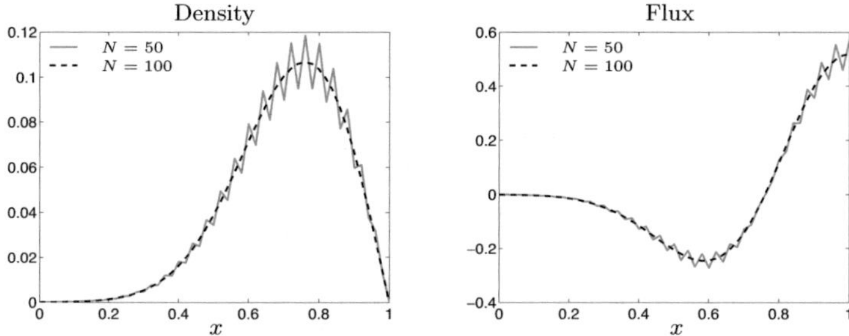

FIGURE 8.21. Lattice Boltzmann solutions for the concentrated initial peak on uniform grids.

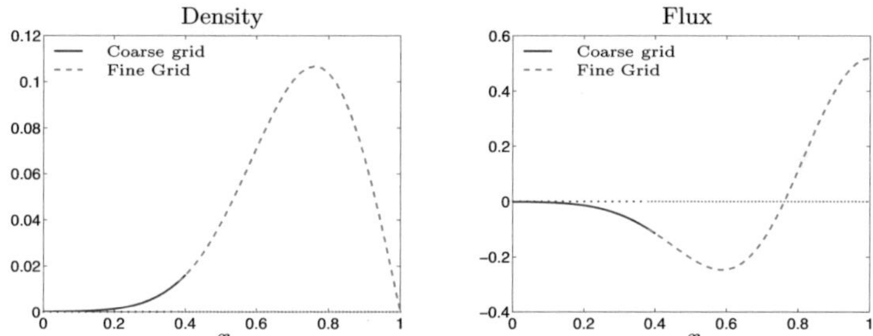

FIGURE 8.22. Lattice Boltzmann solutions for the concentrated initial peak on coupled grids with $h = 1/50$ and $h = 1/100$.

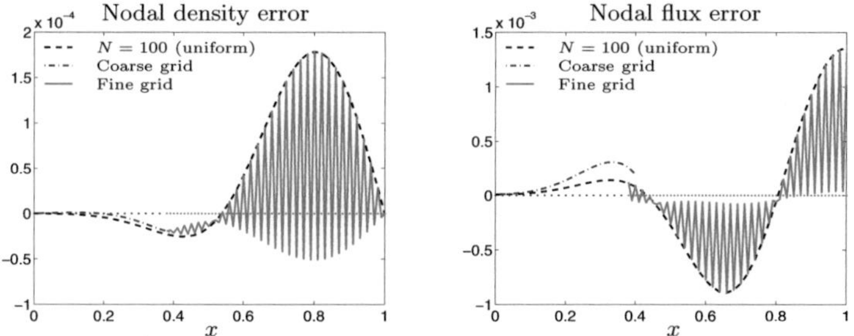

FIGURE 8.23. Errors of the lattice Boltzmann solutions for the concentrated initial peak on coupled grids.

$N = 3200$ that is viewed as a proper approximation of the exact solution. The nodal errors on the fine grid exhibit oscillations. The errors on the uniform grid are envelopes to the errors on the fine grid. In this sense, the solutions on the coupled grids render the optimal results. The error in the flux on the coarse grid is slightly amplified in comparison to the one on the fine grid.

In reasonable applications, the local refinements of the grid have to be adapted to the distribution of the errors. The grid intersections have to be kept away from the maxima of the errors in order to avoid amplifications of the errors. However,

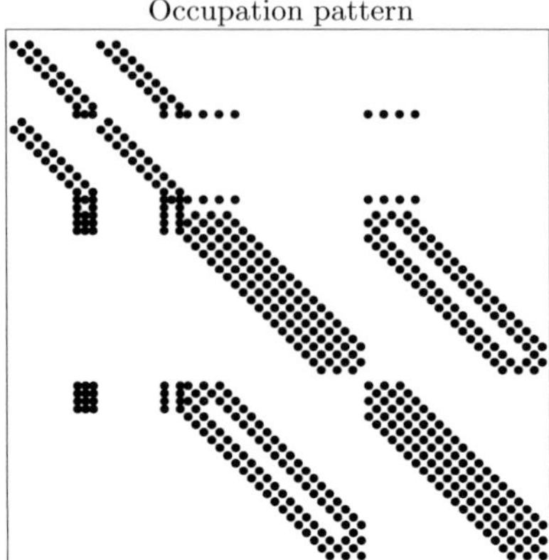

FIGURE 8.24. Occupation pattern for the time evolution matrix on double overlapping VCFD grids with density intersection conditions.

this procedure requires knowledge of the structure of the errors. For most of the problems, this knowledge can only be gained by a posteriori error estimation techniques that are not yet available in the context of lattice Boltzmann methods.

8.7. The Time Evolution Operator on Coupled Grids

On the examined coupled grids, one lattice Boltzmann step on the coarse grid requires four lattice Boltzmann steps on the fine grid. The data have to be exchanged via interpolation steps. In the matrix formulation for the RJ-systems, these steps can be combined and rewritten as a single matrix operation. The occupation pattern of the time evolution matrix for the coupling of double overlapping VCFD grids with density intersection conditions is depicted in Figure 8.24. The nonsymmetric entries stem from the interpolation steps and the data exchange at the intersection.

By using the MATLAB®-routine eig, we examine the eigenvectors of the time evolution operator for double overlapping VCFD grids with density intersection conditions. In the following figures the density component and the flux component of the eigenvectors are plotted on the coarse and on the fine grid. For eigenvalues close to 1, we find low-frequency eigenvectors with a global spread; see Figure 8.25. For increasing frequencies the eigenvalues decay as expected.

Furthermore, we discover low-frequency eigenvectors that only live on the finer grid without any interference onto the coarse grid. In Figure 8.26 one can see that these eigenvectors have vanishing values in the grid points that stem from the unrefined grid. This structure enables the handling of additional information on the finer grid, which finds expression in the improved convergence on the coupled grids.

On the coarse grid low-frequency eigenvectors can be found with eigenvalues close to $1 - 2\omega$; see Figure 8.27. For a relaxation parameter $\omega > 1/2$, this fact leads to fast-decaying oscillations with respect to time.

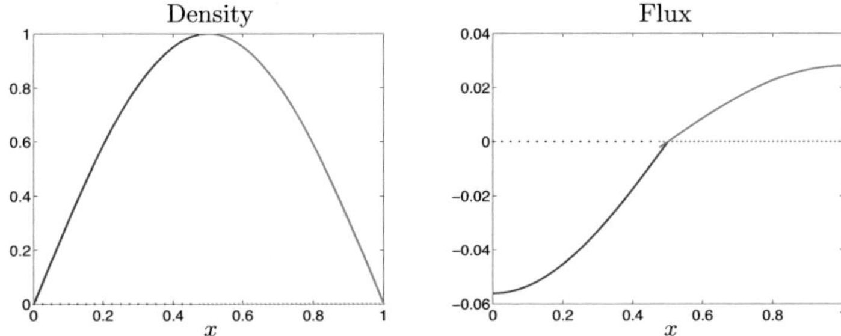

FIGURE 8.25. Low-frequency eigenvector on the global grid for an eigenvalue $\lesssim 1$.

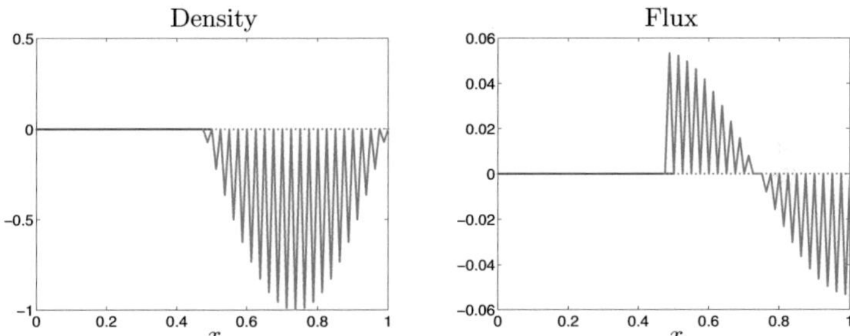

FIGURE 8.26. Low-frequency eigenvector on the fine grid for an eigenvalue $\lesssim 1$ with vanishing values in the global grid points.

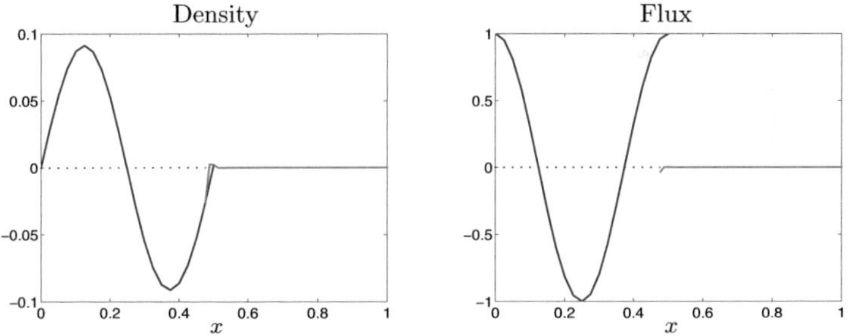

FIGURE 8.27. Low-frequency eigenvector on the coarse grid with an eigenvalue $\approx 1 - 2\omega$.

High-frequency eigenvectors on the coarse grid can be found with eigenvalues close to -1. These eigenvectors are responsible for the oscillations of the solutions for nonsmooth data.

The high-frequnecy oscillations on the fine grid are furnished with eigenvalues close to $(2\omega - 1)^2$. Hence, these eigenvectors are decaying very fast and do not give rise to spurious oscillations with respect to time.

Not all of the eigenvectors show a global behavior or are simply restricted to the coarse or the fine grid. We also find eigenvectors with interferences between the

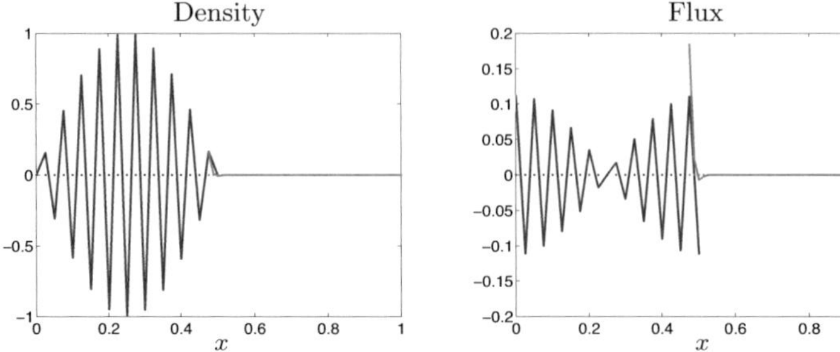

FIGURE 8.28. Highly oscillating eigenvector on the coarse grid with an eigenvalue $\gtrsim -1$.

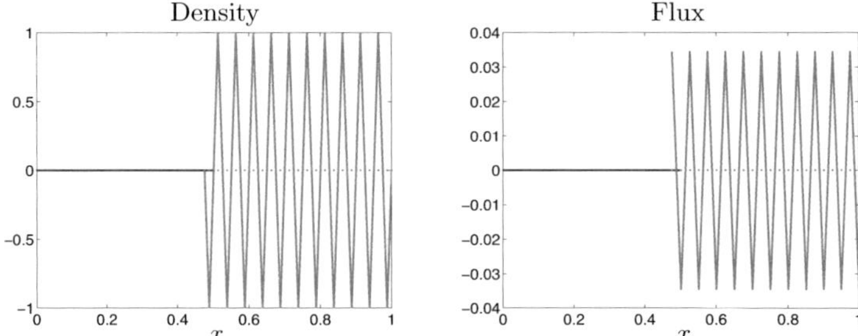

FIGURE 8.29. Highly oscillating eigenvector on the fine grid with an eigenvalue $\approx (2\omega - 1)^2$.

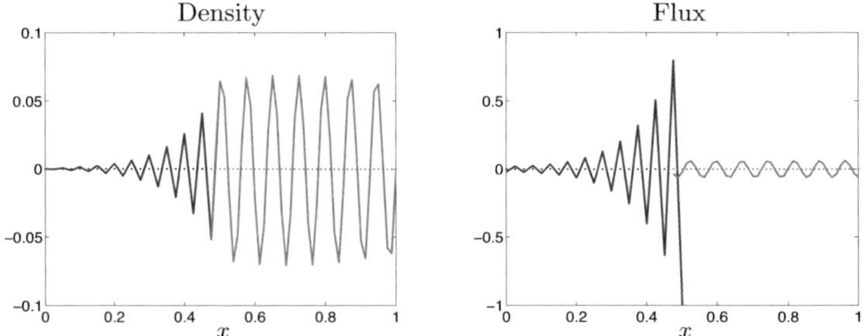

FIGURE 8.30. Eigenvector on the fine grid with transmission to the coarse grid with an eigenvalue $\lesssim 2\omega - 1$.

coupled grids. In Figure 8.30 one of these eigenvectors is plotted. The corresponding eigenvalue is close to $2\omega - 1$.

8.8. LB-Algorithms on a Hierarchy of Refinement Zones

In the applications it may be required to use nested grids with multiple refinement zones, where the basis is a uniform grid with grid size h_C. Let the interval Ω be divided into K subintervals Ω_i, $i = 1, \ldots, K$. Each subinterval Ω_i is uniformly discretized with the grid size h_i and contains at least eight nodes. The grid sizes have to be of the form $h_i = h^C 2^{-p_i}$, where p_i gives the refinement level of the subinterval Ω_i, $p_i = 0, 1, 2, \ldots$, and h^C is the grid size of the coarsest level with $p_i = 0$. Neighboring subintervals are allowed to differ by one level only. The lattice Boltzmann algorithm on this hierarchy of grids can be described as follows:

Let the initial density r_0 be given. Define the local fluxes $j_0^i := -h_i \partial_x r_0/(2\omega)$ in Ω_i for $i = 1, \ldots, K$. Compute the local distribution functions u_0^i and v_0^i by

$$u_0^i := \frac{1}{2}(r_0 + j_0^i), \quad i = 1, \ldots, K,$$

$$v_0^i := \frac{1}{2}(r_0 - j_0^i), \quad i = 1, \ldots, K.$$

The algorithm on the hierarchy of grids then reads:

LB-Algorithm on a hierarchy of refinements:
  i) Start on the intervals Ω_i of the coarsest level with $p_i = 0$.
  ii) Perform one LB-step on all intervals of this level. Interpolate the outflow values at the intersections to the neighboring finer grid in time.
  iii) Goto the next finer level.
  iv) Repeat ii) and iii) until the finest level is reached.
  v) Perform four LB-steps on the finest level. Compute the missing inflow values at the intersections by fitting the interpolated values.
  vi) Goto the next coarser level. Fit the missing inflow values with the information from the finer grid.
  vii) Do one LB-step on all intervals of this level and check:
    • If the number of the total LB-steps done on this level is not divisible by four, then interpolate the data at the intersections and goto the next finer level (if finer level exists, otherwise stay and proceed).
    • If the number of the total LB-steps done on this level is divisible by four, then goto the next coarser level and fit the missing inflow values on the coarse grid.
  viii) Repeat vii) until the coarsest level is reached again. Then goto i).

8.9. Efficiency of Local Refinements

In this closing section we want to examine the efficiency of local refinements with respect to the premise of optimal decay of the errors on the refined grids.

We assume that the error is uniformly distributed in the interval Ω up to a small domain Ω_μ with relative width μ, where the error is amplified by a factor of f. The total operating expense on a uniform grid with N grid points and the grid size h is given by $A_0 = CN^3$ with a constant C, due to the prescribed space-time coupling. For a globally uniform error distribution, the domain Ω_μ has to be refined with a grid size $2^{-p}h$ for a $p \in \mathbb{N}$, where staggered grids have to be used. For the uniform error distribution, we require $f = 4^p$ in the optimal case which leads to a condition for p. The additional computational expense is $A_\mu = C\mu 8^p N^3 = \mu 8^p A_0$. Hence, the total expense for the local refinement is given by

$$A = (1 - \mu)A_0 + A_\mu \approx (1 + \mu 8^p)A_0 = (1 + \mu f^{3/2})A_0.$$

On a uniformly refined grid, the total effort would be $8^p A_0$ to achieve the same local error. Hence, the local refinement reduces the total effort by a factor of approximately μ. For an adequate performance of the adaptive algorithm $\mu f^{3/2} = \mathcal{O}(1)$ is required. The effect for the applications is that only refinements about points should be used.

In the d-dimensional case with $d \in \{2, 3\}$ we consider a rectangle or a cuboid Ω_μ with relative edge lengths μ_i, $i = 1, \ldots, d$, and volume $V_\mu := \prod_{i=1}^d \mu_i$. In Ω_μ

the local error is assumed to be amplified by a factor of f. A local refinement on Ω_μ with grid size $2^{-p}h$ leads to a decay of the local error by a factor of 4^p. The total operating expense on a uniform grid with N gridpoints in each dimension is $A_0 = CN^{2+d}$. The additional effort needed to achieve a globally uniform error distribution by applying local grid refinement is $A_\mu = 2^{(2+d)p}V_\mu A_0 = f^{1+d/2}V_\mu A_0$. Hence we gain the total effort

$$A = (1 - V_\mu)A_0 + A_\mu \approx (1 + V_\mu 2^{(2+d)p})A_0 = (1 + V_\mu f^{1+d/2})A_0.$$

On a uniformly refined grid, the total effort would be $2^{(2+d)p}A_0$ to achieve the same local error. Hence, the total effort is reduced by a factor of approximately V_μ by applying local refinement in Ω_μ. Therefore, appropriate choices are refinements about lines or planes, where V_μ tends to zero.

Summary and Outlook

The topic of this work was the study of lattice Boltzmann methods for a one-dimensional model problem. The analytical examinations were basically split into three major parts.

In the first part, we considered the fluid-dynamic limit equation of the model problem, namely the heat equation on a bounded interval subject to several types of boundary conditions. We gave results on the classical and on the weak existence theory. Furthermore, we proved a priori estimates for the solutions in terms of energy estimates and by following the concept of Fourier solutions. Special care was taken on the regularity of the solutions and the necessary assumptions on the data.

In the second part, the interface between the fluid-dynamic limit equation and the discrete lattice Boltzmann equations was set up by the velocity discrete system. For our model problem, this is the Goldstein-Taylor model which forms a hyperbolic advection system. The parabolic limit of this advection system was worked out by the transformations to singularly perturbed telegraph equations. The presented Fourier solutions clarified the passage from the hyperbolic nature to the parabolic nature. A priori estimates for the solutions were given in terms of energy estimates. With the aid of the a priori estimates we proved convergence of the solutions of the advection system towards solutions of the heat equation in second order with respect to the scaling factor that corresponds to the Knudsen number and the Mach number.

In the third and major part, we derived the lattice Boltzmann schemes as discretizations of the advection system in the diffusion scaling and performed a detailed numerical analysis including numerical experiments. The equations were discretized in a finite difference and in a finite volume context, following the characteristic directions on vertex centered and cell centered grids. An extensive list of boundary conditions was proposed.

For a given viscosity, the lattice Boltzmann schemes are endowed with a prescribed coupling of the time step τ and the grid size h of the form $\gamma\tau = h^2$. Beside the number N of grid points, the only free quantity to choose is the relaxation parameter ω of the collision term. The choice of ω has a direct influence on the distribution of the eigenvalues of the time evolution matrix. However, there is no possibility to keep the eigenvalues away from the value -1, which turned out to be unfavorable in the case of nonsmooth data.

A discrete Fourier analysis of the time evolution matrices in several boundary situations was performed, which led to an insight into the structure of the discrete solutions and the coupling of the flux and the density. A priori estimates for the discrete solutions were found in the finite difference context. These estimates were used to prove second order convergence of the lattice Boltzmann solutions towards the solution of the heat equation in terms of the grid size h for a fixed end time. Special attention was taken at the boundaries, where the residuals show decreased consistency orders.

The numerical results confirmed the obtained theoretical results. Second order convergence was observed for the density and the flux for all presented boundary conditions. The dependence on the parameters and the initial data was checked. For vanishing source terms and a specific choice of the parameters, fourth order convergence for the density was detected.

Oscillations in the lattice Boltzmann solutions were observed for nonsmooth data. This behavior has to be attributed to the eigenvalues close to -1. In this context, discretizations in connection with the explicit Euler method appeared to be superior. Here, the space-time coupling has not to be kept at a fixed ratio.

For the application of nested grids, we provided several grid coupling algorithms. On vertex centered grids, we obtained the desired improvements that enable the reduction of the computational effort in the case of concentrated errors.

In future works, we have to face the question how our stability and convergence results can be extended to nonlinear and multi-dimensional problems. As we have seen, the convergence processes depend on the smoothness of the analytical solution. But for the Navier-Stokes equations, the regularity of the solutions is an unsettled problem.

Furthermore, the question has to be tackled how the lattice Boltzmann methods can be included in a variational framework. An approach by Petrov-Galerkin finite element methods would allow for a posteriori error estimation techniques that are the basis for the introduction of adaptive methods. The space-time coupling and the corresponding high computational costs require reductions.

A variational framework for Boltzmann-type equations with respect to velocities can be found for the semiconductor model in Ref. [**46**] and in Ref. [**43**]. Spectral Galerkin methods employing Hermite polynomials up to arbitrary orders are used and adaptive numerical implementations are introduced. The equations of the Goldstein-Taylor model can be obtained from the semiconductor model by the application of Gaussian quadrature formulas.

Another issue that has to be investigated in future works is the grid coupling in the multi-dimensional case. In higher dimensions, the transformation of the distribution functions and the macroscopic quantities do not form a closed system. Furthermore, the requirement of overlapping grids is a challenge. For our one-dimensional model problem, stability estimates and convergence proofs for the coupled problem on nonuniform grids are still missing. With respect to our results, improved data interpolation formulas have to be found in the case of grid coupling on cell centered grids. Spatial interpolations are also required in the multi-dimensional case.

Grid coupling algorithms are proposed in Ref. [**13**] and Ref. [**14**]. In the two-dimensional case, double overlapping grids are used, since the equilibrium distributions are employed to determine the intersection conditions. Single overlapping grids for a one-dimensional model with three velocities are the objective in Ref. [**42**]. Extensions are provided for the two-dimensional case.

The great advantage of lattice Boltzmann methods is its easy implementation and the applicability to a great variety of problems. With few lines of code reasonable solutions can be found, whereas standard procedures as the finite element method applied to the classical fluid-dynamic equations require much more technical background.

Bibliography

[1] H. W. Alt. *Lineare Funktionalanalysis*. 3. Auflage. Springer, Berlin, 1999.

[2] C. Bardos, F. Golse, and D. Levermore. Fluid dynamic limits of kinetic equations. I: Formal derivations. *J. Stat. Phys.*, 63:323–344, 1991.

[3] C. Bardos, F. Golse, and D. Levermore. Fluid dynamic limits of kinetic equations. II: Convergence proofs for the Boltzmann equation. *Commun. Pure Appl. Math.*, 46:667–753, 1993.

[4] R. Benzi, S. Succi, and M. Vergassola. The lattice Boltzmann equation: theory and applications. *Physics Reports*, 222, No. 3:145–197, 1992.

[5] J. E. Broadwell. Shock structure in a simple discrete velocity gas. *Phys. Fluids*, 7:1243–1247, 1964.

[6] J. E. Broadwell. Study of rarefied shear flow by the discrete velocity method. *J. Fluid Mech.*, 19:401–414, 1964.

[7] R. E. Caflisch and G. C. Papanicolaou. The fluid-dynamical limit of a nonlinear model Boltzmann equation. *Commun. Pure Appl. Math.*, 32:589–616, 1979.

[8] T. Carleman. *Problèmes mathématiques dans la théorie cinetique des gaz*. Publ. Scient. Inst. Mittag-Leffler, Uppsala, 1957.

[9] C. Cercignani. *The Boltzmann equation and its applications*. Springer, New York, 1988.

[10] C. Cercignani. *Mathematical methods in kinetic theory*. 2. ed.. Plenum press, New York, 1990.

[11] S. Chen and G. D. Doolen. Lattice Boltzmann method for fluid flows. *Ann. Rev. Fluid. Mech.*, 30:329–364, 1998.

[12] A. De Masi, R. Esposito, and J. L. Lebowitz. Incompressible Navier Stokes and Euler limits of the Boltzmann equation. *Commun. Pure Appl. Math.*, 42:1189–1214, 1989.

[13] O. Filippova and D. Hänel. Grid refinement for lattice-BGK models. *J. Comput. Phys.*, 147:219–228, 1998.

[14] O. Filippova and D. Hänel. Acceleration of lattice-BGK schemes with grid refinement. *J. Comput. Phys.*, 165:407–427, 2000.

[15] A. Friedman. *Partial differential equations of parabolic type*. Prentice Hall, Englewood Cliffs, 1964.

[16] U. Frisch, D. d'Humières, B. Hasslacher, P. Lallemand, Y. Pomeau, and J.-P. Rivet. Lattice gas hydrodynamics in two and three dimensions. *Complex Systems*, 1:649–707, 1987.

[17] U. Frisch, B. Hasslacher, and Y. Pomeau. Lattice-gas automata for the Navier-Stokes equation. *Phys. Rev. Lett.*, 56:1505–1508, 1986.

[18] E. Gabetta and B. Perthame. Scaling limits for the Ruijgrok-Wu model of the Boltzmann equations. *Math. Methods Appl. Sci.*, 24:949–967, 2001.

[19] R. Gatignol. *Théorie cinétique des gaz à répartition discrète de vitesses*. Lecture Notes in Physics 36, Springer, Berlin, 1975.

[20] S. Goldstein. On diffusion by discontinuous movements, and on the telegraph equation. *Quart. J. Mech. Appl. Math.*, 4:129–156, 1951.

[21] H. Grad. Singular and nonuniform limits of solutions of the Boltzmann equation. *SIAM AMS Proceedings, I., Transport Theory*, pages 296–308, 1969.

[22] C. Grossmann and H.-G Roos. *Numerik partieller Differentialgleichungen*. 2. durchges. Auflage. Teubner, Stuttgart, 1994.

[23] M Hanke-Bourgeois. *Grundlagen der numerischen Mathematik und des wissenschaftlichen Rechnens*. Teubner, Stuttgart, 2002.

[24] X. He and L.-S. Luo. Lattice Boltzmann model for the incompressible Navier-Stokes equation. *J. Stat. Phys.*, 88, No. 3-4:927–944, 1997.

[25] N. J. Higham. *Accuracy and stability of numerical algorithms*. SIAM, Philadelphia, 1996.

[26] W. Hundsdorfer and J. Verwer. *Numerical solution of time-dependent advection-diffusion-reaction equations*. Springer, Berlin, 2003.

[27] S. Hwang. Fluid-dynamic limit for the Ruijgrok-Wu model of the Boltzmann equation. *J. Math. Anal. Appl.*, 315:327–336, 2006.

[28] S. Jin, L. Pareschi, and G. Toscani. Diffusive relaxation schemes for multiscale discrete-velocity kinetic equations. *SIAM J. Numer. Anal.*, 35, No. 6:2405–2439, 1998.

[29] M. Junk. A finite difference approximation of the lattice Boltzmann method. *Numer. Methods Partial Differ. Equations*, 17, No. 4:383–402, 2001.

[30] M. Junk, A. Klar, and L.-S. Luo. Theory of the lattice Boltzmann method: Mathematical analysis of the lattice Boltzmann equation. *To appear in JCP.*

[31] E. Kamke. *Differentialgleichungen, II., Partielle Differentialgleichungen.* 4., wesentlich veränderte Auflage. Akademische Verlagsgesellschaft, Leipzig, 1962.

[32] T. G. Kurtz. Convergence of sequences of semigroups of nonlinear operators with an application to gas kinetics. *Trans. Amer. Math. Soc.*, 186:259–272, 1973.

[33] S. Larsson and V. Thomée. *Partielle Differentialgleichungen und numerische Methoden.* Springer, Berlin, 2005.

[34] P. L. Lions and G. Toscani. Diffusive limits for finite velocity Boltzmann kinetic models. *Rev. Mat. Iberoamericana*, 13, No. 3:473–513, 1997.

[35] H. P. McKean. The central limit theorem for Carleman's equation. *Isr. J. Math.*, 21:54–92, 1975.

[36] T. Platkowski and R. Illner. Discrete velocity models of the Boltzmann equation: A survey on the mathematical aspects of the theory. *SIAM Rev.*, 30, No. 2:213–255, 1988.

[37] A. D. Polyanin and V. F. Zaitsev. *Handbuch der linearen Differentialgleichungen.* Spektrum Akademischer Verlag, Heidelberg, 1996.

[38] F. Poupaud. Diffusion approximation of the linear semiconductor Boltzmann equation: analysis of boundary layers. *Asymptotic Analysis*, 4, No. 4:293–317, 1991.

[39] M. H. Protter and H. F. Weinberger. *Maximum principles in differential equations.* Prentice Hall, Englewood Cliffs, 1967.

[40] A. Quarteroni and A. Valli. *Numerical approximation of partial differential equations.* Springer, Berlin, 1994.

[41] P. A. Raviart and J.-M. Thomas. *Introduction à l'analyse numérique des équations aux dérivées partielles,* 2. Auflage. Masson, Paris, 1988.

[42] M. Rheinländer. A consistent grid coupling method for lattice-Boltzmann schemes. *J. Stat. Phys.*, 121, No. 1-2:49–74, 2005.

[43] C. Ringhofer, C. Schmeisser, and A. Zwirchmayr. Moment methods for the semiconductor Boltzmann equation on bounded position domains. *SIAM J. Numer. Anal.*, 39, No. 3:1078–1095, 2001.

[44] T. W. Ruijgrok and T. T. Wu. A completely solvable model of the nonlinear Boltzmann equation. *Physica A*, 113:401–416, 1982.

[45] F. Salvarani. Diffusion limits for the initial-boundary value problem of the Goldstein-Taylor model. *Rend. Sem. Mat. Univ. Pol. Torino*, 57, No. 3:211–222, 1999.

[46] C. Schmeisser and A. Zwirchmayr. Convergence of moment methods for linear kinetic equations. *SIAM J. Numer. Anal.*, 36, No. 1:74–88, 1998.

[47] W. A. Strauss. *Partielle Differentialgleichungen.* Vieweg, Braunschweig, 1995.

[48] G. I. Taylor. Diffusion by continuous movements. *Proc. London Math. Soc.*, 20:196–212, 1921.

[49] J. Wloka. *Partielle Differentialgleichungen.* Teubner, Stuttgart, 1982.

[50] D. A. Wolf-Gladrow. *Lattice-gas cellular automata and lattice Boltzmann models. An introduction.* Lecture Notes in Mathematics 1725, Springer, Berlin, 2000.

[51] E. Zeidler. *Nonlinear functional analysis and its applications II/A, Linear monotone operators.* Springer, Berlin, 1990.